Recent Advances in Aquatic Food Products Processing

Recent Advances in Aquatic Food Products Processing

Editor

Jingran Bi

MDPI

Basel • Beijing • Wuhan • Barcelona • Belgrade • Novi Sad • Cluj • Manchester

Editor
Jingran Bi
Dalian Polytechnic University
Dalian
China

Editorial Office
MDPI AG
Grosspeteranlage 5
4052 Basel, Switzerland

This is a reprint of articles from the Special Issue published online in the open access journal *Foods* (ISSN 2304-8158) (available at: https://www.mdpi.com/journal/foods/special_issues/883RA0LGZC).

For citation purposes, cite each article independently as indicated on the article page online and as indicated below:

Lastname, A.A.; Lastname, B.B. Article Title. *Journal Name* **Year**, *Volume Number*, Page Range.

ISBN 978-3-7258-2299-7 (Hbk)
ISBN 978-3-7258-2300-0 (PDF)
doi.org/10.3390/books978-3-7258-2300-0

Cover image courtesy of Jingran Bi

Contents

About the Editor

Jingran Bi

Jingran Bi is an associate professor at School of Food Science and Technology, Dalian Polytechnic University, China. She focuses on the research of food quality control and has published over 90 papers, which have been cited over 1000 times with an H-index of 19. She possesses considerable experience in taking charge of national projects and currently oversees a Young Scientists Fund of National Natural Science Foundation of China. Additionally, she serves as a guest editor of Foods (MDPI) and holds a director position at Liaoning Food Science and Technology Society.

foods

MDPI

Editorial

Aquatic Food Products: Processing Technology and Quality Control

Jingran Bi

State Key Laboratory of Marine Food Processing & Safety Control, Dalian Polytechnic University, Dalian 116034, China; bijingran1225@foxmail.com; Tel.: +86-0411-86322020

1. Introduction

Aquatic products have the characteristics of high protein, low fat, and good nutritional balance, and they have become an important source of support to solve world hunger and nutritional deficiencies. More than 900 million people around the world consume most of their daily protein through aquatic products [1], which are gradually becoming an important part of human diet: aquatic product-derived protein intake accounts for 2% of the total animal protein intake per capita [2]. Aquatic products also are an important source of polyunsaturated fatty acids, which have important functions in human health: providing heat, being a carrier of fat-soluble substances, protecting heart health, clearing excess lipids from blood vessels, regulating immune metabolism, etc. [3,4]. In addition, aquatic products have a delicious taste, which is widely welcomed by consumers.

The perishability of aquatic products makes them susceptible to bacterial spoilage throughout the fishing, transportation, processing, and storage stages. Following the spoilage of aquatic products, enzymatic activity within them facilitates the decomposition and degradation of nutrients, leading to the formation of malodorous and toxic substances [5]. The quality and safety of aquatic products are often measured by K value (representing the ratio between the sum of inosine and hypoxanthine and the sum of all other adenosine triphosphate breakdown products), volatile basic nitrogen, biological amine, and other biochemical indexes [6]. Spoilage greatly reduces the edible value and economic value of aquatic products, resulting in a great waste of resources. Therefore, improving the preservation technology of aquatic products and reducing their spoilage rates is one of the problems that the aquatic product processing industry needs to solve. Traditional freshness preservation methods are mainly chemical preservation, including salt, acid, smoke, etc. [7]. Chemical preservation methods, although effective in inhibiting bacterial growth and despite boasting a high level of sterilization efficiency, may potentially compromise the texture, distinctive umami quality, and delicate taste of aquatic products. Additionally, they may leave behind chemical residues, contributing to environmental pollution. With the increasing demand for food safety and health, some chemical and physical preservation methods that destroy nutrients are rejected by consumers. In recent years, extracting natural substances from plants to ensure the quality and safety of food has become a research trend. In addition, it is worth nothing that bacteria are the most important factor causing the deterioration of aquatic products, as they can gradually decompose large molecules such as proteins and lipids in aquatic products into small molecular compounds such as amino acids and fatty acids, leading to the generation of harmful substances such as amines and hydrogen sulfide and to the production of unpleasant odors [8,9]. Currently, as the scholarly exploration into bacterial quorum sensing (QS) systems intensifies, the pivotal role bacterial QS plays in the degradation of aquatic products has been gradually discovered. Researchers have strived to elucidate the molecular mechanisms of the effects of QS signaling molecules on nutrient degradation [10]. As QS can inhibit or slow down the process of aquatic product spoilage, it has provided a new way of aquatic product preservation.

Citation: Bi, J. Aquatic Food Products: Processing Technology and Quality Control. *Foods* **2024**, *13*, 2806. https://doi.org/10.3390/foods13172806

Received: 7 August 2024
Accepted: 26 August 2024
Published: 3 September 2024

Proper processing can prevent spoilage and extend the shelf life of aquatic products while maintaining their nutritional value, texture, and flavor. At present, the commonly used processing methods of aquatic products include hot processing, pickling, high-pressure and ultrasonic methods, and so on [11]. Heating is the most common way of processing aquatic products, and it can easily change the flavor of food and its nutritional composition. The traditional heating method uses radiation to heat the surface of an object and then gradually heats the interior through conduction and convection [12]. Using this method during processing can involve problems such as low heating efficiency, long local heating time, short internal heating time, uneven heating, and others, which leads to the loss of nutrients, bad taste, and a dry, rough texture of aquatic products. In addition to the fresh consumption of aquatic products, curing is also a favorite processing method; it can not only inhibit the multiplication of microorganisms, but also extend the preservation time of aquatic products [13]. High-pressure treatment is a non-heat treatment process that can replace heat treatment. Ultra-high-pressure processing is not affected by the size and shape of aquatic products, because the pressure applied during processing is uniform. Ultra-high-pressure processing can inactivate microorganisms at lower temperatures, ensuring the safety of food and extending its shelf life. Unlike hot processing, ultra-high-pressure processing can break down non-covalent bonds in proteins, squeezing them and changing their molecular conformation, but it has no effect on the covalent bonds of small molecular substances such as amino acids, so it can maintain food flavor and nutrition to a large extent [14,15]. As an advanced food processing technology, ultrasonic processing has broad prospects in food processing and preservation. It is a mild but targeted form of processing that can improve the quality and safety of aquatic products. Although ultrasound treatment alone is not sufficient to inactivate the various harmful enzymes in aquatic products, ultrasound has shown potential for a highly efficient inactivation of enzymes and pathogens compared to mild heat treatment [16]. As an important nutrient component of aquatic products, protein plays an important role in aquatic product quality during processing. The spatial structure of proteins is maintained by secondary bonds (such as hydrogen bonds), which change from their original ordered spatial structure to a disordered spatial structure under the various processing methods [17]. Changes in protein structure can affect the properties of protein in the following ways: (i) physicochemical properties: the tight structure becomes loose, the hydrophobic groups are exposed, the asymmetry of protein increases, and the physicochemical indexes (such as solubility) change [18–20]; (ii) biological activities: function is changed; (iii) biochemical properties: the protein denatures and the crystal structure disappears and can be easily hydrolyzed by protease. The application of diverse processing methods leads to varying degrees of protein alterations, consequently affecting the flavor of aquatic products and their nutritional composition [18–20]. Therefore, it is of great significance to study the influence of aquatic product processing methods on protein properties for the development of superior processing technologies.

This Editorial refers to the Special Issue "Recent Advances in Aquatic Food Products Processing". The Special Issue, which contains nine research articles and one review, highlights traditional processing (such as heating, salting, drying, smoking, natural fermentation), modern processing (such as ultra-high-pressure processing, low-salt fermentation, rapid freezing-thawing, etc.), and byproduct processing, as well as the quality change mechanisms that occur during the processing of aquatic products. To encourage readers to explore this Special Issue, I briefly describe the various contributions in the following paragraphs.

2. An Overview of Published Articles

The paper by Hematyar et al. (contribution 1) reported the relationship between fish welfare and ultimate filet quality. The study evaluated the effects of pre-slaughter handling and stocking density on filet quality in largemouth bass (*Micropterus salmoides*). The fish were divided into three groups and reared at various stocking densities of 35, 50, and 65 kg·m^{-3}, respectively. After feeding for 12 weeks, half of the fish were slaughtered by direct percussion on the head without air exposure (control group), and the other half were subjected to acute antemortem stress through anoxia for 3 min. Blood and filet samples were collected for quality and nutritional evaluation. It was found that anoxia triggered an earlier occurrence of rigor mortis, with a lower initial postmortem pH than that of the control group. Moreover, the stocking density and anoxia might have caused stress in the fish. When the fish were stimulated, their blood cortisol and plasma glucose levels climbed dramatically. Moreover, the stressful rearing environment at high stocking density also contributed to the hypoxia, resulting in the rapid depletion of glycogen reserves and the antioxidant defense. The oxidative enzyme activities in the intensive stocking groups were significantly higher than those of the sparse stocking group. Furthermore, low stocking density and pre-slaughter handling without anoxia also exhibited a slowing effect on the lipid oxidation and protein oxidation process by metal catalyzation, which was beneficial to the textural parameters of filets. In conclusion, fish welfare plays an essential role in the improvement of filet quality.

Aussanasuwannakul et al. (contribution 2) revealed crucial elements of online market success for traditional fermented aquatic foods. In detail, this work evaluated the nutritional composition, storage stability, sensory quality, and packaging efficiency of roasted pickled fish powder (RPFP), underlining the balance between the characteristics of traditional food and consumers' health requests. Based on the commercial benchmark, the protein content was measured at 40.17%, and the fat content was 10.60%. Compared to the commercial benchmark, the developed RPFP exhibited a higher protein content in the herbal flavor product, with a value of 28.97%, and a superior fat content in the spicy flavor products, at 19.51%. Moreover, the herbal flavor product was rich in dietary fiber, at 14.23%, and the intense heat of the spicy flavor product could effectively comply with both the nutritional and specific taste expectations of customers. Moreover, through storage stability analysis, including the determination of microbiological, pH level, and water activities, packaging forms such as traditional polypropylene cups and laminated aluminum foil stand-up pouches were evaluated. It was demonstrated that packaging efficiency could not only extend the shelf life of products thanks to excellent barrier properties, but also improve brand performance. Furthermore, the novel method of estimating both the nutritional and sensory dimensions of RPFP products for online market success provided invaluable insights into enhancing the status of traditional food in the field of e-commerce.

Wang et al. (contribution 3) focused on the regulation of quorum sensing (QS) in the microbial coculture system. *Hafnia alvei* (*H. alvei*) and *Pseudomonas fluorescens* (*P. fluorescens*) are the spoilage microorganisms specific to aquatic foods. The authors found that the biofilm contents, extracellular polysaccharides, and biogenic amines in the coculture system of *H. alvei* and *P. fluorescens* were significantly decreased as the *luxI* gene of *H. alvei* was knocked out, implying that QS participated in the dual-species interactions. Moreover, transcriptomics was applied to illustrate the regulatory mechanisms at the molecular biological level. In the coculture system of *H. alvei* ΔluxI and *P. fluorescens*, the transcriptional levels of genes associated with chemotaxis, flagellar assembly, and two-component system pathways were dramatically down-regulated. In addition, in the coculture system, 732 of the genes identified in *P. fluorescens*, mainly related to biofilm formation, ATP-binding cassette transporters, and amino acid metabolism, were significantly differentially expressed. These results demonstrated that the disruption of QS in *H. alvei* could effectively weaken spoilage development in the microbial coculture system, which offers a new insight into the inhibition of *H. alvei* and *P. fluorescens* and into the preservation of aquatic products.

Zhang et al. (contribution 4) developed a rapid non-contact cutting position identification method based on a constructed linear laser data acquisition system. This method eliminated the reliance on manual subjective experience, consequently making it suitable for large-scale production lines. The 3D point cloud information of the surface contour of the fish body was nondestructively collected via line laser scanning technology. Using principal component analysis (PCA), the characteristic variables of the ventrodorsal boundary line were extracted, which reduced the time-consuming computation and enhanced the recognition ability to identify the cutting position. Then, the fish head identification models were established using Least Squares Support Vector Machines (LS-SVMs), Particle Swarm Optimization-Back Propagation (PSO-BP) networks, and Long and Short-Term Memory (LSTM) neural networks, respectively. The paper found that the LSTM model was superior, with minimal error between the predicted and actual values and with satisfactory reliability. Further, from the viewpoint of fish morphology, this identification technology based on the ventral-dorsal demarcation line is reasonable and therefore has strong potential in the development of intelligent fish processing and equipment to meet the decapitation needs of salmon and tuna and other large carp-shaped fish.

Li et al. (contribution 5) proposed an amino-carboxymethyl chitosan (ACC)/dialdehyde starch (DAS) film to effectively load and slow-release ε-polylysine (ε-PL), which could be applied in the aquatic preservation field. The content of amino groups on chitosan was 0.83 ± 0.02 mmol/g after modification by carboxymethylation and amination, and the aldehyde content of starch was $89.8 \pm 0.07\%$ after oxidation by periodate. The ACC/DAS film was prepared by the Schiff base reaction between the amino group on ACC and the aldehyde group on DAS, which built a firm three-dimensional network skeleton structure. Due to occurrence of the Schiff base reaction, imide bonds (-NCH-) were formed, and the two long-chain polysaccharides were cross-linked, thus disrupting the crystal arrangement. Swelling analysis found that the ACC/DAS porous structure had a strong water-holding capacity. However, the over cross-linking network prepared by ACC/DAS with a -NH2: CHO molar ratio of 1:0.6 displayed a low flexibility, which restricted the penetration of water molecules. The loading ratio of ε-PL on ACC/DAS was 99.2%. Combined with the antibacterial activities of ε-PL and ACC, ACC/DAS/ε-PL exhibited an excellent long-time inhibition capability against *S. aureus* and *E. coli*, which makes it a good candidate for active food packaging.

Li et al. (contribution 6) explored the food safety of shrimp paste fermented by a new starter culture, *Tetragenococcus muriaticus* TS (*T. muriaticus* TS). Shrimp paste, rich in nutrients and with a unique flavor, is popular in Southeast Asian and Chinese coastal areas. During the traditional fermentation process of shrimp paste, biogenic amines, which are a hazardous food factor, are always generated by microorganisms via the decarboxylated reduction of amino acids. The autochthonic salt-tolerant *T. muriaticus* TS exhibited a positive amine oxidase activity, which could degrade biogenic amines in vitro. According to high-throughput sequencing data, *T. muriaticus* TS could disrupt the bacterial structure and interspecific correlations, and especially suppress the growth of bioamine-producing bacteria. Consequently, after fermenting for 35 days, putrescine, cadaverine, and histamine concentrations in shrimp paste were significantly reduced by 19.20%, 14.01%, and 28.62%, respectively. Meanwhile, *T. muriaticus* TS also effectively inhibited the total volatile base nitrogen and improved the amino acid nitrogen concentrations. Through HS-SPME-GC-MS analysis, it was found that the odor of the shrimp paste fermented by *T. muriaticus* TS was improved, and pyrazines were enhanced while amines were weakened. Overall, *T. muriaticus* TS is an effective starter culture to prepare shrimp paste with high flavor quality and safety.

Zhu et al. (contribution 7) studied the influence of drying methods on pufferfish filet quality. Hot air drying (HAD), a common aquatic food product processing method, always effectively improves the drying speed. Cold air drying (CAD) is an appropriate approach to retain sensory quality and nutrition. Cold and hot air combined drying (CHACD) may combine the advantages of the two and avoid their disadvantages. The study found that during drying, mobile water molecules were reduced via evaporation, and the immobilized water content of CHACD was between that of HAD and CAD. CHACD also enhanced the textural attributes of the filets, such as their springiness, chewiness, and toughness, owing to their intact, well-arranged, and well-defined microstructure. Although drying inevitably causes lipid oxidation and protein degradation, CHACD could effectively alleviate this problem. All in all, CHACD exhibits advantages over CAD and HAD, such as an accelerated drying time, reduced lipid oxidation, improved protein stability, and tightened tissue structure, providing a novel drying method for pufferfish processing.

Qian et al. (contribution 8) worked on modified atmosphere packaging (MAP), which is an eco-friendly and convenient physical technique to prolong the shelf life of food. In their paper, the effects of MAP's gas composition, including carbon dioxide (CO_2), nitrogen (N_2), and oxygen (O_2), on the quality of salmon were studied. The study found that MAP with a CO_2 concentration of over 40% could effectively inhibit the growth of mesophilic bacteria. However, excessive CO_2 ($\geq$80%) might facilitate proteolysis, thus providing nutrients for bacterial growth, resulting in an increase in the total mesophilic bacterial count and the accumulation of TVB-N, as well as in a decrease in hardness, water-holding capacity, and myofibrillar protein content. Furthermore, a MAP environment with an $O_{2\ content}$ higher than 20% probably promotes protein degradation, consequently affecting muscle texture. According to electronic nose analysis, 60% CO_2/20% O_2/20% N_2 and 60% CO_2/10% O_2/30% N_2 were most suitable to retain the original odor characteristics of salmon. In summary, 60% CO_2/10% O_2/30% N_2 was the optimal MAP gas composition for the preservation of salmon by inhibiting bacteria growth and resisting protein degradation.

Chen et al. (contribution 9) described the processing quality of crabs during thermal processing. Along with thermal processing, protein degradation is closely related to the nutrition, texture, and flavor profiles of aquatic products. For example, free amino acids (FAAs) are the end products of protein degradation that provide the umami, sweet, and bitter tastes of heated aquatic foods. It was found that bitter amino acid content in crabs did not undergo any significant change during steaming, while sweet (Thr, Gly, Ala, and Pro) and umami amino acids (Asp and Glu) climbed to the maximum value after 10 cmin of steaming and then declined. Furthermore, flavor nucleotides such as guanosine monophosphate, inosine monophosphate, and adenosine monophosphate also exhibited a synergistic effect with free amino acids on the umami and sweetness profiles of crabs during steaming. Combined with electronic tongue analysis and sensory assessment, the recommended steaming time for crab was found to be 10–15 min. These findings provide a theoretical foundation for the large-scale standardized processing of crabs.

Finally, Fan et al. (contribution 10) reviewed the characteristics, mechanism, and inhibition of endogenous proteases in sea cucumber (*Apostichopus japonicas*). The autolysis of sea cucumber is always triggered by ultraviolet irradiation, mechanical injury, ambient temperature, etc. During autolysis, endogenous proteases play an essential role in damaging collagen fibers or the interfibrillar proteoglycan bridge, resulting in tissue self-dissolution, which causes a decrease in edible suitability and economic loss. Therefore, it is necessary to systematically understand the action mechanism and inhibition methods of endogenous proteases. In the paper, the external trigger factors of sea cucumber autolysis and the changes in protein composition and mechanical properties during autolysis were described in detail. Furthermore, the classification, characteristics, possible mechanisms, and natural inhibitors of endogenous proteases were specifically reviewed. Additionally, it was pointed out that molecular biological mechanisms, optimization processes, and applications of effective protease inhibitors in the sustainable production of sea cucumber should be the focus of future research.

3. Conclusions

This compilation of articles devoted to describing the sensory quality, protein degradation, and hazard generation and inhibition in the process of aquatic product storage, preservation, fermentation, and hot processing provides a theoretical basis for the growth of the aquatic product industry.

With the continuous development of the global aquatic product market, the improvement of aquatic product processing technology has become the main research direction. The current processing technology of aquatic products mainly includes defishment, prefabrication, sterilization, and preservation, etc. These processing technologies can effectively improve the quality of aquatic products, improve their taste, extend their shelf life, and maintain their nutritional value. Meanwhile, the focus of research on aquatic product processing technology has gradually shifted from a single method to a multi-technology composite application to better preserve the nutrients in and taste of food. The quality improvement process of aquatic products needs to integrate a variety of high-tech processing technologies, and the exploration, development, and large-scale application of new processing technologies are the necessary conditions for its future development.

Funding: This article was supported by the National Natural Science Foundation of China (32102050), Liaoning provincial education department basic scientific research projects (LJK MZ20220875), and Open Foundation of State Key Laboratory of Marine Food Processing & Safety Control (SKL202406).

Conflicts of Interest: The authors declare no conflicts of interest.

List of Contributions:

1. Hematyar, N.; Rahimnejad, S.; Gorakh Waghmare, S.; Malinovskyi, O.; Policar, T. Effects of Stocking Density and Pre-Slaughter Handling on the Fillet Quality of Largemouth Bass (*Micropterus salmoides*): Implications for Fish Welfare. *Foods* **2024**, *13*, 1477.
2. Aussanasuwannakul, A.; Butsuwan, P. Evaluating Microbiological Safety, Sensory Quality, and Packaging for Online Market Success of Roasted Pickled Fish Powder. *Foods* **2024**, *13*, 861.
3. Wang, Y.; Li, X.; Zhang, G.; Bi, J.; Hou, H. Transcriptome Reveals Regulation of Quorum Sensing of *Hafnia alvei* H4 on the Coculture System of *Hafnia alvei* H4 and *Pseudomonas fluorescens* ATCC13525. *Foods* **2024**, *13*, 336.
4. Zhang, X.; Gong, Z.; Liang, X.; Sun, W.; Ma, J.; Wang, H. Line Laser Scanning Combined with Machine Learning for Fish Head Cutting Position Identification. *Foods* **2023**, *12*, 4518.
5. Li, Y.; Qiu, Y.; Hou, H.; Zhang, G.; Hao, H.; Bi, J. The Preparation and Properties of Amino-Carboxymethyl Chitosan-Based Antibacterial Hydrogel Loaded with ε-Polylysine. *Foods* **2023**, *12*, 3807.
6. Li, X.; Zhang, Y.; Ma, X.; Zhang, G.; Hou, H. Effects of a Novel Starter Culture on Quality Improvement and Putrescine, Cadaverine, and Histamine Inhibition of Fermented Shrimp Paste. *Foods* **2023**, *12*, 2833.
7. Zhu, Y.; Chen, X.; Qiao, K.; Chen, B.; Xu, M.; Cai, S.; Shi, W.; Liu, Z. Combined Effects of Cold and Hot Air Drying on Physicochemical Properties of Semi-Dried Takifugu obscurus Fillets. *Foods* **2023**, *12*, 1649.
8. Qian, Y.-F.; Liu, C.-C.; Zhang, J.-J.; Ertbjerg, P.; Yang, S.-P. Effects of Modified Atmosphere Packaging with Varied CO_2 and O_2 Concentrations on the Texture, Protein, and Odor Characteristics of Salmon during Cold Storage. *Foods* **2022**, *11*, 3560.
9. Chen, Q.; Zhang, Y.; Jing, L.; Xiao, N.; Wu, X.; Shi, W. Changes in Protein Degradation and Non-Volatile Flavor Substances of Swimming Crab (*Portunus trituberculatus*) during Steaming. *Foods* **2022**, *11*, 3502.
10. Fan, X.; Wu, K.; Tian, X.; Benjakul, S.; Li, Y.; Sang, X.; Zhao, Q.; Zhang, J. Endogenous Proteases in Sea Cucumber (*Apostichopus japonicas*): Deterioration and Prevention during Handling, Processing, and Preservation. *Foods* **2024**, *13*, 2153.

References

1. Serra, V.; Pastorelli, G.; Tedesco, D.E.A.; Turin, L.; Guerrini, A. Alternative protein sources in aquafeed: Current scenario and future perspectives. *Vet. Anim. Sci.* **2024**, *25*, 100381. [CrossRef]
2. Wang, M.; Cheng, Y.; Li, X.; Nian, L.; Yuan, B.; Cheng, S.; Wang, S.; Cao, C. Effects of microgels fabricated by microfluidic on the stability, antioxidant, and immunoenhancing activities of aquatic protein. *J. Future Foods* **2025**, *5*, 57–67. [CrossRef]
3. Chen, J.-N.; Zhang, Y.-Y.; Huang, X.; Wang, H.-P.; Dong, X.; Zhu, B.; Qin, L. Analysis of Lipid Molecule Profiling and Conversion Pathway in Mandarin Fish (*Siniperca chuatsi*) during Fermentation via Untargeted Lipidomics. *J. Agric. Food Chem.* **2023**, *71*, 8673–8684. [CrossRef] [PubMed]
4. Li, B.; Xu, L.; He, R.; Li, Y.; Li, G.; Deng, Y.; Wang, Z.; Li, F.; Li, T.; Rong, S. Association of aquatic food consumption, long-chain polyunsaturated n-3 fatty acid intake, and blood mercury levels with cognitive function in middle-aged and older adults. *Clin. Nutr.* **2024**, *43*, 1635–1642. [CrossRef]
5. Chu, Y.; Wang, J.; Xie, J. Foodomics in aquatic products quality assessment during storage: An advanced and reliable approach. *Food Biosci.* **2024**, *58*, 103734. [CrossRef]
6. Ding, T.; Li, Y. Biogenic amines are important indices for characterizing the freshness and hygienic quality of aquatic products: A review. *LWT* **2024**, *194*, 115793. [CrossRef]
7. Wei, X.; Zhang, M.; Chen, K.; Huang, M.; Mujumdar, A.S.; Yang, C. Intelligent detection and control of quality deterioration of fresh aquatic products in the supply chain: A review. *Comput. Electron. Agric.* **2024**, *218*, 108720. [CrossRef]

8. Zhou, Y.; Zhang, Y.; Liang, J.; Hong, H.; Luo, Y.; Li, B.; Tan, Y. From formation to solutions: Off-flavors and innovative removal strategies for farmed freshwater fish. *Trends Food Sci. Technol.* **2024**, *144*, 104318. [CrossRef]
9. Zhao, L.; Zhang, M.; Wang, H. Inhibition of the fishy odor from boiled crab meatballs during storage via novel combination of radio frequency and carbon dots. *Food Control* **2022**, *136*, 108843. [CrossRef]
10. Ge, Z.; Du, X.; Liu, J.; Zhu, J.; Hao, H.; Bi, J.; Hou, H.; Zhang, G. Benzyl isothiocyanate suppresses biofilms and virulence factors as a quorum sensing inhibitor in Pseudomonas fluorescens. *LWT* **2024**, *204*, 116387. [CrossRef]
11. Siewe, F.B.; Makebe, C.W.; Claudette Bakisu Muala, W.; Laya, A.; Raymond Nkongho, N.; Meliko, M.O.; Kudre, T.G.; Bhaskar, N. Advances in processing, reaction pathways, stabilisation and food applications of natural seafood flavourings. *Food Biosci.* **2024**, *58*, 103627. [CrossRef]
12. Ismail, M.I.; Yunus, N.A.; Hashim, H. Integration of solar heating systems for low-temperature heat demand in food processing industry—A review. *Renew. Sustain. Energy Rev.* **2021**, *147*, 111192. [CrossRef]
13. Wang, J.; Huang, X.-H.; Zhang, Y.-Y.; Li, S.; Dong, X.; Qin, L. Effect of sodium salt on meat products and reduction sodium strategies—A review. *Meat Sci.* **2023**, *205*, 109296. [CrossRef] [PubMed]
14. Li, N.; Wang, Y.; Tan, Z.; Xu, Y.; Liu, X.; Liu, Y.; Zhou, D.; Li, D. Effect of ultra-high pressure heat-assisted technology combined with L-cysteine on the color of ready-to-eat shrimp during storage. *Food Chem.* **2024**, *460*, 140634. [CrossRef] [PubMed]
15. Zhu, C.; Zeng, X.; Chen, L.; Liu, M.; Zheng, M.; Liu, J.; Liu, H. Changes in quality characteristics based on protein oxidation and microbial action of ultra-high pressure-treated grass carp (*Ctenopharyngodon idella*) fillets during magnetic field storage. *Food Chem.* **2024**, *434*, 137464. [CrossRef] [PubMed]
16. Zheng, X.; Zou, B.; Zhang, J.; Cai, W.; Na, X.; Du, M.; Zhu, B.; Wu, C. Recent advances of ultrasound-assisted technology on aquatic protein processing: Extraction, modification, and freezing/thawing-induced oxidation. *Trends Food Sci. Technol.* **2024**, *144*, 104309. [CrossRef]
17. Arakawa, T.; Tomioka, Y.; Akuta, T.; Shiraki, K. The contrasting roles of co-solvents in protein formulations and food products. *Biophys. Chem.* **2024**, *312*, 107282. [CrossRef] [PubMed]
18. Zhou, J.; Feng, Q.; Fu, H.; Ren, K.; Shang, W.; Li, C.; Zhang, X.; Mai, T.T.N.; He, Y. Current trends and perspectives on aquatic-derived protein: A focus on structure-technofunctional properties relationship and application for food preservation. *Trends Food Sci. Technol.* **2024**, *151*, 104651. [CrossRef]
19. Huang, X.; Wang, H.; Tu, Z. A comprehensive review of the control and utilization of aquatic animal products by autolysis-based processes: Mechanism, process, factors, and application. *Food Res. Int.* **2023**, *164*, 112325. [CrossRef] [PubMed]
20. Jiao, X.; Yang, H.; Li, X.; Cao, H.; Zhang, N.; Yan, B.; Hu, B.; Huang, J.; Zhao, J.; Zhang, H.; et al. Green and sustainable microwave processing of surimi seafood: A review of protein component interactions, mechanisms, and industrial applications. *Trends Food Sci. Technol.* **2024**, *143*, 104266. [CrossRef]

foods

MDPI

Article

Effects of Stocking Density and Pre-Slaughter Handling on the Fillet Quality of Largemouth Bass (*Micropterus salmoides*): Implications for Fish Welfare

Nima Hematyar [1,*], Samad Rahimnejad [1,2], Swapnil Gorakh Waghmare [1], Oleksandr Malinovskyi [1] and Tomas Policar [1]

[1] Faculty of Fisheries and Protection of Waters, South Bohemian Research Center of Aquaculture and Biodiversity of Hydrocenoses, Research Institute of Fish Culture and Hydrobiology, University of South Bohemia in Ceske Budejovice, Zátiší 728/II, 389 25 Vodňany, Czech Republic; s.rahimnejad@um.es (S.R.); swaghmare@mcw.edu (S.G.W.); omalinovskyi@frov.jcu.cz (O.M.); policar@frov.jcu.cz (T.P.)

[2] Immunobiology for Aquaculture Group, Department of Cell Biology and Histology, Faculty of Biology, Regional Campus of International Excellence "Campus Mare Nostrum", University of Murcia, 30100 Murcia, Spain

* Correspondence: hematyar@frov.jcu.cz

Abstract: There is currently insufficient acknowledgment of the relationship between fish welfare and ultimate fillet quality. The purpose of this study was to assess the impacts of pre-slaughter handling and stocking density as fish welfare markers on fillet quality of largemouth bass (*Micropterus salmoides*). Fish from three stocking densities of 35, 50, and 65 kg·m^{-3} were reared in a recirculating aquaculture system (RAS) for 12 weeks and received commercial feed. Ultimately, the fish were either stunned with percussion on the head (control group) or subjected to air exposure for 3 min (anoxia group) before stunning and subsequent collection of blood and fillet samples. Western blot analysis revealed the degradation of actin in both groups. Additionally, higher oxidation progress and lower hardness and pH were observed in anoxia compared to the control group. We observed higher hardness at 35 kg·m^{-3} in anoxia compared to 50 and 65 km^{-3}. The initial hardness values at 35, 50, and 65 km^{-3} were 1073, 841, and 813 (g) respectively in the anoxia group. Furthermore, the anoxia and control groups had rigor mortis after 6 and 10 h, respectively. Cortisol and glucose levels, and oxidative enzymes activity were higher in anoxia than in the control group. In conclusion, oxidation induced by anoxia likely plays a crucial role as a promoter of the quality deterioration of largemouth bass fillets.

Keywords: welfare; anoxia; oxidation development; blood biochemistry; antioxidant capacity

Citation: Hematyar, N.; Rahimnejad, S.; Gorakh Waghmare, S.; Malinovskyi, O.; Policar, T. Effects of Stocking Density and Pre-Slaughter Handling on the Fillet Quality of Largemouth Bass (*Micropterus salmoides*): Implications for Fish Welfare. *Foods* **2024**, *13*, 1477. https://doi.org/10.3390/foods13101477

Academic Editor: Jingran Bi

Received: 8 April 2024
Revised: 30 April 2024
Accepted: 2 May 2024
Published: 10 May 2024

1. Introduction

Largemouth bass (*Micropterus salmoides*) is one of the fish species well-suited for aquaculture production [1]. Due to its exceptional fillet quality, willingness to consume commercial feed, and rapid growth rate, it is a good fit for intensive aquaculture [2,3].

Seafood products quality can be assessed by various criteria such as nutritional quality, food safety standards, and sensory analysis [4]. The freshness of fish fillet is influenced by different processing conditions, a crucial aspect for the fishing industry.

Throughout pre-slaughter handling, animals may undergo physical and psychological stressors, leading to heightened anxiety, increased heart rate, respiration, and irritability. Overconsumption of nutrients and water triggers hormonal secretion, resulting in changes to biochemical indicators in the plasma and energy metabolism, adversely impacting animal welfare, production yield, and meat quality [5,6]. There has been a growing concern for animal welfare like stunning methods and stocking density, which are now recognized as critical elements for ensuring food safety and fillet quality. The interval preceding slaughter, during which fish are removed from the water, induces stress in the fish. Inadequate slaughter methods can either expedite or prolong the resolution of rigor mortis in fish, contingent upon their stress level. The 3-min anoxia approach aligns with the guidelines of [7]. The stocking density during aquaculture and slaughter procedures are key aspects of fish welfare, which could also affect the quality of fish fillets. However, assessing fish welfare proves challenging given the lack of consensus regarding pain receptors and anxiety in fish. Additionally, there is insufficient research aimed at establishing a correlation between fillet quality and fish welfare [8]. Furthermore, factors such as handling and rearing conditions including transportation, water temperature and quality, and stocking density serve as indicators of fish welfare, which in turn affect the quality of fish fillets [9]. Increasing stocking density can enhance fish productivity; however, it is associated with reduced fish welfare. Given the influence of stocking density on fillet quality, assessing fillet quality can offer a valuable understanding of both fish welfare and the economic conditions of fish rearing, thereby enhancing the quality of the final product.

Poor slaughtering procedures can elevate stress levels in fish, hastening the onset of rigor mortis [10]. The stress that animals undergo before slaughter can elevate lactic acid production and diminish water-holding capacity (WHC), promoting protein degradation, lipid oxidation, and microbial growth [11]. The rapid decline in pH after slaughtering causes the denaturation of myosin protein and alters its solubility to an insoluble state.

Seafood products are highly susceptible to microbial and enzymatic activity during postmortem conditions [12]. The quality of seafood products may be negatively impacted by protein and lipid oxidation, as along with endogenous enzymatic activity, leading to a decrease in their value [13]. Previous studies have examined the effect of slaughtering stress on the onset of lipid oxidation in African catfish (*Clarias gariepinus*) and gilthead seabream (*Sparus aurata*) [14,15]. In marine organisms, various environmental stressors trigger the generation of reactive oxygen species (ROS); with stressors stimulating their production. This can lead to an enhanced antioxidant capacity in fish when exposed to abnormal culture conditions. The in-vivo systems possess cellular defense mechanisms such as glutathione peroxidase (GPx) and superoxide dismutase (SOD) to remove the accumulation of ROS. For instance, glutathione peroxidase catalyzes peroxide reduction by oxidizing glutathione (GSH) to oxidized glutathione (GSSG).

The purpose of this study was to investigate the effects of fish welfare indices, such as stocking density and pre-slaughter procedures, on the quality of fillet and protein profile of largemouth bass obtained from fish intensively reared under different constant stocking densities. Additionally, the study aimed to establish a relationship between pre-slaughter conditions and fillet quality.

2. Materials and Methods

2.1. Ethical Statement

The University of South Bohemia authorized all animal procedures and licenses were provided to Nos. 58672/2020-MZE-18134 and Nos. 33446/2020-MZE-18134 under the NAZV QK22020144 project.

2.2. Fish Rearing under Different Densities

Largemouth bass (*Micropterus salmoides*) juveniles with a total length (TL) of 35–45 mm and body weight (BW) of 0.37–0.42 g was obtained from the Fishery School in the Czech Republic. These fish were harvested from a traditional pond used for juvenile fish production in early July. The fish were acclimated to the recirculating aquaculture system conditions for four weeks and fed on a commercial feed (Europa 15, 2–3 mm size pellets, Skretting) according to [16]. Also, the optimal environmental conditions described by [3] were applied. The water temperature during the experiment was 25.5 ± 0.8 °C. The experimental fish (TL 235 ± 21.7 mm and BW 163 ± 55.4 g) were reared in the same RAS under three different constant densities (D) of 35 (D35), 50 (D50), and 65 (D65) $kg{\cdot}m^{-3}$, presenting three experimental groups. The fish were cultured in a total of nine tanks (each density group had three replications) for three months to reach a marketable size of TL = 388 ± 39.7 mm and

BW = 351 ± 72.3 g. Prior to sampling, the fish underwent a fasting period and were given a 48-h purification period in tanks with a temperature of 26 °C and 96% oxygen saturation.

2.3. *Slaughtering Method*

After fasting, 54 experimental fish from each density group (18 fish per replicate) were randomly captured and subjected to an anoxia test by keeping the fish out of water for 3 min [7]. In this way, three experimental anoxia (AN) groups were created: D35AN, D50AN, and D65AN. The fish were subjected to acute antemortem stress through anoxia and percussion on the head, a common practice in fish farming before slaughter. The same number of fish (n = 54) from each stocking density group were killed by direct percussion on the head without air exposure. These three groups were considered control (C) groups (D35C, D50C, and D65C). All the groups (D35AN; D50AN; D65AN; D35C; D50C, and D65C) were handled under the same conditions to avoid additional stress.

2.4. *Sample Collection*

After slaughter, the fish were cleaned in cold water and divided into 108 fillets, with two fillets from each fish, resulting in 54 right and 54 left fillets per group. The fillets collected from each fish were individually packed in plastic bags to avoid drying the fillet surface and stored at 4 °C. Fillet quality was assessed immediately and after 24 and 48 h of refrigerator storage as follows. In total, six left-hand fillets were randomly selected from each experimental group for hardness analysis. On the other hand, six randomly selected right-hand fillets were used for TBARS and proteomics analysis. For analysis of antioxidant capacity (catalase, superoxide dismutase, and glutathione peroxidase), three right-hand fillets were taken from each group immediately after slaughter. All flesh samples were frozen with liquid nitrogen and kept at −80 °C until analysis [17].

In addition, in total, 162 whole fish (27 fish per group) were sampled separately to determine the rigor index and pH. As part of the experiment, three fish were randomly chosen from each group and placed on ice for various durations, namely 0, 6, 9, 12, 24, 36, 48, 60, and 72 h. In this regard, we used the same fish for each time point to measure the pH and rigor index.

After administering anesthesia with 0.03 mL/L clove oil, blood samples were collected from the caudal vein of 18 fish (3 from each group) right after they were slaughtered and before they were filleted. Heparinized syringes were used for this procedure. The plasma samples were used for evaluation of the stress response factors such as glucose and cortisol concentrations.

2.5. *Analytical Procedures*

2.5.1. pH

The pH levels of fish were measured at different time intervals (0, 6, 9, 12, 24, 36, 48, 60, and 72 h) using a pH probe (Testo 206, Lenzkirch, Germany). Three fish per group were tested and the probe was placed behind the head into the higher mass of the fillet. The statistical bars were not included in this graph to ignore complications. However, the p-value was added as a Supplementary Materials file.

2.5.2. Rigor Index

Rigor mortis development was assessed using the tail drop method [16] using 3 whole fish from each group during 72 h of post slaughtering at intervals of 0, 6, 9, 12, 24, 36, 48, 60, and 72 h. The entire fish was stored on ice while measuring the rigor index. The rigor index (Ir) was determined using the following formula: $Ir = [(Lo - Lt)/Lo] \times 100$. Here, L is the vertical drop (in cm) of the tail when half of the fish fork length was placed on the edge of a table. This measurement was taken as a function of time. The initial tail drop is Lo, while measurements throughout the experiment (t = 0–72 h) are represented by Lt. T = 0, indicating the time immediately after slaughter. Statistical bars were excluded to avoid complications. However, the p-value was included as a Supplementary Materials file.

2.5.3. Analysis of Glucose and Cortisol Concentrations

The blood samples collected for biochemical analysis were placed in a microcentrifuge (MPW 55, MPW Instruments, Poland, Ohio) and centrifuged at 1500× *g* for 10 min. The resulting plasma was separated and transferred to Eppendorf tubes, which were kept on ice. The tubes were then stored at −80 °C for subsequent analysis of glucose (GLU) and cortisol (COR) concentrations using a VETTEST 8008 hematology analyzer (IDEXX Laboratories Inc., Westbrook, ME, USA).

2.5.4. Antioxidant Capacity

The catalase activity (CAT; EC 1.11.1.6) was measured by spectrophotometric analysis of H_2O_2 decomposition at 240 nm (n = 6) [18]. The total activity of superoxide dismutase (SOD; EC 1.15.1.1) was measured using a spectrophotometer at a wavelength of 560 nm as described by [19]. The measurement of glutathione peroxidase (GPX; EC 1.11.1.9) involved estimating the reduction in nicotinamide adenine dinucleotide phosphate (NADPH) over time by using reduced glutathione (GSH) as a co-substrate. This was further converted to glutathione disulfide (GSSG) to become GPX. The reduction rate of NADPH was measured by its absorbance at 340 nm [20].

2.5.5. Hardness Analysis

The instrumental hardness analysis was conducted on six left fillets from each group using a texture analyzer called TA-XT Plus, manufactured by Stable Micro Systems, Godalming, UK. The fillet was flattened to half of its initial thickness by using a flat-ended cylinder (10 mm diameter, P/10 type). The cylinder was applied perpendicular to the muscle fibers, below the dorsal fin. The pressure was applied at a speed of 2 mm/s. Hardness was defined based on the greatest force measured during the first compression, represented in grams. To avoid confusion, the statistical bars have been excluded from the graph. However, the *p*-value was added as a Supplementary Materials File.

2.5.6. Thiobarbituric Acid Reactive Substances (TBARS)

The measurement of lipid oxidation was conducted through the thiobarbituric acid reactive substances (TBARS) method, according to [21]. Six right fillets in each group were used for the semi-frozen samples. These samples were chopped, and connective tissues and detectable fat were eliminated. Approximately 1 g of muscle tissue was extracted from each subsample by using an Ultra Turrax from Janke and Kunkel, Staufen, Germany. Then, the samples were homogenized with 9.1 mL (0.61 mol·L^{-1}) of trichloroacetic acid (TCA) solution and 0.2 mL (0.09 mol·L^{-1}) of butylated hydroxytoluene (BHT) in methanol at a speed of approximately 14,000 rpm for 3 sets of 20 s each. The homogenate was passed through Munktell paper (Munktell Filter AB, Grycksbo, Sweden) and filtered. Then, two times, 1.5 mL of the filtrate was moved to new tubes. To the first sample (test sample), 1.5 mL of a solution containing 0.02 mol/L of thiobarbituric acid (TBA) was added, while to the second sample (sample blank), 1.5 mL of water was added. The reaction was allowed to proceed in darkness for 15–20 h at room temperature (20 °C). After that, the reaction complex was detected using a UV–visual spectrophotometer (Specord 210; Analytik, Jena, Germany) at a wavelength of 530 nm, against the sample blank. The amount of TBARS was expressed in terms of malondialdehyde (MDA) (µg/g). In this graph, statistical bars were not included to avoid complications. However, the *p*-value was added as a Supplementary Materials File.

2.5.7. Proteomics Sample Preparation

Muscle Protein Extraction

To minimize protein degradation, almost 100 mg of frozen fish muscle from six right fillets in each group was cut and weighed at −20 °C. The muscle tissue was ground in 50 mM phosphate buffered saline (PBS) solution (500 µL). This solution was prepared with a 0.01 M phosphate buffer concentration and a 0.154 M sodium chloride concentration at pH 7.4. The crude extracts were then transferred to an Eppendorf tube.

SDS–PAGE

Standard protocols conducted sodium dodecyl sulfate-polyacrylamide gel electrophoresis (SDS-PAGE) [22]. For each sample, 20 µL of the solution was mixed with Laemmli sample buffer to attain a final protein concentration of 2 $\mu g \cdot \mu L^{-1}$. The mixture was then heated for 2 min at a temperature of 95 °C. After heating, the samples were loaded onto a 10% Criterion Tris-glycine Gel from Bio-Rad, located in Hercules, CA, USA. The samples were then subjected to electrophoresis by applying a constant electrical potential of 200 V. The protein ladder utilized was the Spectra Multicolor Broad Range (15–220 kDa) from Thermo Scientific, Rockford, IL, USA. The gel was then electrophoresed and stained with 0.5% Coomassie Brilliant Blue G-250 (Bull Korean Chem Soc. 2002, Eschenstr, Germany).

Immunoblotting

For immunoblotting, the DNPH reaction was performed directly on the protein homogenate obtained by centrifugation at 12,600× *g* for 3 min. The supernatant was utilized for analysis after adjusting the protein concentration to 10 mg/mL using a BCA kit from Pierce, Rockford, IL. Protein carbonyls were treated with a mixture of 20 µL of sample (1:2), 12% SDS, 10% TFA, and 10 mM DNPH and incubated at room temperature for 30 min. The reaction was halted by adding 40 µL of neutralization buffer (1:2) containing 2 M Tris-base, 30% glycerol, and 20 mM dithioerythritol (DTE) before being separated through SDS-PAGE. The experiment involved centrifuging the samples for 3 min at a force of 12,600× *g*, followed by loading them onto a 10% Tris-glycine gel (Bio-Rad, Hercules, CA, USA) for SDS-PAGE. The gels were then transferred onto 0.2 µm polyvinylidene difluoride (PVDF) membranes using a Trans-Blot SD semidry transfer cell (Bio-Rad Laboratories, Hercules, CA, USA) at 0.35 A and max 50 V for 60 min. The next step involved blocking the membranes with 5% skim milk in Tris-buffered saline and incubating them overnight at +4 °C with anti-DNP (Sigma Aldrich, Darmstadt, Germany), produced in rabbit, at a 1:16,000 dilution in Tris-buffered saline (TBS). After being washed in TBS, the membranes were incubated with the secondary antibody, horseradish peroxidase-conjugated swine anti-rabbit (DAKO Denmark A/S, Glostrup, Denmark), at a 1:8000 dilution, followed by another wash in TBS. Finally, the blot was developed using an ECL ± kit (Bio-Rad Laboratories, Hercules, CA, USA) and the images were analyzed using Image Lab software 6.1 (Molecular Imager Chemi Doc XRS+, Bio-Rad Laboratories, USA).

2.6. Statistical Analysis

The statistical evaluation was conducted using the Statistica CZ 12 software package version 4.2.2, with one-factor ANOVA followed by Tukey's comparison test ($p < 0.05$). The data are presented as mean ± SD. To ensure normal distribution and homogeneity of variance, the data were verified using Levene's test. General linear models were used to analyze the data and assess the effects of the pre-slaughter handling (anoxia and control) and time, as well as their interactions. A study was conducted to compare the storage efficiencies of rigor index, pH, hardness, and lipid oxidation (TBARS) after the death of fish. A three-way ANOVA was used to analyze the data using the R statistical program version 4.2.2. The study compared the effects of stocking densities (35, 50, and 65 $kg \cdot m^{-3}$) and pre-slaughter procedures (anoxia and control) on the aforementioned factors. For blood samples and antioxidant capacity enzyme analysis, a two-way ANOVA was conducted using the R statistical program version 4.2.2. The stocking densities (35, 50, and 65 $kg \cdot m^{-3}$) and pre-slaughter procedures (anoxia and control) were the two factors considered for the analysis. Before the ANOVA analysis, the normality and homogeneity of the data were checked. A three-way ANOVA p-value table for the data variables was included in Supplementary Materials Table S1.

3. Results

3.1. pH

The pH in filets of the control groups (C) was markedly higher ($p < 0.05$) compared to anoxic (AN) filets regardless of fish stocking density (Figure 1). Furthermore, the pH decreased noticeably in both the control (C) and anoxic (AN) groups, by elapsing time. Moreover, pH was affected by the stocking density in both anoxic (AN) and control (C) groups during 72 h of refrigerator storage.

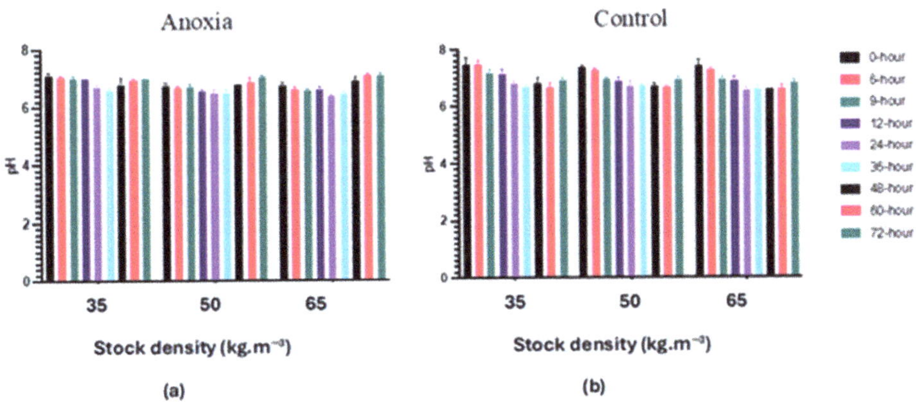

Figure 1. Changes of pH in largemouth bass fillet in anoxia (**a**) and control (**b**) groups during 72 h of refrigerated storage on ice by concerning three different fish stock densities (35, 50, and 65 kg·m^{-3}) storage on ice (mean ± SD; $n = 3$).

3.2. Rigor Mortis Index

The rigor index data following a 72-h stunning period and fish preservation at 4 °C are presented in Figure 2. Anoxia (AN) and control (C) groups had rigor mortis after 6 and 10 h, respectively. According to the results, rigor mortis started earlier in the fillets from fish of the anoxia groups (AN) compared to the control (C) without any effect of culture density. Furthermore, fish exposed to anoxia reached full rigor earlier than the control at all stocking densities. However, the development of the rigor index was unaffected by the fish stocking density.

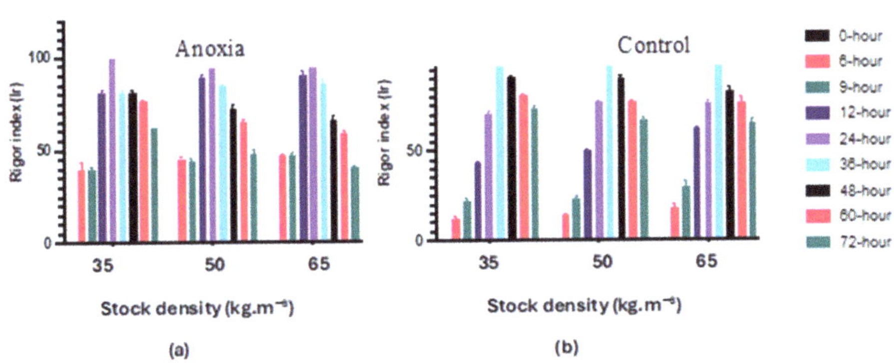

Figure 2. Changes of rigor index in largemouth bass fillets in anoxia (**a**) and control (**b**) groups during 72 h refrigerated storage on ice by concerning 3 different fish stock densities (35, 50 and 65 kg·m^{-3}) (mean ± SD; $n = 3$).

3.3. Blood Glucose and Cortisol

Fish exposed to anoxia showed slightly higher plasma cortisol levels (Figure 3) but they were not significant. Moreover, stocking density did not show a significant impact on the cortisol level. Additionally, glucose concentration (Figure 4) was not affected by different stocking densities and its concentration was significantly higher in fish exposed to anoxia groups (AN) than in the control groups (C).

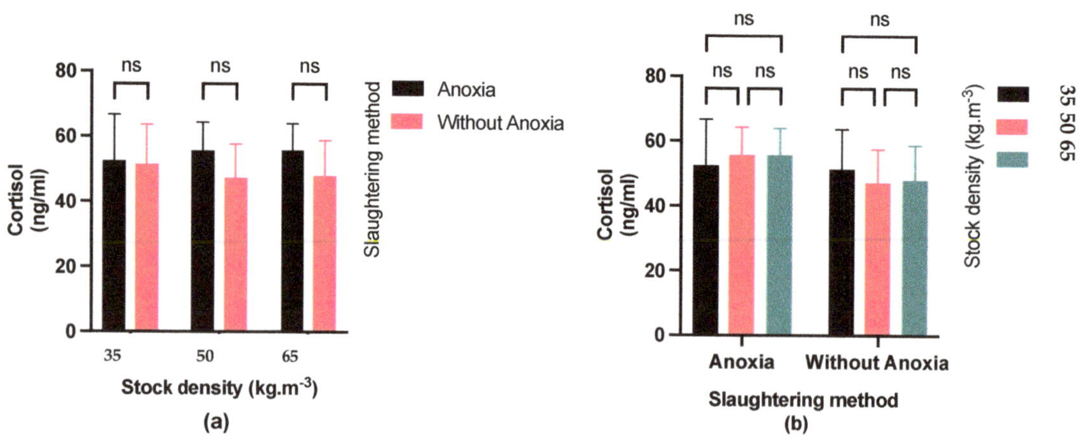

Figure 3. (**a**) Effect of the pre-slaughter procedures (anoxia and without anoxia) in the different fish densities (35, 50, and 65 kg·m^{-3}) on cortisol levels during ($p < 0.05$) multiple comparisons test); (**b**) Effect of stocking densities (35, 50, and 65 kg·m^{-3}) in different pre-slaughter procedures (anoxia and without anoxia groups) on cortisol level; ($p < 0.05$, Tukey's multiple comparisons test). Bar graph showing mean and standard deviation (ns: non-significant).

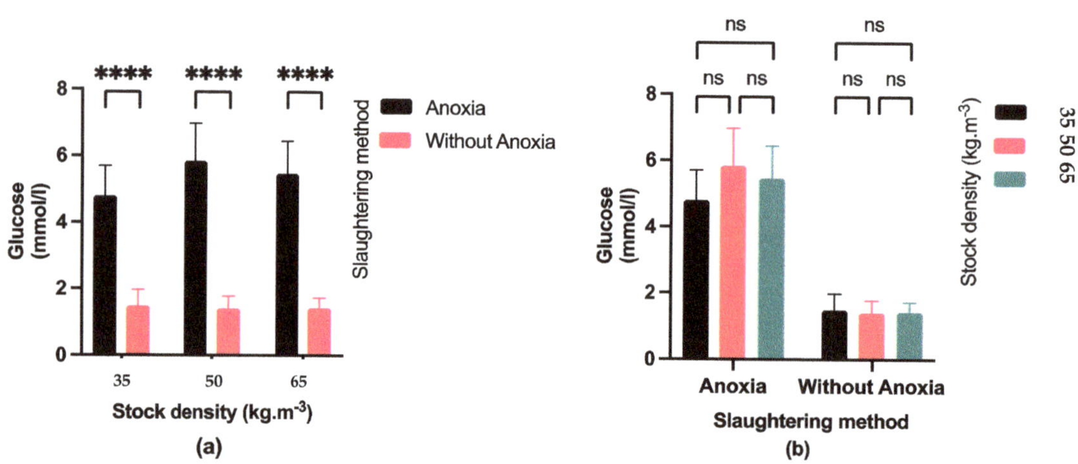

Figure 4. (**a**) Effect of pre-slaughter procedures (anoxia and without anoxia) in different stocking densities (35, 50, and 65 kg·m^{-3}) on glucose level; ($p < 0.05$, multiple comparisons test); (**b**) Effect of stocking densities (35, 50 and 65 kg·m^{-3}) in different pre-slaughter procedures (anoxia and without anoxia groups) on glucose level; ($p < 0.05$, Tukey's multiple comparisons test). Bar graph showing mean and standard deviation (ns: non-significant; **** $p < 0.0001$).

3.4. Antioxidant Enzymes Activity

Fish exposed to anoxia (AN) displayed significantly higher catalase (CAT) activity (Figure 5) compared to the control groups (C). Fish exposed to anoxia showed higher superoxide dismutase (SOD) activity than the control only at a stocking density of 65 kg·m^{-3} (Figure 6). At stocking densities of 50 and 65 kg·m^{-3}, fish exposed to anoxia (AN) had significantly higher glutathione peroxidase (GPX) activity in comparison to the control group (C) (Figure 7). No significant effects were observed on CAT and GPX activities due to stocking density, while SOD activity showed a marked increase at stocking densities of 50 and 65 kg·m^{-3} in comparison to the lowest stocking density.

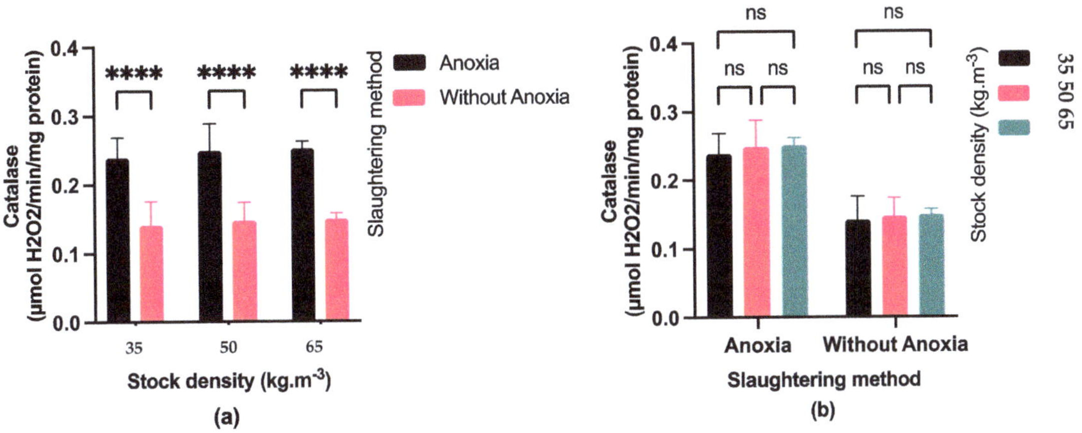

Figure 5. (**a**) Effect of pre-slaughter procedures (anoxia and without anoxia) under different stocking densities (35, 50 and 65 kg·m^{-3}) on catalase activity; ($p < 0.05$, multiple comparisons test); (**b**) Effect of stocking densities (35, 50 and 65 kg·m^{-3}) in different pre-slaughter procedures (anoxia and without anoxia groups) on catalase activity; ($p < 0.05$, Tukey's multiple comparisons test). Bar graph showing mean and standard deviation (ns: non-significant; **** $p < 0.0001$).

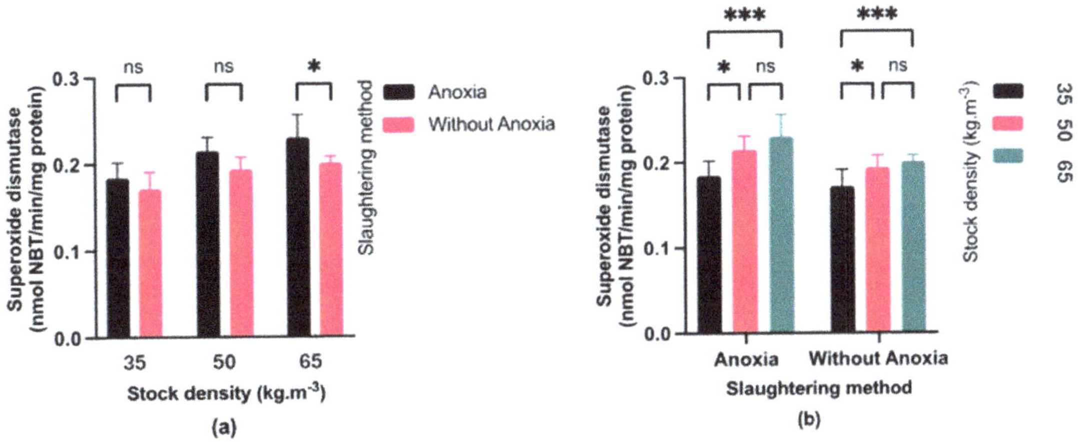

Figure 6. (**a**) Effect of pre-slaughter procedures (anoxia and without anoxia) under different stocking densities (35, 50, and 65 kg·m^{-3}) on superoxide dismutase activity; ($p < 0.05$, multiple comparisons test); (**b**) Effect of stocking densities (35, 50 and 65 kg·m^{-3}) in different pre-slaughter procedures (anoxia and without anoxia groups) on superoxide dismutase activity; ($p < 0.05$, Tukey's multiple comparisons test). Bar graph showing mean and standard deviation (ns: non-significant; * $p < 0.05$; *** $p < 0.001$).

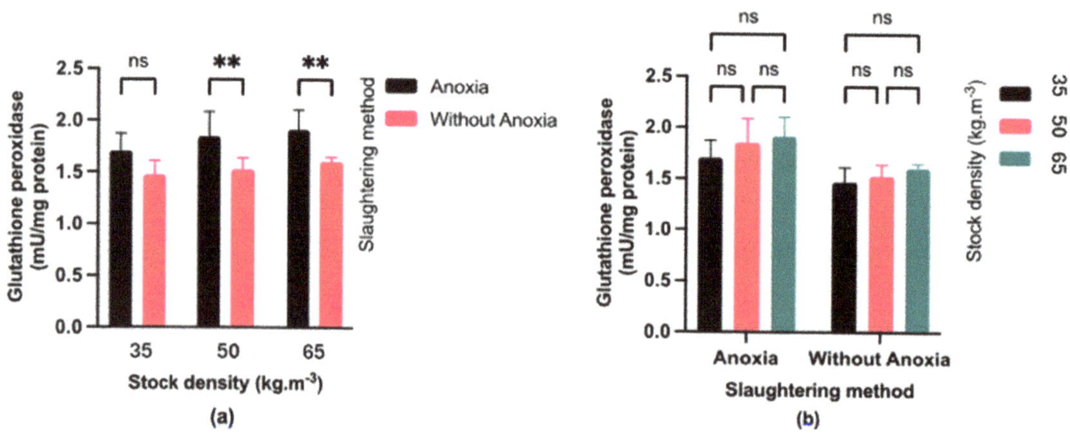

Figure 7. (**a**) Effect of pre-slaughter procedures (anoxia and without anoxia groups) on glutathione peroxidase activity during different stocking densities (35, 50 and 65 kg·m^{-3}); ($p < 0.05$, multiple comparisons test); (**b**) Effect of stocking densities (35, 50 and 65 kg·m^{-3}) on glutathione peroxidase activity in different pre-slaughter procedures (anoxia and without anoxia groups); ($p < 0.05$, Tukey's multiple comparisons test). Bar graph showing mean and standard deviation (ns: non-significant; ** $p < 0.01$).

3.5. Hardness

The hardness of the control fillets (Group C) was significantly higher ($p < 0.05$) than the anoxic fillets (Group AN) after 48 h of storage (Figure 8). Furthermore, we observed a significant declining trend in all groups during the storage time. As the stocking density increased, the fillet hardness decreased gradually, especially in the anoxia (AN) group.

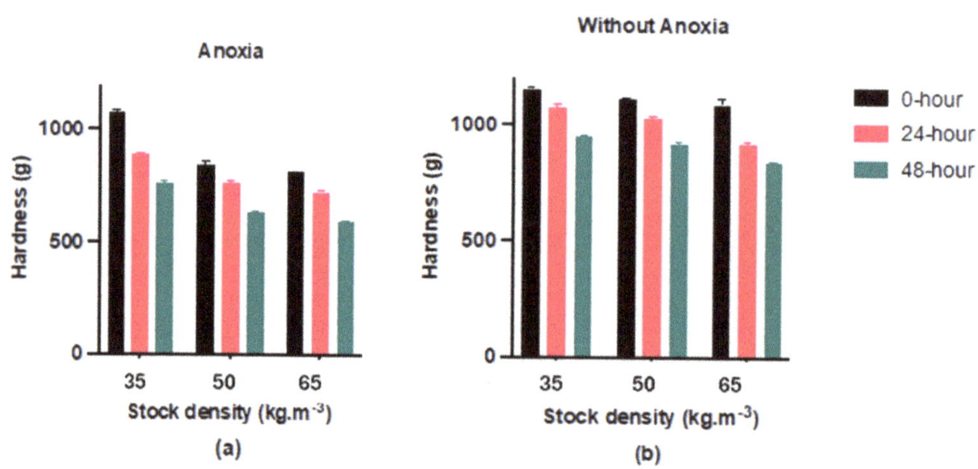

Figure 8. Hardness changes in largemouth bass fillet in anoxia (**a**) and without anoxia (**b**) groups during 48 h refrigerated storage at +4 °C with intervals 24 h by concerning three different fish stock densities (35, 50, and 65 kg·m^{-3}) (mean ± S.D., $n = 6$).

Regarding the effect of pre-slaughter handling, the control group showed a higher level of hardness compared to the anoxia group. This study confirmed that the control group had less protein and lipid oxidation development, as demonstrated by the Western blot and TBARs assay. Therefore, the higher oxidation development in the anoxia (AN) group could be related to the lower hardness.

3.6. Thiobarbituric Acid Reactive Substances (TBARS)

MDA concentration was significantly higher in the muscles of fish exposed to anoxia (AN) compared to the control groups (C) (Figure 9). Meanwhile, in fish exposed to anoxia (AN), the MDA concentration increased significantly after 12 h and then plateaued. Regarding the control group (C), the MDA concentration increased considerably after 24 h. Furthermore, MDA increased significantly during 48 h of storage at 4 °C in all treatments. The oxidation development was affected by different stocking densities in the fish exposed to anoxia (AN), where higher densities promoted lipid oxidation development.

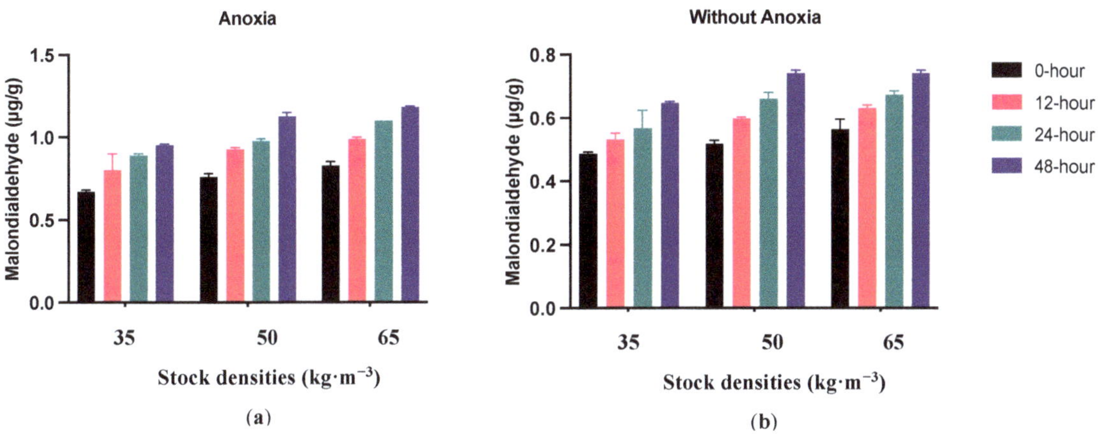

Figure 9. TBARS changes in largemouth bass fillet after anoxia (**a**) and without anoxia (**b**) groups during 48 h of refrigerated storage at +4 °C with by concerning three different fish stock densities (35, 50 and 65 kg·m^{-3}) (mean ± S.D., n = 6).

3.7. Protein Changes during Postmortem

Several bands between 15 and 220 kDa were found in the protein profile of largemouth bass fillets (Figure 10A). The bands may be attributed to specific components such as myosin heavy chain (MHC) at 200 kDa, nebulin (107 kDa), actin (43 kDa), troponin (30 kDa), and myosin light chain components (25–15 kDa) [23]. Consistent protein patterns were observed among all groups except for the control group at 35 kg·m^{-3}, which displayed an intense MHC on SDS-PAGE. The protein carbonyl groups with respect to Western blot (immunoblot) among all groups during 48 h storage at 4 °C (Figure 10B) showed an increasing trend in band intensity by elapsing time at approximately 43 kDa, which indicates actin protein degradation. Additionally, less oxidized carbonyl was observed in the control group (C) than in the group subjected to anoxia (AN). Furthermore, protein carbonyl oxidation was affected by density. Regardless of the stunning method, we observed a more intense band by increasing the stock density.

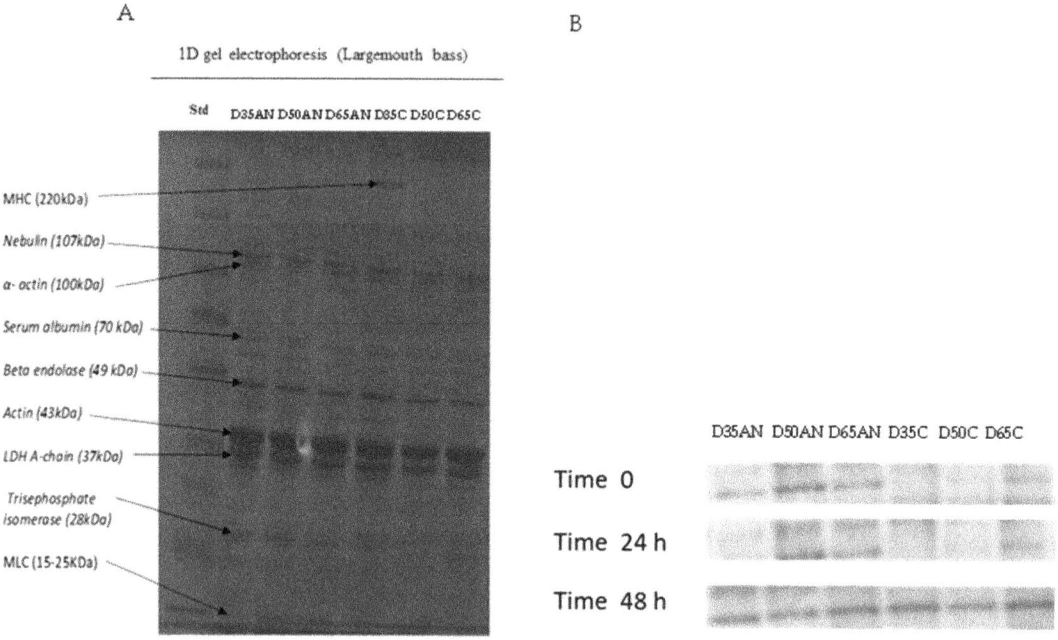

Figure 10. (**A**) SDS-polyacrylamide gel electrophoresis and (**B**) immunoblotting against protein carbonyl in largemouth bass (*Micropterus salmoides*) refrigerated storage fillet under +4 °C from fish groups cultured under three different fish densities, 35; 50; and 65 kg·m^{-3}, subjected to anoxia and control conditions before the slaughtering. SD35C, SD50C, SD65C (C = Control), SD35AN, SD50AN, and SD65AN (AN = Anoxia).

4. Discussion

Fish exposed to anoxia struggled more than control fish, according to our observations made during the slaughter period. This activity triggered earlier rigor mortis in anoxic fish as well as lower initial postmortem pH [24]. The result of our study is in accordance with the previous studies [25,26]. In contrast with our findings, ref. [10] did not observe any differences in pH during 3- and 6-min anoxia in Nile tilapias. Additionally, stress during slaughter can result in a higher ultimate pH or a faster pH decline after death [27]. The animal's pre-slaughter reactions can change the muscle's energy metabolism, significantly impacting the final muscle quality. For instance, lower pH and muscle energy are associated with exhaustion before slaughtering [28]. In addition, the reduction in muscle pH might be associated with the generation of H^+ ions resulting from ATP degradation [29]. The pH decreased significantly until 24 h due to lactic acid formation and then, after 36 h, increased slightly owing to ammonia formation. Higher stocking density, on the other hand, is linked to more stress during rearing, which may cause lower levels of muscle glycogen and higher ultimate pH [27]. Fish subjected to high densities may exhibit increased struggle, leading to a faster depletion of glycogen stores. Glycogen plays a key role in fish muscle function, and its depletion can impact postmortem pH regulation. One possible reason for the observed phenomenon could be linked to the chronic stress experienced by the sampled fish. This stress could have been caused by various factors, including poor water quality and high stocking densities. Therefore, higher stocking density in both groups showed lower initial muscle pH compared to the lower density.

According to our results, the onset of rigor mortis started after 6 and 12 h in the fish exposed to anoxia and in the control, respectively. The results indicate that rigor mortis started earlier under anoxic conditions [30]. Ref. [31] revealed that in fish exposed to stressful conditions, rigor mortis starts earlier compared to unstressed fish. The delay of rigor mortis onset in control fish compared to anoxic fish depends on ATP decline. Because of limited glycogen reserves in fish muscle exposed to anoxia, postmortem analysis started earlier than that in the control group. Most likely, fish are exhausted when exposed to anoxic conditions, leading them to enter rigor mortis more rapidly [8,32]. The observed drop in pH, linked directly to the greater rigor mortis loss, suggests a clear correlation between these factors and the final fillet quality.

Assessment of fish stress commonly involves measuring cortisol levels and blood glucose concentration. When fish are stimulated, they produce hormones like cortisol, which increases the synthesis of glucose [33]. According to our results, exposing largemouth bass to anoxia led to raised blood cortisol levels. Moreover, a similar increasing tendency was observed for the plasma glucose level. This finding is in line with the results obtained in Channel catfish and Asian seabass [34,35]. An alteration in blood cortisol levels is considered the principal response to stress in the body [36]. The findings of our study indicated that regardless of stocking density, anoxia might induce stress in fish [36]. Cortisol activates the central nervous system by enhancing glucose synthesis via various metabolic pathways, such as the breakdown of muscle and liver glycogen reserves. In this context, anoxia has been observed to cause stress and physiological alterations in largemouth bass, as evidenced by changes in blood glucose and cortisol levels. During anoxia, fish undergo oxygen deprivation, leading to a metabolic shift from aerobic to anaerobic pathways. Fish resort to anaerobic glycolysis to produce energy since they cannot engage in aerobic respiration in the absence of oxygen. This process involves breaking down glycogen into glucose to produce ATP for cellular processes. As a result, blood glucose levels in fish can increase during anoxia since glycogen stores are mobilized to meet the energy demands of tissues [37,38]. As a result, the highest blood glucose level in the fish decreased, leading to a decrease in its utilization for biochemical reactions in the body after 3 min of pre-slaughter anoxia. Increased glucose production can assist fish in coping with the energy demands caused by stress. Fish release cortisol, a stress hormone, in response to various stressors, including anoxia. Cortisol helps regulate metabolism and energy mobilization during stress responses [37]. The fish reared at higher density may experience an increase in serum glucose and cortisol levels due to energy metabolism through gluconeogenesis and glycogenolysis pathways [38]. This augmentation in energy would help the fish cope with stressful situations, such as high stocking density. Nonetheless, raising stocking densities did not result in elevated blood cortisol and glucose levels, probably suggesting that fish in the current study acclimated to the density during the rearing period.

Rapid pH declines after death led to the build-up of muscle ROS and increased levels of antioxidant enzymes such as SOD, CAT, and GSH-Px [39]. Due to the activity of the antioxidant defense system during adaptation time, we observed that the SOD, CAT, and GPx activities increased under anoxic conditions in comparison to the control group. This condition implies that the antioxidant defense system may play a vital role in facilitating an adaptive mechanism to withstand anoxia. The activation of the antioxidant defense system may also be linked to the molecular mechanisms involved in oxygen detection and the associated transduction pathways that govern intermediate metabolism during anoxia [40].

Moreover, elevated stress levels during rearing were associated with increased stocking densities, leading to higher SOD activity at 50 and 65 $kg{\cdot}m^{-3}$ stocking densities compared to 35 kg·m−3 stocking densities. In agreement with our findings, [41] indicated higher SOD and MDA levels in fillets of blunt snout bream (*Megalobrama amblycephala*) reared at higher stocking density. Higher stocking density in aquaculture can lead to hypoxia [42], which triggers antioxidant defense. However, in our study stocking density did not affect CAT and GPX due to the lower level of antioxidant defense.

The reduction in hardness observed across all groups following storage is likely attributed to enzymatic protein degradation, as documented in prior studies [15,43,44]. Our study showed that lower stocking densities of 35 and 50 $kg{\cdot}m^{-3}$ produced firmer fillets compared to the higher density of 65 $kg{\cdot}m^{-3}$, suggesting that fish welfare during rearing impacts the ultimate quality of fillets [15]. However, ref. [45] reported no significant difference in hardness across all stocking densities. The present study found that the hardness decreased in all samples, which was consistent with previous research [46]. We hypothesize that fish reared under stocking densities of 35 and 50 (kg·m^{-3}) may experience greater opportunities for vigorous swimming, a factor that could favorably influence the texture. Thus, increasing stocking density is contrary to fish welfare and has an adverse influence on fillet quality. According to [47], pre-slaughter stressors like crowding, noisy environments, and chasing led to an increase in the expression and activity of cathepsin B and L in farmed Atlantic salmon fillets. This increase in cathepsin activity can result in faster degradation of muscle proteins and negatively impact the texture of the fillets. Lower muscle hardness negatively impacts the economic value of fish by making them unsuitable for industrial processes or human consumption [48]. It has been observed that the fish stocked at higher densities displayed lower flesh quality. This occurred as a result of alterations in muscle structure, including a reduction in the area of muscular bundles and an increase in intramuscular connective tissue [34]. It is suggested that high stocking densities have negative effects on the quality of largemouth bass flesh.

In the current study, the consequence of anoxia was the acceleration of lipid oxidation, as confirmed by the elevated MDA concentration in the fillets of fish exposed to anoxia. Our speculation is that this could be associated with the excessive build-up of MDA. MDA can cause cytotoxicity and protein denaturation by cross-linking with proteins, nucleic acids, and amino phospholipids [49]. Fish exposed to anoxia demonstrated higher levels of lipid oxidation compared to the control group. In agreement with our result, ref. [50] reported a relationship between pre-slaughter stress and lipid oxidation in rainbow trout (*Oncorhynchus mykiss*). This could be associated with inadequate ATP production and a rapid decrease in pH levels [51,52]. Moreover, the higher pH level in anoxia fillets when compared to control fillets could serve as an alternate reason for reducing the formation of met-Hb and Hb-deoxygenation, while also minimizing the progression of lipid oxidation. Likewise, the higher lipid oxidation level under anoxic conditions in this experiment could be associated with the higher level of superoxide radicals or hydroxyl radicals.

Additionally, lipid oxidation development started earlier in anoxia than in the control group. In this regard, the results indicated that pre-slaughtering stressful situations can speed up the development of oxidation [53]. The higher presence of lipid-oxygenated products in the exposed fish fillet led to the progression of lipid oxidation during post-mortem when compared to the control group. It might be one of the reasons that lipid oxidation started earlier in exposed fish than in the control group. Thus, severe stressful circumstances or stress duration may render the adaptive response ineffective, increasing lipid oxidation.

Moreover, lipid oxidation was affected by stocking density in the anoxia group. [41] found that fillets of blunt-snout bream (*Megalobrama amblycephala*) reared at higher stocking density had higher levels of SOD and MDA, which is consistent with our findings. Higher aquaculture stocking densities have the potential to induce hypoxia, which activates the antioxidant defense system [42].

During the refrigerator storage period, the protein patterns were analyzed in all groups by using SDS-PAGE to identify any protein alterations. We confirmed the impact of stocking density on protein profile by indicating MHC at a density of 35 kg·m^{-3} in the control group. Proteolysis is likely responsible for the susceptibility of MHC due to the formation of cross-linked proteins [54]. It appears that MHC was negatively impacted by high stocking density and anoxia in this study. According to [15], stocking density has a significant impact on the quality of African catfish fillets during storage. The SDS-PAGE result indicated the degradation of MHC in all groups except 35 kg·m^{-3}.

Fish fillets exposed to anoxia had higher protein carbonyls than control fillets after 48 h of storage, as confirmed by Western blotting. The exposure of fish to anoxia resulted in higher protein carbonyls than control fillets. Protein carbonyls were more intense in anoxia fillets compared to control fillets during 48 h storage, which may be due to actin protein oxidation. Protein oxidation caused by metal-catalyzed cleavages is a significant form of oxidative damage during storage time [55]. During the postmortem period, protein degradation mainly occurs due to calpain activity [56]. Moreover, higher levels of lipid oxidation products can lead to increased protein oxidation (through metal catalyzation) in anoxia fillets, which may result in poorer textural parameters (hardness) compared to control fillets with less oxidized protein. It is reasonable to assume that even a slight degradation of actin can weaken the myofibrillar lattice, which can affect the texture of meat. The extent of the effect depends on where the thin filament's actin is degraded. In agreement with our result, [15] reported higher protein oxidation confirmed by Western blot during postmortem storage of African catfish. We assume that at higher stocking densities, MDA can combine with the nucleophilic groups of proteins, leading to protein denaturation. On the other hand, when taking the stocking density into account, we noticed a lesser protein oxidation at lower stocking densities. According to [57], exercise results in more rigid connective tissue leading to a firmer texture. We propose that fish stocked at the density of 35 kg/m^{-3} may have greater opportunities for intense swimming, resulting in better texture and lower protein oxidation.

5. Conclusions

Recent research has confirmed that anoxia and stocking density significantly impact the ultimate fillet quality. Due to the increased stress from the fish exposed to anoxia, higher levels of lipid oxidation in their fillets were observed when compared to the control group. In this investigation, the influence of stress on the progression of protein oxidation was confirmed by Western blot analysis. In addition, the control group showed lower levels of antioxidant defense, blood cortisol, and glucose, as well as greater levels of hardness and pH than the anoxia group. We observed lower hardness and higher oxidation progress in the higher stocking density of the anoxic group. The present study confirmed the role of pre-slaughtering handling on the final fillet quality.

Supplementary Materials: The following supporting information can be downloaded at: https://www.mdpi.com/article/10.3390/foods13101477/s1, Table S1: Three-Way ANOVA *p*-value table for the data variables.

Author Contributions: Conceptualization, N.H. and T.P.; data curation, N.H., formal analysis, N.H., S.R. and S.G.W.; funding acquisition, T.P.; investigation, T.P.; methodology, T.P., O.M. and N.H.; project administration, T.P.; supervision, T.P.; writing—original draft, N.H.; writing—review and editing; N.H., S.R. and T.P. All authors have read and agreed to the published version of the manuscript.

Funding: The study was financially supported by the Ministry of Agriculture of the Czech Republic, NAZV project QK22020144.

Institutional Review Board Statement: The University of South Bohemia authorized all animal procedures, and licenses were provided to Nos. 58672/2020-MZE-18134 and Nos. 33446/2020-MZE-18134 under the NAZV QK22020144 project.

Informed Consent Statement: Not applicable.

Data Availability Statement: The original contributions presented in the study are included in the article and Supplementary Materials, further inquiries can be directed to the corresponding author.

Acknowledgments: The Ministry of Agriculture of the Czech Republic financially supported the study under the NAZV project QK22020144. Samad Rahimnejad was supported financially by a Maria Zambrano contract of University of Murcia (UMU) within the framework of the programme for the requailification of the Spanish university system (Ministry of Universities) during the period 2021–2023 funded by the European Union-Next GenerationEU.

Conflicts of Interest: The authors declare no real or perceived conflicts of interest.

Abbreviations

SD35C, SD50C, SD65C (C = Control), SD35AN, SD50AN and SD65AN (AN = Anoxia). (SDS–PAGE), sodium dodecyl sulfate–polyacrylamide gel electrophoresis; PVDF, polyvinylidene difluoride;

WHC, water holding capacity; TBARS, thiobarbituric acid reactive substances; RAS, recirculating aquaculture systems; SDS, sodium dodecyl sulfate; TFA, trifluoroacetic acid; DTE, dithioerythritol; DNPH, 2,4-dinitrophenylhydrazine; MDA, Malondialdehyde; ATP, adenosine triphosphate.

References

1. Wang, D.; Yao, H.; Li, Y.-H.; Xu, Y.-J.; Ma, X.-F.; Wang, H.-P. Global diversity and genetic landscape of natural populations and hatchery stocks of largemouth bass micropterus salmoides across American and Asian regions. *Sci. Rep.* **2019**, *9*, 16697. [CrossRef] [PubMed]
2. Wang, Y.; Ni, J.; Nie, Z.; Gao, J.; Sun, Y.; Shao, N.; Li, Q.; Hu, J.; Xu, P.; Xu, G. Effects of stocking density on growth, serum parameters, antioxidant status, liver and intestine histology and gene expression of largemouth bass (Micropterus salmoides) farmed in the in-pond raceway system. *Aquac. Res.* **2020**, *51*, 5228–5240. [CrossRef]
3. Pěnka, T.; Malinovskyi, O.; Křišťan, J.; Imentai, A.; Policar, T. Effect of density and mixed culture of largemouth bass (*Micropterus salmoides*) and pikeperch (*Sander lucioperca*) on growth, survival and feed conversion rate in intensive culture. *Czech J. Anim. Sci.* **2021**, *66*, 428–440. [CrossRef]
4. Olafsdottir, G.; Nesvadba, P.; Di Natale, C.; Careche, M.; Oehlenschläger, J.; Tryggvadóttir, S.a.V.; Schubring, R.; Kroeger, M.; Heia, K.; Esaiassen, M.; et al. Multisensor for fish quality determination. *Trends Food Sci. Technol.* **2004**, *15*, 86–93. [CrossRef]
5. Adzitey, F. Effect of pre-slaughter animal handling on carcass and meat quality. *Int. Food Res. J.* **2011**, *18*, 485–491.
6. Warner, R.; Ferguson, D.; Cottrell, J.; Knee, B.W. Acute stress induced by the preslaughter use of electric prodders causes tougher beef meat. *Aust. J. Exp. Agric.* **2007**, *47*, 782–788. [CrossRef]
7. Castro, P.; Lewandowski, V.; Souza, M.; Coradini, M.; Alexandre, A.; Sary, C.; Ribeiro, R. Effect of different periods of pre-slaughter stress on the quality of the Nile tilapia meat. *Food Sci. Technol.* **2016**, *37*, 52–58. [CrossRef]
8. Daskalova, A. Farmed fish welfare: Stress, post-mortem muscle metabolism, and stress-related meat quality changes. *Int. Aquat. Res.* **2019**, *11*, 113–124. [CrossRef]
9. Gatica, M.C.; Monti, G.; Knowles, T.G.; Gallo, C. Effects of crowding on blood constituents and flesh quality variables in Atlantic salmon (*Salmo salar* L.). *Arch. Med. Vet.* **2010**, *42*, 187–193. [CrossRef]
10. Zuanazzi, J.S.G.; Lara, J.A.F.d.; Goes, E.S.d.R.; Almeida, F.L.A.d.; Oliveira, C.A.L.d.; Ribeiro, R.P. Anoxia stress and effect on flesh quality and gene expression of tilapia. *Food Sci. Technol.* **2019**, *39*, 195–202. [CrossRef]
11. Tornberg, E.; Wahlgren, M.; Brøndum, J.; Engelsen, S. Pre-rigor conditions in beef under varying temperature- and pH-falls studied with rigometer, NMR and NIR. *Food Chem.* **2000**, *69*, 407–418. [CrossRef]
12. Fu, L.; Zhang, S.; Zhou, J.; Liu, C.; Lin, J.; Wang, Y. Alterations of Protein Expression in the Muscle of Pacific White Shrimp (*Litopenaeus vannamei*) Contribute to Postmortem Changes. *J. Shellfish Res.* **2014**, *33*, 815–823. [CrossRef]
13. Mei, J.; Ma, X.; Xie, J. Review on Natural Preservatives for Extending Fish Shelf Life. *Foods* **2019**, *8*, 490. [CrossRef] [PubMed]
14. Tejada, M.; Huidobro, A. Quality of farmed gilthead seabream (*Sparus aurata*) during ice storage related to the slaughter method and gutting. *Eur. Food Res. Technol.* **2002**, *215*, 1–7. [CrossRef]
15. Hematyar, N.; Imentai, A.; Křišťan, J.; Gorakh Waghmare, S.; Policar, T. Considering Two Aspects of Fish Welfare on African Catfish (*Clarias gariepinus*) Fillet throughout Postmortem Condition: Efficiency and Mechanisms. *Foods* **2022**, *11*, 4090. [CrossRef] [PubMed]
16. Stara, A.; Kristan, J.; Zuskova, E.; Velisek, J. Effect of chronic exposure to prometryne on oxidative stress and antioxidant response in common carp (*Cyprinus carpio* L.). *Pestic. Biochem. Physiol.* **2013**, *105*, 18–23. [CrossRef] [PubMed]
17. Bito, M. Studies on rigor mortis of fish. I. Defference in the mode of rigor mortis among some varieties of fish by modified Cutting's method. *Bull. Tokai Reg. Fish. Res. Lab.* **1983**, *109*, 89–96.
18. Beers, R.F.; Sizer, I.W. A spectrophotometric method for measuring the breakdown of hydrogen peroxide by catalase. *J. Biol. Chem.* **1952**, *195*, 133–140. [CrossRef] [PubMed]
19. Marklund, S.; Marklund, G. Involvement of the superoxide anion radical in the autoxidation of pyrogallol and a convenient assay for superoxide dismutase. *Eur. J. Biochem.* **1974**, *47*, 469–474. [CrossRef]
20. Flohé, L.; Günzler, W.A. Assays of glutathione peroxidase. In *Methods in Enzymology*; Elsevier: Amsterdam, The Netherlands, 1984; Volume 105, pp. 114–120.
21. Miller, B.C.; Lau, H.W.; Tyler, N.E.; Cottam, G.L. Liver Composition and Lipid-Metabolism in NZB/W F1 Female Mice Fed Dehydroisoandrosterone. *Biochim. Biophys. Acta* **1988**, *962*, 25–36. [CrossRef]
22. Laemmli, U.K.; Eiserling, F.A. Studies on Morphopoiesis of Head of Phage T-even V. formation of polyheads. *Mol. Gen. Genet.* **1968**, *101*, 333–345. [CrossRef]
23. Seki, N.; Watanabe, T. Changes in morphological and biochemical properties of the myofibrils from carp muscle during postmortem storage [*Cyprinus carpio*]. *Bull.-Jpn. Soc. Sci. Fish.* **1982**, *48*, 517–524. [CrossRef]
24. Terlouw, C.; Picard, B.; Deiss, V.; Berri, C.; Hocquette, J.-F.; Lebret, B.; Lefèvre, F.; Hamill, R.; Gagaoua, M. Understanding the Determination of Meat Quality Using Biochemical Characteristics of the Muscle: Stress at Slaughter and Other Missing Keys. *Foods* **2021**, *10*, 84. [CrossRef]
25. Hultmann, L.; Phu, T.M.; Tobiassen, T.; Aas-Hansen, Ø.; Rustad, T. Effects of pre-slaughter stress on proteolytic enzyme activities and muscle quality of farmed Atlantic cod (*Gadus morhua*). *Food Chem.* **2012**, *134*, 1399–1408. [CrossRef] [PubMed]

26. Kristoffersen, S.; Tobiassen, T.; Steinsund, V.; Olsen, R.L. Slaughter stress, postmortem muscle pH and rigor development in farmed Atlantic cod (*Gadus morhua* L.). *Int. J. Food Sci. Technol.* **2006**, *41*, 861–864. [CrossRef]
27. Barrasso, R.; Ceci, E.; Tufarelli, V.; Casalino, G.; Luposella, F.; Fustinoni, F.; Dimuccio, M.M.; Bozzo, G. Religious slaughtering: Implications on pH and temperature of bovine carcasses. *Saudi J. Biol. Sci.* **2022**, *29*, 2396–2401. [CrossRef]
28. Lefevre, F.; Cos, I.; Pottinger, T.G.; Bugeon, J. Selection for stress responsiveness and slaughter stress affect flesh quality in pan-size rainbow trout, *Oncorhynchus mykiss*. *Aquaculture* **2016**, *464*, 654–664. [CrossRef]
29. Strasburg, G.; Xiong, Y.L.; Chiang, W. Physiology and chemistry of edible muscle tissues. In *Fennema's Food Chemistry*; CRC Press: Boca Raton, FL, USA, 2007; pp. 935–986.
30. Wilkinson, R.J.; Paton, N.; Porter, M.J.R. The effects of pre-harvest stress and harvest method on the stress response, rigor onset, muscle pH and drip loss in barramundi (Lates calcarifer). *Aquaculture* **2008**, *282*, 26–32. [CrossRef]
31. Roth, B.; Moeller, D.; Veland, J.O.; Imsland, A.; Slinde, E. The Effect of Stunning Methods on Rigor Mortis and Texture Properties of Atlantic Salmon (*Salmo salar*). *J. Food Sci.* **2002**, *67*, 1462–1466. [CrossRef]
32. Wu, H.; Arai, M.; Ohnuki, H.; Yoshiura, Y.; Endo, H. Development of a Flow Injection Biosensor System Enables Glucose and Cortisol Simultaneous Measurement for the Evaluation of Fish Stress. In *Electrochemical Society Meeting Abstracts*; The Electrochemical Society, Inc.: Pennington, NJ, USA, 2016.
33. Sadhu, N.; Sharma, S.K.; Joseph, S.; Dube, P.; Philipose, K. Chronic stress due to high stocking density in open sea cage farming induces variation in biochemical and immunological functions in Asian seabass (*Lates calcarifer*, Bloch). *Fish Physiol. Biochem.* **2014**, *40*, 1105–1113. [CrossRef]
34. Refaey, M.M.; Li, D.; Tian, X.; Zhang, Z.; Zhang, X.; Li, L.; Tang, R. High stocking density alters growth performance, blood biochemistry, intestinal histology, and muscle quality of channel catfish Ictalurus punctatus. *Aquaculture* **2018**, *492*, 73–81. [CrossRef]
35. Ellis, T.; Yildiz, H.Y.; López-Olmeda, J.; Spedicato, M.T.; Tort, L.; Øverli, Ø.; Martins, C.I.M. Cortisol and finfish welfare. *Fish Physiol. Biochem.* **2012**, *38*, 163–188. [CrossRef] [PubMed]
36. O'Connor, E.A.; Pottinger, T.G.; Sneddon, L.U. The effects of acute and chronic hypoxia on cortisol, glucose and lactate concentrations in different populations of three-spined stickleback. *Fish Physiol. Biochem.* **2011**, *37*, 461–469. [CrossRef] [PubMed]
37. Zhang, T.; Zhang, L.; Yin, T.; You, J.; Liu, R.; Huang, Q.; Shi, L.; Wang, L.; Liao, T.; Wang, W.; et al. Recent understanding of stress response on muscle quality of fish: From the perspective of industrial chain. *Trends Food Sci. Technol.* **2023**, *140*, 104145. [CrossRef]
38. Roque, A.; Gras, N.; Rey-Planellas, S.; Fatsini, E.; Pallisera, J.; Duncan, N.; Muñoz, I.; Velarde, A.; Hernandez, M.D. The feasibility of using gas mixture to stun seabream (Sparus aurata) before slaughtering in aquaculture production. *Aquaculture* **2021**, *545*, 737168. [CrossRef]
39. Macedo, A.B.; Moraes, L.H.R.; Mizobuti, D.S.; Fogaça, A.R.; Moraes, F.d.S.R.; Hermes, T.d.A.; Pertille, A.; Minatel, E. Low-Level Laser Therapy (LLLT) in Dystrophin-Deficient Muscle Cells: Effects on Regeneration Capacity, Inflammation Response and Oxidative Stress. *PLoS ONE* **2015**, *10*, e0128567. [CrossRef]
40. Hochachka, P.W.; Land, S.C.; Buck, L.T. Oxygen sensing and signal transduction in metabolic defense against hypoxia: Lessons from vertebrate facultative anaerobes. *Comp. Biochem. Physiol. Part A Physiol.* **1997**, *118*, 23–29. [CrossRef] [PubMed]
41. Yang, Z.; Xu, G.; Ge, X.; Liu, B.; Xu, P.; Song, C.; Zhou, Q.; Zhang, H.; Zhang, W.; Shan, F.; et al. The effects of crowding stress on the growth, physiological response, and gene expression of the Nrf2-Keap1 signaling pathway in blunt snout bream (Megalobrama amblycephala) reared under in-pond raceway conditions. *Comp. Biochem. Physiol. Part A Mol. Integr. Physiol.* **2019**, *231*, 19–29. [CrossRef] [PubMed]
42. Green, B.W.; Rawles, S.D.; Fuller, S.A.; Beck, B.H.; McEntire, M.E. Hypoxia affects performance traits and body composition of juvenile hybrid striped bass (*Morone chrysops* × *M. saxatilis*). *Aquac. Res.* **2016**, *47*, 2266–2275. [CrossRef]
43. Yu, D.; Li, P.; Xu, Y.; Jiang, Q.; Xia, W. Physicochemical, microbiological, and sensory attributes of chitosan-coated grass carp (*Ctenopharyngodon idellus*) fillets stored at 4 °C. *Int. J. Food Prop.* **2017**, *20*, 390–401. [CrossRef]
44. Li, H.; Xu, B. *Advanced Materials Researches and Application*; Trans Tech Publication Ltd.: Zürich, Switzerland, 2013.
45. Çagiltay, F.; Erkan, N.; Ulusoy, Ş.; Selcuk, A.; Özden, Ö. *Effects of Stock Density on Texture-Colour Quality and Chemical Composition of Rainbow Trout (*Oncorhynchus mykiss*)*; AquaDocs: Edmond, OK, USA, 2015.
46. Sigholt, T.; Erikson, U.; Rustad, T.; Johansen, S.; Nordtvedt, T.S.; Seland, A. Handling Stress and Storage Temperature Affect Meat Quality of Farmed-raised Atlantic Salmon (*Salmo salar*). *J. Food Sci.* **1997**, *62*, 898–905. [CrossRef]
47. Bahuaud, D.; Mørkøre, T.; Østbye, T.K.; Veiseth-Kent, E.; Thomassen, M.S.; Ofstad, R. Muscle structure responses and lysosomal cathepsins B and L in farmed Atlantic salmon (*Salmo salar* L.) pre- and post-rigor fillets exposed to short and long-term crowding stress. *Food Chem.* **2010**, *118*, 602–615. [CrossRef]
48. Bjørnevik, M.; Solbakken, V. Preslaughter stress and subsequent effect on flesh quality in farmed cod. *Aquac. Res.* **2010**, *41*, e467–e474. [CrossRef]
49. Trenzado, C.E.; Morales, A.E.; Palma, J.M.; de la Higuera, M. Blood antioxidant defenses and hematological adjustments in crowded/uncrowded rainbow trout (*Oncorhynchus mykiss*) fed on diets with different levels of antioxidant vitamins and HUFA. *Comp. Biochem. Physiol. Part C Toxicol. Pharmacol.* **2009**, *149*, 440–447. [CrossRef] [PubMed]
50. Secci, G.; Serra, A.; Concollato, A.; Conte, G.; Mele, M.; Olsen, R.E.; Parisi, G. Carbon monoxide as stunning/killing method on farmed Atlantic salmon (*Salmo salar*): Effects on lipid and cholesterol oxidation. *J. Sci. Food Agric.* **2016**, *96*, 2426–2432. [CrossRef]

51. Bayır, M.; Bayır, A.; Aras, N.M. A comparison of the effect of long-term starvation on responses to low-temperature stress by juvenile rainbow (*Oncorhynchus mykiss*) and brown (*Salmo trutta*) trout reveal different responses in the two species. *Mar. Freshw. Behav. Physiol.* **2014**, *47*, 239–251. [CrossRef]
52. Pérez-Sánchez, J.; Borrel, M.; Bermejo-Nogales, A.; Benedito-Palos, L.; Saera-Vila, A.; Calduch-Giner, J.A.; Kaushik, S. Dietary oils mediate cortisol kinetics and the hepatic mRNA expression profile of stress-responsive genes in gilthead sea bream (*Sparus aurata*) exposed to crowding stress. Implications on energy homeostasis and stress susceptibility. *Comp. Biochem. Physiol. Part D Genom. Proteom.* **2013**, *8*, 123–130. [CrossRef] [PubMed]
53. Lushchak, V.I.; Lushchak, L.P.; Mota, A.A.; Hermes-Lima, M. Oxidative stress and antioxidant defenses in goldfish *Carassius auratus* during anoxia and reoxygenation. *Am. J. Physiol.-Regul. Integr. Comp. Physiol.* **2001**, *280*, R100–R107. [CrossRef]
54. Soladoye, O.P.; Juárez, M.L.; Aalhus, J.L.; Shand, P.; Estévez, M. Protein Oxidation in Processed Meat: Mechanisms and Potential Implications on Human Health. *Compr. Rev. Food Sci. Food Saf.* **2015**, *14*, 106–122. [CrossRef]
55. Hematyar, N.; Rustad, T.; Sampels, S.; Kastrup Dalsgaard, T. Relationship between lipid and protein oxidation in fish. *Aquac. Res.* **2019**, *50*, 1393–1403. [CrossRef]
56. Liu, P.; Zhang, Z.; Guo, X.; Zhu, X.; Mao, X.; Guo, X.; Deng, X.; Zhang, J. μ-Calpain oxidation and proteolytic changes on myofibrillar proteins from Coregonus Peled in vitro. *Food Chem.* **2021**, *361*, 130100. [CrossRef] [PubMed]
57. Bugeon, J.; Lefevre, F.; Fauconneau, B. Fillet texture and muscle structure in brown trout (*Salmo trutta*) subjected to long-term exercise. *Aquac. Res.* **2003**, *34*, 1287–1295. [CrossRef]

foods

MDPI

Article

Evaluating Microbiological Safety, Sensory Quality, and Packaging for Online Market Success of Roasted Pickled Fish Powder

Aunchalee Aussanasuwannakul [1,]* and Pisut Butsuwan [2]

1 Department of Food Chemistry and Physics, Institute of Food Research and Product Development, Kasetsart University, Bangkok 10903, Thailand
2 Department of Food Processing and Preservation, Institute of Food Research and Product Development, Kasetsart University, Bangkok 10903, Thailand; ifrpisut@ku.ac.th
* Correspondence: aunchalee.a@ku.th or ifrala@ku.ac.th; Tel.: +660-29428629-35

Abstract: This study explores the pivotal roles of microbiological safety, sensory qualities, packaging efficiency, and consumer preferences in determining the success of roasted pickled fish powder (RPFP) variants in the online marketplace. The comparison of the nutritional composition of the developed RPFP variants with a commercial benchmark reveals significant differences: the protein content in the herbal flavor variant is found to be 28.97%, which is lower than the 40.17% found in the commercial benchmark, while the fat content in the spicy flavor variant is measured at 19.51%, exceeding the 10.60% present in the benchmark. Additionally, the herbal flavor boasts a superior dietary fiber content of 14.23%, highlighting the intricate relationship between nutritional content and sensory attributes, which is critical in online retail, where physical product evaluation is not possible. Our comprehensive approach, evaluating both nutritional and sensory dimensions, introduces a novel perspective to the adaptation of traditional food products for e-commerce, addressing a gap in the literature. Despite this study's limitations, including a focused market analysis and constrained sample size, our findings offer valuable insights into enhancing product quality and integrity in the digital marketplace, positioning RPFP for competitive success while suggesting directions for future research.

Keywords: roasted pickled fish powder; sensory evaluation; just-about-right; check-all-that-apply; microbiological safety; traditional food products; consumer preferences; packaging optimization; digital marketplace; shelf-life extension

Citation: Aussanasuwannakul, A.; Butsuwan, P. Evaluating Microbiological Safety, Sensory Quality, and Packaging for Online Market Success of Roasted Pickled Fish Powder. *Foods* **2024**, *13*, 861. https://doi.org/10.3390/foods13060861

Academic Editor: Jingran Bi

Received: 13 February 2024
Revised: 8 March 2024
Accepted: 11 March 2024
Published: 12 March 2024

1. Introduction

The integration of traditional food products into the digital marketplace represents a significant evolution within the food processing and preservation domain, marking a dynamic confluence of heritage culinary practices and contemporary consumer expectations. Our study, focusing on the development of value-added roasted pickled fish powder (RPFP), strives to harmonize traditional preservation methods with the demands of today's online consumer preferences. The digital transformation of community-based food enterprises highlights the urgent need to adapt product offerings to cater to a wider and more diverse consumer base, thus enhancing their economic viability and contributing to sustainable local economies [1–5].

Despite significant advances in food technology and e-commerce, there exists a notable gap in our understanding of how innovative food products, especially those emanating from community settings, align with the intricacies of an online-centric market. To bridge this gap, an integrated approach that encompasses a digital presence, insights into consumer behavior, and technological adaptation is essential [6,7]. This study is poised to elevate RPFP to meet high standards of chemical and nutritional quality, ensuring its resonance with the expectations of the online marketplace and leveraging the food product development process through stages crucial for digital market alignment [8–10].

Moreover, this research seeks to bolster the economic prospects of community enterprises and empower local producers, offering a model for successful online integration. Scientifically, it aims to enrich the dialogue on food science and product development, illuminating how traditional foods can be reimagined for the digital market. This endeavor gains particular relevance against the backdrop of Thailand's digital transformation, characterized by high internet penetration and burgeoning online marketplaces, providing a fertile ground for projects aiming to empower women through digital and economic means [11,12].

Plaa-som, a traditional Thai fermented fish dish, serves as a pertinent backdrop for this study, showcasing the versatility and cultural significance of traditional food products. Research into plaa-som's fermentation process, nutritional enhancements, and adaptability to health-conscious trends lays a foundational understanding for RPFP's potential in online markets [13–19]. By detailing the methodologies underpinning RPFP's development, this study aims to offer insights into the multifaceted approach required for the successful online retailing of traditional food products.

2. Materials and Methods

2.1. *Roasted Flavored Pickled Fish (Plaa-som) Powder Preparation*

Fresh pearl spot (*Probarbus jullieni*) fish, ranging from 700 to 1200 g, were procured from a local market and immediately transported in an icebox to the food processing facility of the Freshwater Fish Processing Community Enterprise (Aunt Uan Village No. 3) in Sam Wa East Subdistrict, Khlong Sam Wa District, Bangkok, Thailand. Processing commenced within an hour of purchase. The preparation of the pickled fish (plaa-som) followed traditional methods, as outlined by Onsurathum et al. [20], with subsequent flavoring and dehydration steps adapted from Begum et al. [21]. First, the fish (2 kg) was cleaned under running tap water, gutted, and mixed with glutinous rice, minced garlic, and table salt. This mixture was then divided, sealed in plastic bags, and left to ferment at room temperature for 3 days, emphasizing the importance of hygiene to control sensory quality variation due to the natural fermentation process [22]. Post-fermentation, the fish was steamed for 20 min, seasoned with tamarind juice and salt, fried to reduce moisture, and finally roasted to achieve a moisture content below 10%. The ingredients of each flavor variant of the roasted pickled fish powder are detailed in Table 1.

Following preparation, 50 g samples of roasted pickled fish powder, while still hot (75–80 °C), were packaged into either 10×15 cm^2 laminated foil stand-up pouches made from a laminated film consisting of polyethylene terephthalate, aluminum, and polyethylene with a zip lock or polypropylene cups, and these were immediately sealed, labeled, and stored at room temperature (27 °C) for further analysis. The packaging materials were supplied by TK Supplies and Business Solutions Co., Ltd., Bangkok, Thailand.

2.2. *Storage Stability*

Following storage under ambient conditions (25 °C, 60% relative humidity), the samples underwent qualitative shelf-life testing, including microbiological quality assessments in accordance with local regulations [23] and hedonic testing. In this study, we compared two recipes (herbal vs. spicy flavor) across two time points (0 and 2 months), as well as measuring water activity, moisture content, and pH. The examination was solely centered on the pH and water activity data of the herbal-flavored RPFPs packaged in plastic cups and aluminum pouches over a duration of two months. A commercial benchmark was not utilized for this part of the study, as the aim was to investigate the inherent stability of the newly developed RPFP variants under different storage conditions. This approach was selected to isolate and understand the effects of packaging and time on the product's stability without an external comparison to commercial products.

Table 1. Ingredients of roasted pickled fish powder variants.

Code	Product	Brand	Sample	Ingredients
152	Commercial benchmark	Kamnan Chul farm		Pickled fish (36.90%), fried garlic (16%), fried shallot (16%), granulated sugar (9%), tamarind juice (8%), chili powder (2.8%), kaffir lime leaves (1.6%), seasoning powder (0.7%)
243	Roasted pickled fish (herbal flavor)	Developed product		Pickled fish (60%), dried bird's eye chili (15%), fried shallot (10%), fried herd mix (garlic, galangal, lemon grass) (5%), tamarind juice (5%), salt (4%), seasoning powder (1%)
697	Roasted pickled fish (spicy flavor)	Developed product		Pickled fish (60%), dried bird's eye chili (15%), fried shallot (10%), fried garlic (5%), tamarind juice (5%), salt (4%), seasoning powder (1%)

2.3. Proximate Composition and Nutritional Value

The nutritional composition, including moisture, crude protein, crude lipid, and ash contents, was analyzed in triplicate following AOAC standard procedures [24], with the methods of Sullivan and Carpenter [25] employed for a nutritional labeling analysis.

In our nutritional evaluation, detailed methodologies were employed to ascertain the nutritional composition and potential health benefits of the developed roasted pickled fish powder (RPFP) variants. The assessment included the determination of the moisture content, protein, fat, ash, total carbohydrates, and energy value using standard procedures as outlined by AOAC International. The protein content was measured using the Kjeldahl method, while the fat content was determined through Soxhlet extraction. The ash content was assessed via incineration in a muffle furnace, and carbohydrates were calculated by difference, subtracting the sum of moisture, protein, fat, and ash from 100. The energy value was estimated using the Atwater system, multiplying protein and carbohydrates by 4 kcal/g and fat by 9 kcal/g.

Furthermore, specific nutritional markers, such as dietary fiber, sodium, and calcium, were quantified using established analytical techniques. Dietary fiber was analyzed using the enzymatic–gravimetric method, and sodium and calcium levels were measured via atomic absorption spectrophotometry, ensuring a comprehensive evaluation of the RPFPs' nutritional profile. These methods adhere to the rigorous standards set by AOAC International, providing a reliable basis for the nutritional characterization of our RPFP variants.

2.4. *pH and Water Activity Determination*

pH measurements were conducted using a digital pH meter after blending the samples with distilled water. The water activity of the samples was analyzed using a water activity meter (Model LabMaster-aw; Novasina, Zurich, Switzerland).

2.5. *Microbiological Properties*

2.5.1. Yeast and Mold Determination

Yeast and mold in the samples were analyzed following the dilution plating technique described in the FDA's Bacteriological Analytical Manual (BAM) [26]. First, the samples were prepared by adding 0.1% peptone water to create a 10^{-1} dilution and homogenized. For plating, the spread-plate method was employed using dichloran rose Bengal chloramphenicol agar. Each dilution was plated in triplicate. The plates were incubated in the dark at 25 °C. After a 5-day incubation period, plates containing 10–150 colonies were selected for counting. Colony forming units (CFU) per gram or milliliter were calculated based on average colony counts, rounding off to two significant figures. This enumeration process aids in accurately quantifying yeast and mold presence in food samples, which is crucial for ensuring food safety and quality.

2.5.2. *Escherichia coli* Determination

The determination of *E. coli* followed the method described in the FDA's BAM [27]. Briefly, the method involved preparing a sample, transferring it to Lauryl tryptose broth for presumptive testing, and incubating it at 35 °C for 24–48 h. Positive results proceeded to BGLB broth, with incubation under the same conditions. Confirmed tests used EC broth, incubated at 44.5 °C for 24 h, with further checks at 48 h. *E. coli* was then isolated using eosin methylene blue agar and identified via colony characteristics and biochemical tests. *E. coli* was quantified using the most probable number estimation method.

2.5.3. *Staphylococcus aureus* Determination

S. aureus was determined in accordance with the FDA's BAM (CFU/g) [28]. After sample preparation, the sample was diluted, plated on Baird–Parker agar, and incubated at 35–37 °C for 45–48 h. Colonies typical of *S. aureus* were counted and confirmed via coagulase and other ancillary tests. The number of *S. aureus* per gram of food was reported based on colony counts. This method is suitable for foods where more than 100 *S. aureus* cells per gram are expected.

2.5.4. *Clostridium perfringens* Determination

C. perfringens was determined according to the BAM of the FDA [29]. First, a sample (25 g) was homogenized in 225 mL of peptone dilution fluid, followed by serial dilutions and plating on tryptose–sulfite–cycloserine agar. The plates were incubated anaerobically at 35 °C for 24 h. Presumptive *C. perfringens* colonies, which appear black with a white precipitate, were confirmed using further biochemical tests. The count of *C. perfringens* was calculated per gram of sample based on the colony count. This procedure allows for the accurate detection and quantification of *C. perfringens* in food samples.

2.5.5. *Bacillus cereus* Determination

B. cereus was determined according to the method outlined in the FDA's BAM [30]. The sample was homogenized, followed by plating serial dilutions on mannitol egg-yolk polymyxin agar or Bacara agar. These plates were incubated at 30 °C for 24–48 h. Presumptive *B. cereus* colonies, typically characterized by pink colonies with a lecithinase reaction, were counted. Confirmation involved Gram staining and biochemical tests. The number of *B. cereus* per gram of food was calculated based on colony counts.

2.5.6. *Salmonella* spp. Determination

The method for determining *Salmonella* spp. was in accordance with ISO 6570-1:2017/Amd.1:2020 [31]. It involved the pre-enrichment of a specific amount of sample in a non-selective broth, selective enrichment, and isolation on differential media. This method ensures the precise detection of *Salmonella* through a series of steps that include incubation at specific temperatures and durations (24 ± 2 h at 35 °C). For the confirmation of presumptive *Salmonella* colonies, biochemical and serological tests were conducted, with a count of *Salmonella* per gram or milliliter (25 g) of sample obtained.

2.6. Sensory Evaluation and Consumer Study

All subjects provided informed consent for inclusion before participating in the study. The study was conducted in accordance with the international guidelines for human research protection, and the methodology was approved by the Research Ethics Committee of Kasetsart University (COE No. COE67/013). Sensory profiling and a consumer analysis of the three RPFP samples were carried out by 60 untrained volunteer panelists (aged 18 years or over) who were habitual consumers of flavored fish products. The panelists were employees of the Institute of Food Research and Product Development, Kasetsart University, Bangkok, Thailand, and they had previously been recruited for and familiarized with the Sensory and Consumer Research Unit's sensory evaluation protocol. All the panelists completed a questionnaire. The aim of the questionnaire was to ensure a comprehensive understanding of consumer responses. Participant selection was demographically diverse and followed the methodology of Monteiro et al. [32]. The first part of the questionnaire included six short questions on gender, age, education, income, and frequency of exposure to the product. Age category was defined as Generations Z (younger than 25 years), Y (26–43 years), and X (44–58 years), which are related to online shopping behavior [33,34]. The second part involved a sensory evaluation and consumer study of the developed roasted pickled fish samples, with the herbal and spicy flavors evaluated and compared against a commercial benchmark (Kamnan Chul Farm, Rai Nai Chul Kunwong Co., Ltd., Phetchabun, Thailand). The sensory evaluation and consumer study detailed in Sections 2.6.1–2.6.4 encompassed comprehensive analyses. These analyses aimed to assess the sensory attributes of appearance, taste, color, texture, odor, and overall acceptability. Furthermore, consumer preferences were evaluated to understand the acceptance of and purchase intent for the developed samples, in comparison with a commercial benchmark. The study employed structured sensory evaluation methods and consumer surveys to gather insights into the sensory appeal and market potential of the product variants, providing a basis for refining product development and marketing strategies for the online market. The sample was removed from a freshly opened package. Each sample (5 g of RPFP) was coded with three random digits. The sample was presented to the panelists in a clear, press-seal plastic cup to prevent it from absorbing moisture. The sensory evaluation was conducted as a full crossover, where each panelist evaluated each sample in a sequential monadic presentation. Furthermore, the presentation order of the sample product across panelists was balanced, and the assignment of the presentation order to the panelists was randomized.

2.6.1. Acceptance Test

For acceptance testing and acceptability scaling, the consumer appeal of the products was assessed using a rating scale for the degree of liking or disliking. The acceptance scales included the traditional 9-point hedonic scale and just about right (JAR) scale [35].

Liking or Hedonic Test

The 9-point hedonic scale (1: dislike extremely to 9: like extremely) was used for determining the likability or acceptance of appearance, flavor, and texture and overall acceptance.

Just About Right (JAR) Scale

The 7-point JAR scale was employed to determine the optimal intensity levels of key sensory attributes, including color, sour, salty, spicy, flavor, and texture. These attributes were rated from 1 = "too little" to 5 = "JAR" and 7 = "too much".

2.6.2. Check All That Apply (CATA) Analysis

On the CATA questionnaire, developed in accordance with the literature [36], the participants selected sensory attributes that they considered relevant from a predetermined list of descriptors. The descriptors used were created from a focus group discussion (FGD), which was conducted in line with that outlined by Nurazizah et al. [37]. The focus group included participants with relevant food product experience collaboratively generating sensory descriptors to understand consumer perceptions of RPFP. Table 2 shows the terms identified by the FGD, together with sensory attributes (no. 1–7) and product positioning (no. 8–14). The CATA question was as follows: "Please check all the attributes that describe the RPFP you have just tried".

Table 2. Sensory attributes and non-sensory (product positioning) attributes based on the results of the FGD.

No.	Attribute	Description
1	Tangy	A zesty, acidic quality typical in fermented foods
2	Salty	The flavor attribute from sodium ions, essential in flavoring and preservation
3	Spicy	A heat sensation from capsaicinoids or similar compounds
4	Light and airy	Textural characteristic of a low-density, porous food matrix
5	Crumbly roasted granules	A brittle, uneven texture resulting from roasting and the Maillard reaction
6	Aromatic with herbal notes	A complex scent profile from volatile compounds in herbs
7	Intensely flavorful	A strong, multidimensional taste and aroma profile
8	100% roasted pickled fish	Product of lacto-fermented fish and thermal processing
9	Non-greasy	Absence of excess surface lipids, enhancing mouthfeel
10	Bone free	A uniform texture due to the removal of calcified elements
11	No MSG, no preservatives	The quality of having no added MSG or synthetic preservatives
12	Protein packed	A high protein level, significant for nutrition labeling
13	Shelf stability	Extended preservation of product quality and freshness
14	Packaging integrity	Use of spoilage-preventing technologies in packaging

2.6.3. Consumer Purchase Intent

Structured questionnaires were administered to assess the likelihood of consumers purchasing the RPFPs, combining methodologies from Patel et al. [38] and Murillo et al. [39], with consumer purchasing intent evaluated under various hypothetical purchase scenarios. After completing the CATA questionnaire, the subjects were asked the following question: "If the product had a package size of 50 g in a sealed aluminum bag and a price of 59 baht (USD 1.66) on an online store like Shopee, would you buy it?" There were five possible answers: "definitely would not buy", "probably would not buy", "may or may not buy", "probably would buy", and "definitely would buy".

2.6.4. Consumer Preference

In the consumer preference test, the participants were asked to compare two of the developed products against each other with the aim of product improvement and parity testing. After they had tried two samples, the first question that they were asked was as follows: "Which sample did you prefer overall?" The possible responses were "preferred the first sample", "preferred the second sample", and "no preference/both the same". They were then asked an open-ended question: "Why did you prefer that sample?" The serving sequences were randomized across the participants, with an equal number of participants receiving either the first sample or the second sample first.

2.7. Statistical Analysis

To compare the means of nutritional composition and consumer liking data among RPFPs, an analysis of variance (ANOVA) was performed, followed by post hoc Tukey's tests, using XLSTAT 2023.3.0 (1415) software for Macintosh 14.2 [40]. The results are presented as mean ± standard deviation, with a significance threshold set at $p \leq 0.05$. All analyses were replicated at least three times to ensure reliability. XLSTAT's CATA data analysis tool was used to automate the analysis of the CATA data, including Cochran's Q test, a correspondence analysis (CA), a principal coordinate analysis (PCoA), and a consumer clustering analysis (hierarchical cluster analysis). XLSTAT's CATATIS applied to the JAR data resulted in multivariate and penalty analyses. The CATATIS method for the multivariate analysis of the JAR data resulted in an attribute/product biplot and a penalty analysis.

Descriptive statistics were used to calculate the mean scores and standard deviations for the sensory attributes assessed via the hedonic scale at 0 and 2 months of storage. Paired t-tests contrasted these scores to test the hypothesis of no significant sensory change due to storage. Significant differences were quantified using Cohen's d to illuminate the magnitude of changes. This analysis was replicated thrice for reliability. Data visualization employed box plots to compare sensory attributes and illustrate consumer perceptions' variability and central tendency over time. Histograms and Kernel Density Estimate (KDE) curves visualized the overall satisfaction scores, indicating the range of the consumer experience with the products.

3. Results

3.1. Proximate Composition and Nutritional Value

We present a detailed analysis of the proximate compositions and nutritional values of the herbal- and spicy-flavored RPFPs in comparison to those of a commercial benchmark. This analysis, aimed at developing a value-added product for online sales, emphasizes moisture reduction and optimized packaging to extend shelf life.

Table 3 provides a comparative analysis of the proximate and nutritional compositions of the RPFPs in the two developed flavors (herbal and spicy) and those of the commercial benchmark. The analysis included moisture, protein, fat, ash, total carbohydrate, total energy, and other nutritional parameters. The results revealed distinct differences in the nutritional composition between the developed products and the commercial benchmark. Notably, both developed products exhibited unique profiles in terms of their protein, fat, dietary fiber, sodium, and calcium contents, which are critical for their nutritional and health benefits. The comparative analysis revealed that the protein content of the herbal flavor (28.97%) was higher than that of the spicy flavor (14.48%) but that the protein contents of both were lower than the protein content of the commercial benchmark (40.17%). Thus, although the developed products are rich in proteins, the commercial benchmark is superior in terms of protein content. The fat content of the spicy flavor (19.51%) was higher than that of both the herbal flavor (16.28%) and the commercial benchmark (10.60%), indicating a richer, more energy-dense product. The comparative analysis revealed that the herbal flavor was notably superior in dietary fiber (14.23%) and calcium (422.81 mg/100 g) content compared to the spicy flavor, which exhibited lower dietary fiber (8.22%) and significantly lower calcium (103.20 mg/100 g). This highlights the herbal flavor as potentially offering more health benefits, particularly for bone health and digestive wellness. The sodium content of the herbal flavor (1366 mg/100 g) was higher than that of the spicy flavor (954.88 mg/100 g), which may be a consideration for sodium-restricted diets.

Table 3. Comparative nutritional compositions of the RPFPs (herbal and spicy variants vs. a commercial benchmark).

Analysis Item	Herbal Flavor	Spicy Flavor	Commercial Benchmark
Moisture (%)	15.49 ± 0.12 ^{c}	13.72 ± 0.09 ^{b}	11.56 ± 0.11 ^{a}
Protein (%) (factor 6.25)	28.97 ± 0.13 ^{b}	14.48 ± 0.38 ^{a}	40.17 ± 0.72 ^{c}
Fat (%)	16.28 ± 0.16 ^{b}	19.51 ± 0.88 ^{c}	10.60 ± 0.08 ^{a}
Ash (%)	6.74 ± 0.13 ^{b}	4.03 ± 0.17 ^{a}	8.60 ± 0.15 ^{c}
Total carbohydrate (%)	32.52 ± 0.06 ^{b}	48.26 ± 0.40 ^{c}	29.07 ± 0.16 ^{a}
Total energy (kcal/100 g)	392.48 ± 3.06 ^{b}	426.55 ± 1.98 ^{c}	372.36 ± 3.87 ^{a}
Energy from fat (Kcal/100 g)	146.52 ± 1.37 ^{b}	175.59 ± 2.96 ^{c}	95.40 ± 1.88 ^{a}
Saturated fat (%)	6.16 ± 0.03 ^{a}	7.72 ± 0.10 ^{b}	-
Cholesterol (mg/100 g)	87.87 ± 0.75 ^{b}	42.07 ± 0.50 ^{a}	-
Dietary fiber (%)	14.23 ± 0.04 ^{b}	8.22 ± 0.07 ^{a}	-
Sugars (%)	5.63 ± 0.10 ^{a}	7.14 ± 0.15 ^{b}	-
Sodium (mg/100 g)	1366 ± 4.73 ^{b}	954.88 ± 1.54 ^{a}	-
Vitamin A (beta carotene) (µg/100 g)	646.56 ± 4.68 ^{b}	565.78 ± 6.85 ^{a}	-
Vitamin B1 (mg/100 g)	0.12 ± 0.01 ^{a}	0.11 ± 0.00 ^{a}	-
Vitamin B2 (mg/100)	0.12 ± 0.01 ^{a}	0.10 ± 0.00 ^{a}	-
Calcium (mg/100 g)	422.81 ± 4.68 ^{b}	103.20 ± 2.32 ^{a}	-
Iron (mg/100 g)	4.18 ± 0.03 ^{a}	4.62 ± 0.09 ^{a}	-

$^{a–c}$ Means that entries within the same column that have the same superscript or no superscript are not significantly different ($p > 0.05$; $n = 3$).

3.2. Storage Stability

In the context of enhancing the storage stability of RPFP, with a focus on extending the shelf life for online marketplace viability, this study incorporated a comprehensive analysis of packaging improvements, microbiological safety, and sensory attributes. Such an analysis is vital for ensuring product quality and consumer satisfaction, aligning with the methods described for processing and packaging RPFP.

Selecting the right packaging for the developed RPFPs is crucial for ensuring their quality and safety when sold online. Table 4 provides a comparison of the previously used polypropylene cups with the newly adopted laminated aluminum foil stand-up pouches. Although polypropylene cups might seem a budget-friendly option, their poor barrier properties and vulnerability to shipping damage make them unsuitable for online sales of RPFP. Laminated aluminum foil stand-up pouches, despite their slightly higher upfront cost, offer the following significant advantages:

(1) Preserve fresh flavors and aromas, with customers receiving the product at its peak, maximizing satisfaction;
(2) Guarantee safe consumption, with a robust barrier minimizing spoilage risk, ensuring product safety;
(3) Protect against bumps during transit, arriving in pristine condition;
(4) Enhance brand image, with premium packaging reflecting the quality and care put into the product.

The RPFP in a laminated bag was further used in analyses of storage stability indices comparing the two recipes.

Table 5 presents the microbiological safety and stability of the herbal-flavored RPFPs over time. The table details the microbial analyses of the RPFPs packaged in plastic cups and aluminum pouches, assessing the products at the initial (0 months) and final (2 months) time points of the study. The results confirm the microbiological stability of the herbal flavor in both types of packaging. The pH and water activity values in Table 6, which are critical for assessing physical and chemical stability, remained stable across both product flavors and time points, indicating no significant deterioration in product quality. The results confirmed compliance with Thai Department of Medical Sciences (DMSC) standards, showcasing the products' safety across storage times and packaging types.

Table 4. Comparison of packaging options for RPFP for online sales: polypropylene cup versus laminated aluminum pouch.

Feature	Polypropylene Cup (PP *)	Laminated Aluminum Pouch (PET/AL/PE **)
Image	 	 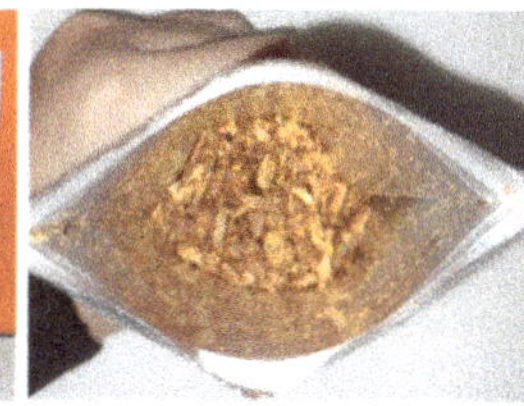
Weight	4.57 g	4.34 g
Dimensions	Diameter = 7 cm; length = 5.5 cm	Width = 10 cm; length = 15 cm
Barrier Properties	Poor; susceptible to moisture, oxygen, and light ingress	Excellent; blocks moisture, oxygen, light, and odors
Shelf Life	Short; prone to flavor degradation and spoilage	Extended; maintains freshness and quality for months
Physical Protection	Vulnerable to dents, punctures, and breakage	Puncture-resistant and sturdy; withstands shipping rigors
Tamper Evident	No; broken seals easily occur	Secure; resealable zip-lock closure prevents contamination
Online Suitability	Not ideal; leakage and breakage risk during shipping	Ideal; protects product during transit and storage
Sustainability	Typically not recyclable or biodegradable	Some pouches are recyclable or contain biodegradable materials
Cost	Lower initial cost	Higher initial cost, but longer shelf life may offset expense

* Polypropylene commonly used in food packaging for its flexibility and resistance; ** A multi-layer film comprising polyethylene terephthalate (PET) for structural integrity, aluminum (AL) as a barrier layer, and polyethylene (PE) for sealing and flexibility.

Table 5. Comprehensive microbiological analysis of herbal-flavored RPFPs across storage conditions and times.

	Criteria	Plastic Cup 0 Months	Plastic Cup 2 Months	AL Pouch 0 Months	AL Pouch 2 Months
Yeasts and molds (CFU/g)	<100	<10	<10	<10	<10
Escherichia coli (MPN/g)	<3	<3	<3	<3	<3
Staphylococcus aureus (CFU/g)	<10	<10	<10	<10	<10
Clostridium perfringens (CFU/g)	<100	<10	<10	<10	<10
Bacillus cereus (CFU/g)	<1000	<10	<10	70	<10
Salmonella spp. (/25 g)	N.D. *	N.D.	N.D.	N.D.	N.D.

* N.D. = Not detected.

The box plots in Figure 1 (top) illustrate the sensory attribute scores of the herbal-flavored RPFPs at 0 and 2 months of storage, showing medians, quartiles, and outliers. Statistical tests revealed no significant differences in appearance, color, odor, taste, texture, or overall liking between storage times, indicating that the sensory attributes remain unaffected for up to 2 months. The histogram and KDE curve in Figure 1 (bottom) highlight the consistency of consumer perceptions of overall quality, with an analysis confirming panel agreement on sensory evaluations across storage times, demonstrating reliable consumer panel performance.

Table 6. Stability of pH and water activity of RPFP: a comparative analysis of flavors over time.

	Herbal Flavor		Spicy Flavor	
Parameter	**0 Months**	**2 Months**	**0 Months**	**2 Months**
pH	4.67	4.65	4.33	4.32
water activity	0.67	0.67	0.598	0.60

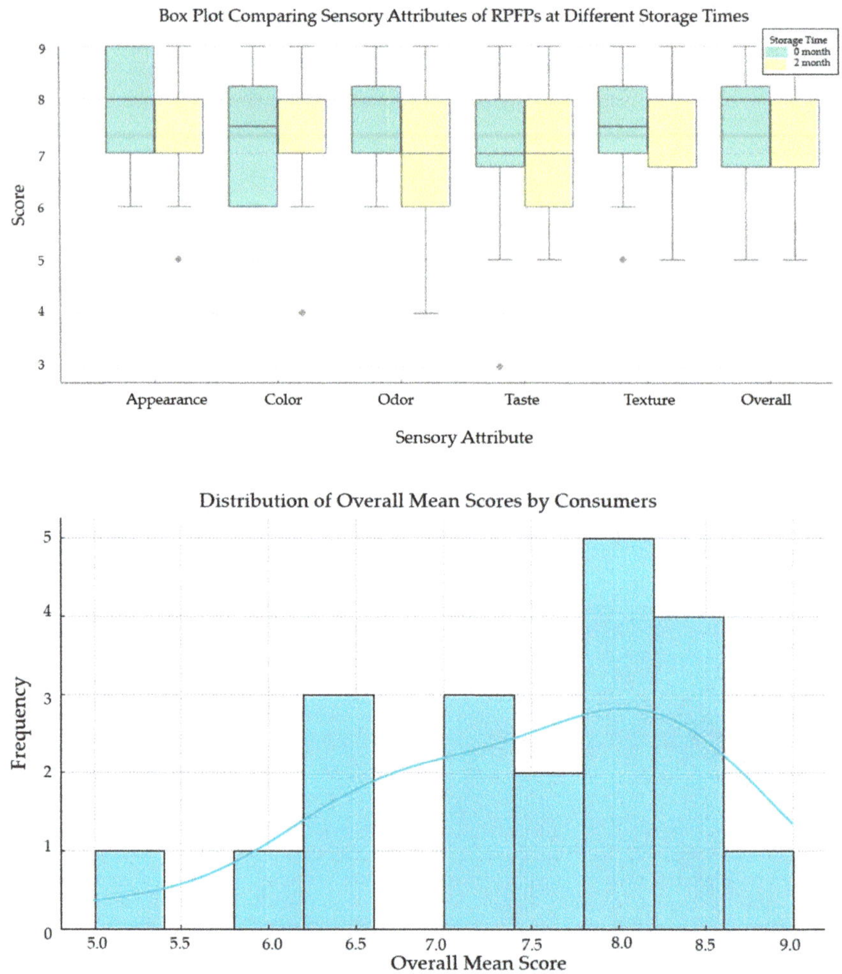

Figure 1. (**Top**) Box plots comparing sensory attributes (appearance, color, odor, taste, texture, and overall liking) of herbal-flavored RPFPs at 0 and 2 months of storage, displaying consumer ratings' variability and central tendency on a 9-point hedonic scale. (**Bottom**) Distribution of overall mean scores by consumers, with a histogram and KDE curve showcasing the central tendency and spread of overall satisfaction scores, highlighting the range of consumer experiences with the products over time.

Both the herbal and spicy flavors exhibited robust storage stability, with no discernible difference in their microbiological safety or sensory attributes, suggesting that the product formulations are well suited to long-term storage. The microbiological and sensory evaluation indicated that both types of packaging (plastic cup vs. aluminum pouch) effectively maintained product quality. However, the transition to aluminum pouches is justified by their superior barrier properties. Furthermore, the aluminum pouch's added benefits of consumer convenience and enhanced physical protection during shipping offer a clear advantage for online sales. In relation to the storage time (0 vs. 2 months), the absence of significant changes in microbiological safety, pH, water activity, and sensory properties over 2 months underscored the excellent storage stability of the RPFP, affirming its suitability as a ready-to-eat product with a prolonged shelf life.

This comprehensive evaluation of storage stability, incorporating microbiological safety, pH, water activity, and sensory attributes, underscores RPFP's potential as a stable and high-quality product suitable for the online marketplace. The findings provide a solid foundation for the further development and optimization of packaging and storage strategies to enhance product appeal and consumer satisfaction.

3.3. Sensory Evaluation and Consumer Study

3.3.1. Demographic Characteristics of Participants

Figure 2 depicts the demographic composition of the participants who took part in the sensory evaluation of the RPFPs. According to the study inclusion criteria, all participants had to be at least 18 years of age. The participants were categorized into different groups to ensure a comprehensive analysis of consumer preferences. The group categorization included gender, age (generation), educational attainment, income level, experience of online shopping, and frequency of online food purchases. Contrary to an equal distribution across these demographics, our findings revealed a diverse composition:

- Gender: The participant pool comprised a significantly higher proportion of females than males: 78% versus 22%.
- Generation: The breakdown by generation showed a predominance of Generation Y (53%), followed by Generation X (32%), with Generation Z (15%) being the least represented.
- Education: Educational levels varied, with 50% holding a bachelor's degree, 25% having an undergraduate level of education, 17% possessing doctoral qualifications, and 8% having a master's degree.
- Income (revenue): The majority, 68%, reported earning less than THB 30,000 (approximately USD 845, based on an exchange rate of THB 35.518 per USD on 7 February 2024). Those with incomes ranging from THB 30,000 to 50,000 accounted for 23%, which translates to approximately USD 845 to USD 1407. A small fraction, 5%, earned above THB 50,000 (approximately USD 1407 or more).
- Online shopping experience: The majority (87%) of the participants had experience shopping online, with just 13% not having such experience.
- Frequency of online food shopping: In total, 49% of the participants seldom shopped for food online, whereas 20% did so once a week, 18% once a month, and 13% never did so.

This detailed demographic breakdown not only underlines the diversity of the participant group but also emphasizes the study's focus on gathering insights from a broad spectrum of consumers who engage with the product category in varying capacities.

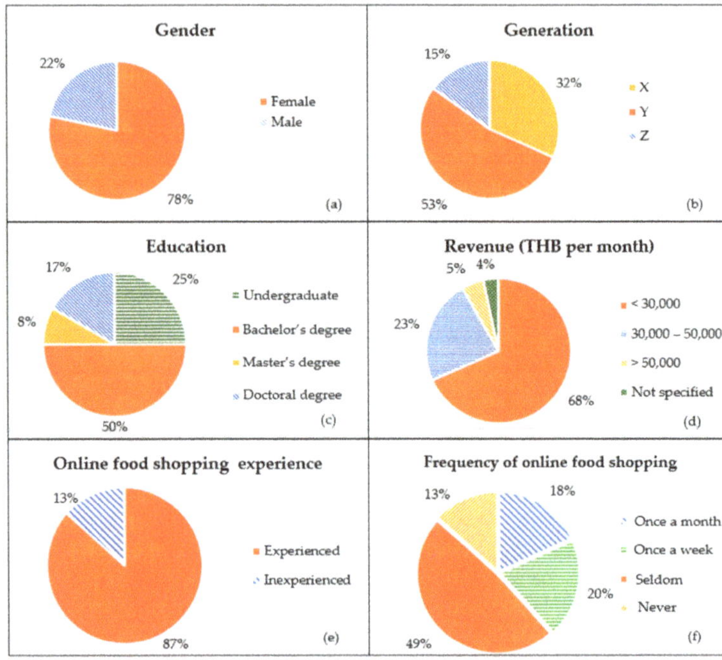

Figure 2. Distribution of the demographic characteristics of the participants (n = 60) in the sensory evaluation of RPFP. The pie charts illustrate the percentage distribution across various categories, including (**a**) gender, (**b**) generational groups (X, Y, Z), (**c**) education levels, (**d**) monthly income (in THB), (**e**) online food shopping experience, and (**f**) frequency of online food shopping.

3.3.2. Liking

In the sensory evaluation segment focusing on consumer liking for RPFP comprising two product prototypes (herbal flavor: sample code 243; spicy flavor: sample code 697) and a commercial benchmark (sample code: 152), an ANOVA was employed to analyze the preference patterns among these offerings. The results of the ANOVA (F = 7.527, Pr > F = 0.001) highlighted statistically significant differences in the liking scores among the tested products. Thus, the findings indicate that preferences for at least one of the prototypes markedly differed from the preferences for the commercial standard, underscoring distinct consumer perceptions tied to each product's sensory profile.

Further examination of these preferences was facilitated by mean charts, which detailed the least square means of the liking scores, effectively revealing the varying degrees of consumer appeal that each product commanded. The commercial benchmark emerged as the most favored product, indicating its broader consumer appeal over the newly introduced herbal and spicy variations.

The differences in preferences are depicted in a bar graph, serving as a representation of the data. As indicated in the bar graph, the commercial benchmark (sample 152) achieved a score of 0.300 ± 0.095, surpassing the scores for the herbal (sample 243) and spicy (sample 697) prototypes, both of which registered a score of −0.150 ± 0.095. This graphical representation not only facilitates an intuitive understanding of the products' relative standings but also highlights the nuanced consumer inclinations that differentiate between the familiar and the novel (Figure 3).

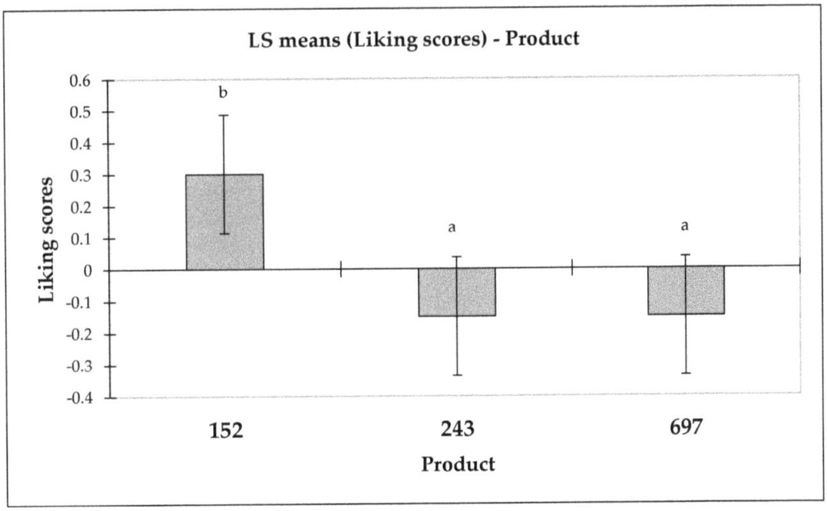

Figure 3. Comparative analysis of consumer liking scores for the RPFP variants, detailing the least square means, along with standard errors for the liking scores of sample 152 (commercial benchmark), sample 243 (herbal flavor), and sample 697 (spicy flavor). Distinctions in statistical significance are denoted by the different letters positioned above the respective bars.

3.3.3. JAR Analysis

In the JAR analysis, consumer perceptions regarding various sensory attributes of the RPFP samples, encompassing color, sourness, saltiness, spiciness, flavor, and texture intensity, were evaluated. The consumer responses were grouped into three distinct classifications: too little, JAR, and too much. Figure 4 shows the results of the JAR analysis and penalty analysis of the sensory attributes of the RPFPs.

The evaluation of the color attribute revealed that 87% of the participants deemed the herbal-flavored product (sample code: 243) to possess optimal intensity, indicating a favorable perception that contrasted with that of the commercial benchmark (sample code: 152), which some participants found overly vibrant, and that of the spicy variant (sample code: 697), criticized for its inadequate color intensity. In the assessment of sourness, the benchmark was considered JAR by 67% of the participants, a higher proportion than the herbal (243) and spicy (697) samples, which received JAR ratings from 57% and 52% of the participants, respectively. These findings point to variation in consumer acceptance of sourness levels, with a notable segment finding the sourness intensity of the benchmark insufficient, while the herbal and spicy variants were perceived by some as excessively sour.

The distribution of the JAR ratings for saltiness demonstrated relative uniformity across all samples, with perceived adequacy ranging from 68 to 77%. However, the attribute of spiciness was particularly divisive for the spicy sample (697), with 88% of consumers finding its spiciness overly intense, in stark contrast to the more moderate perceptions of the herbal (sample code: 243) and benchmark (152) samples.

In terms of flavor, both the benchmark (152) and herbal (243) samples were found to have JAR intensity by 72% and 70% of consumers, respectively. The spicy sample (697), however, was considered to have an appropriate flavor intensity by only 42% of consumers, with a significant portion finding it too mild. Texture was identified as a challenge primarily for the spicy sample, with 80% of consumers indicating an excessively intense texture, unlike the benchmark sample, which was found JAR by 82% of the participants.

In Figure 4 (right side), the penalty analysis is depicted, showing the drop in overall liking on a 9-point hedonic scale against the percentage of consumers who rated sensory attributes as "too low" or "too high". This analysis determined the impact of deviations from the ideal intensity of sensory attributes on overall liking. It revealed that attributes perceived as too low in intensity incurred greater penalties to overall product liking. For example, an insufficient flavor intensity resulted in a significant deduction of 1.5 points on the 9-point hedonic scale, as reported by approximately 45% of consumers. This was particularly critical for the spicy sample (697), where overwhelming spiciness was predicted to reduce the hedonic score by 0.3 points, corroborating the findings from the JAR analysis. Conversely, although the JAR analysis indicated texture and sourness as problematic for the herbal (243) and spicy (697) samples, only the herbal variant was noted to receive a significant penalty for these attributes being overly intense. In contrast, the benchmark (152) sample exhibited significant reductions in overall liking due to insufficient texture and sourness, while saltiness intensity was a key factor for both prototypes, with nearly 20% of consumers indicating its inadequacy, leading to an approximate deduction of 0.9 points on the hedonic scale.

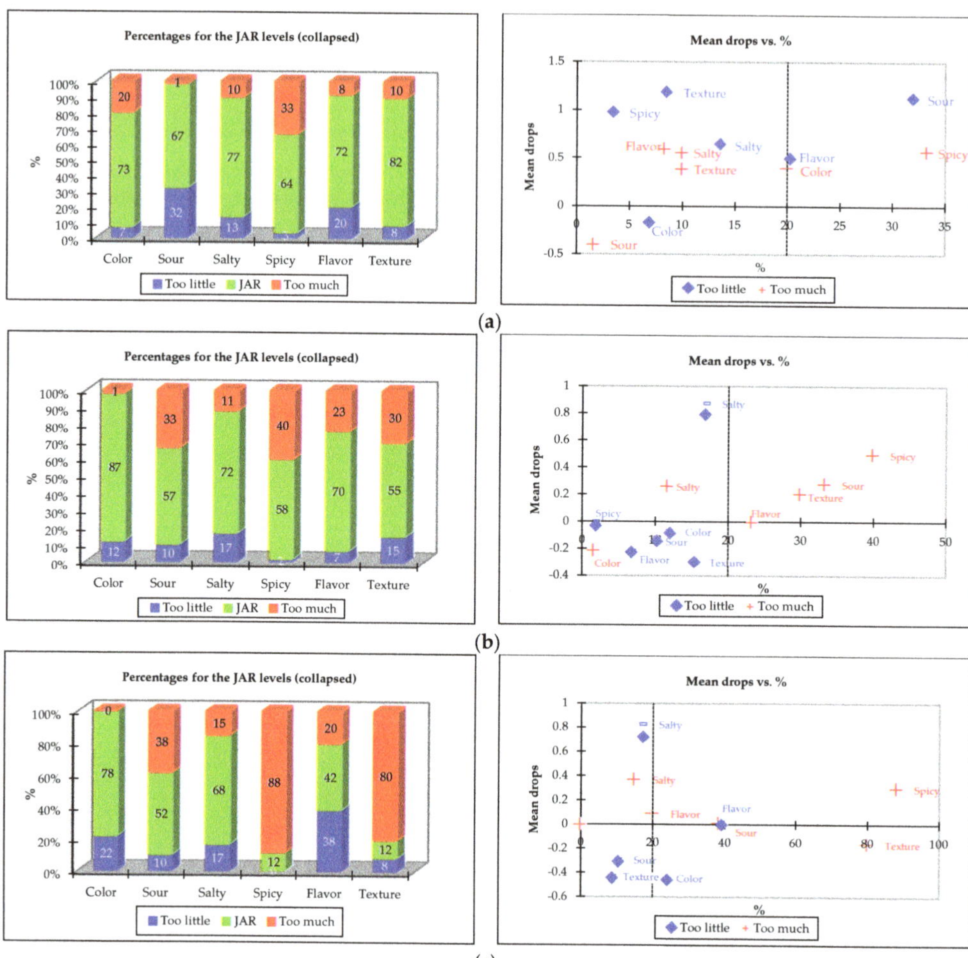

Figure 4. Distribution of the JAR ratings for the sensory attributes of the RPFP samples (**left side**), depicting consumer responses for color, sourness, saltiness, spiciness, flavor, and texture intensity, categorized as too little, JAR, and too much for each product sample, and the penalty analysis on the 9-point hedonic scale for each sensory attribute being too low or too high in intensity (**right side**): (**a**) commercial benchmark (152), (**b**) herbal flavor (243), and (**c**) spicy flavor (697).

3.3.4. CATA Analysis

The CATA methodology was utilized to evaluate the sensory attributes of the RPFP samples comprising the two prototypes designated as "herbal flavor" (sample code: 243) and "spicy flavor" (sample code: 697), as well as a "commercial benchmark" (sample code: 152). The participants selected attributes from a predefined list that accurately reflected the sensory characteristics of each sample. The attributes considered were as follows: "tangy", "salty", "spicy", "light and airy", "crumbly roasted granules", "aromatic with herbal notes", "intensely flavorful", "100% roasted pickled fish", "non-greasy", "bone free", "no MSG, no preservatives", "protein packed", "shelf stability", and "packaging integrity".

Cochran's Q Test

Cochran's Q test revealed statistically significant differences in the perception of these attributes across the samples, highlighting their distinct sensory profiles (Table 7). For example, the "spicy" flavor was predominantly associated with "tangy" and "spicy" attributes, distinguishing it from the other samples. Meanwhile, the "herbal" flavor was noted for its "aromatic with herbal notes" and "light and airy" texture, and the "commercial benchmark" was often linked to "crumbly roasted granules" and "intensely flavorful" qualities, indicating the unique sensory profile of the benchmark compared to that of the prototypes.

Table 7. Cochran's Q test results for the sensory attributes of the RPFP samples.

Attributes	*p*-Values	152	243	697
Tangy	<0.0001	0.200 [a]	0.533 [b]	0.533 [b]
Salty	0.895	0.317 [a]	0.300 [a]	0.300 [a]
Spicy	<0.0001	0.450 [a]	0.600 [a]	0.833 [b]
Light and airy	0.029	0.383 [a]	0.250 [a]	0.250 [a]
Crumbly roasted granules	<0.0001	0.567 [b]	0.317 [a]	0.317 [a]
Aromatic with herbal notes	0.00	0.367 [b]	0.433 [b]	0.150 [a]
Intensely flavorful	<0.0001	0.700 [c]	0.433 [b]	0.150 [a]
100% roasted pickled fish	0.002	0.400 [a]	0.567 [b]	0.567 [b]
Non-greasy	0.549	0.567 [a]	0.517 [a]	0.517 [a]
Bone free	0.135	0.650 [a]	0.583 [a]	0.583 [a]
No MSG, no preservatives	0.513	0.250 [a]	0.217 [a]	0.217 [a]
Protein packed	0.819	0.183 [a]	0.167 [a]	0.167 [a]
Shelf stability	0.368	0.450 [a]	0.417 [a]	0.417 [a]
Packaging integrity	0.069	0.333 [a]	0.267 [a]	0.267 [a]

The table presents the *p*-values for each attribute, with multiple pairwise comparisons using the critical difference (Sheskin) procedure. [a–c] Significant differences are indicated by different letters for each attribute across the product samples (sample codes: 243, 697, and 152 for the herbal flavor, spicy flavor, and benchmark, respectively).

CA

A subsequent CA provided a visual depiction of how each sample correlated with specific sensory attributes, illustrating the unique positioning of each product variant within the sensory space (Figure 5). The results of the CA underscored the alignment of the commercial benchmark with "crumbly roasted granules" and "intensely flavorful" attributes, setting it apart from the prototypes, which were more closely related to the attributes "spicy", "tangy, "and "aromatic with herbal notes".

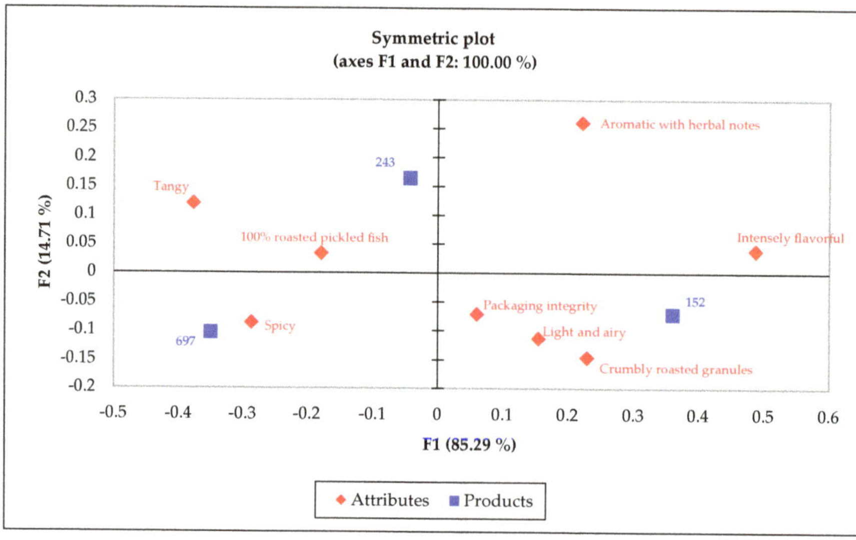

Figure 5. Biplot from CA, exhibiting relationship between CATA attributes and RPFP samples.

PCoA

The PCoA further dissected the relationships between the sensory attributes and consumer perceptions, demonstrating that attributes such as "protein packed", "shelf stability", and "packaging integrity", along with "100% roasted pickled fish" and "non-greasy", played a significant role in shaping overall liking (Figure 6). This analysis captured a substantial portion of the variability in sensory perceptions across the samples.

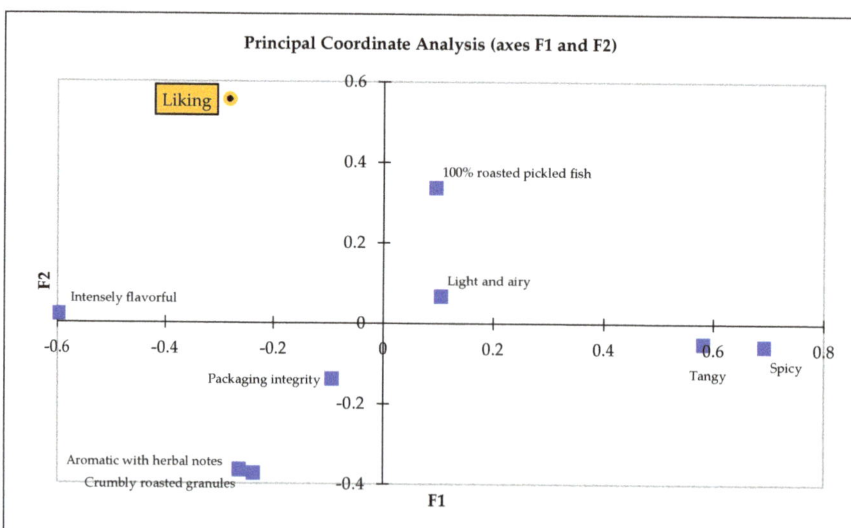

Figure 6. PCoA two-dimensional map depicting interplay between sensory attributes and consumer liking.

Consumer Clustering Analysis

A consumer clustering analysis of the sensory perception of the RPFP samples was systematically conducted using a hierarchical cluster analysis and the CLUSCATA method. This methodological approach allowed for the segmentation of consumers into distinct groups, each reflecting a unique set of sensory experiences with the product samples. The analysis, visualized through a dendrogram in Figure 7, delineated two primary consumer classes, each characterized by distinctive perceptual profiles toward the attributes of the fish powder samples.

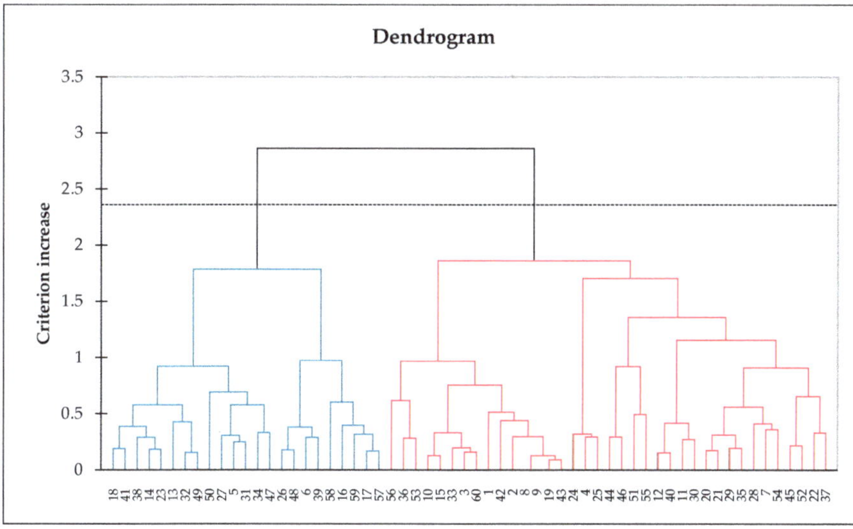

Figure 7. Dendrogram from CLUSCATA analysis, showing segmentation of consumer groups into two main classes based on RPFP sensory perceptions. Clusters are differentiated by color: blue indicates preliminary groupings, red denotes distinct classes as marked by the dotted line.

The exploration of these classes through biplots provided further insights into the perceptual differences between them. Class 1, marked by a homogeneity score of 0.572, displayed a relatively uniform perception of the products, contrasting with Class 2, which had a homogeneity score of 0.495. The biplots revealed that Class 1 was closely aligned with the sensory attributes identified in the CA, with samples 243 and 697 being distinctly separated from sample 152 on the first axis, indicating shared attributes, such as "tangy", "spicy", "100% roasted pickled fish", "shelf stability", "no MSG, no preservatives", and "bone free". Conversely, Class 2 demonstrated a clear differentiation in perception, particularly distinguishing sample 697 from samples 243 and 152, highlighting unique attribute associations for each group.

The biplots for Class 1 and Class 2 detail the variance explained on the first and second axes and illustrate the sensory perception differences among the samples within each class (Figure 8). Class 1′s biplot emphasizes the shared sensory characteristics between the prototypes, and Class 2′s biplot underscores the distinct sensory profiles that set the spicy flavor apart.

Complementing the clustering analysis, demographic insights were obtained to provide a comprehensive understanding of each cluster's composition (Table 8). Cluster 1 predominantly consisted of females (77%), with Generation Y (50%) accounting for a significant proportion, followed by Generations X (23%) and Z (27%). This cluster also showed higher educational attainment, with the majority holding bachelor's degrees (69%) and most earning below THB 30,000 (approximately USD 845, based on an exchange rate of THB 35.518 per USD on 7 February 2024) per month (73%). Their online food shopping experiences varied, with a notable portion shopping seldom (46%) or weekly (19%).

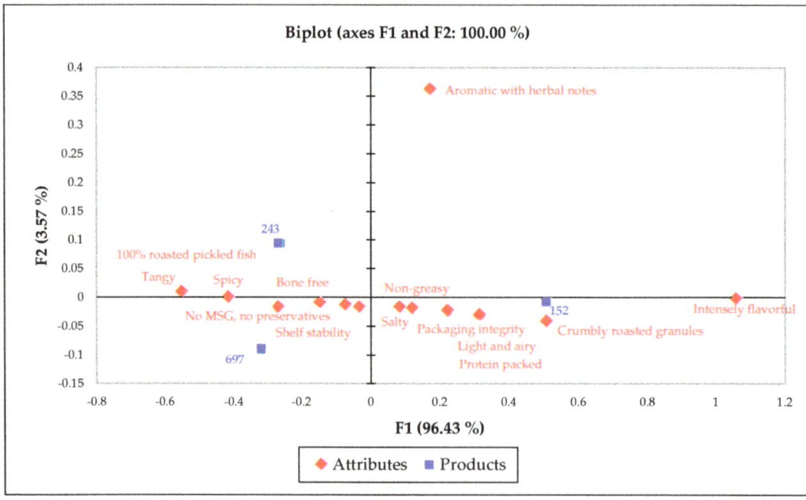

(a)

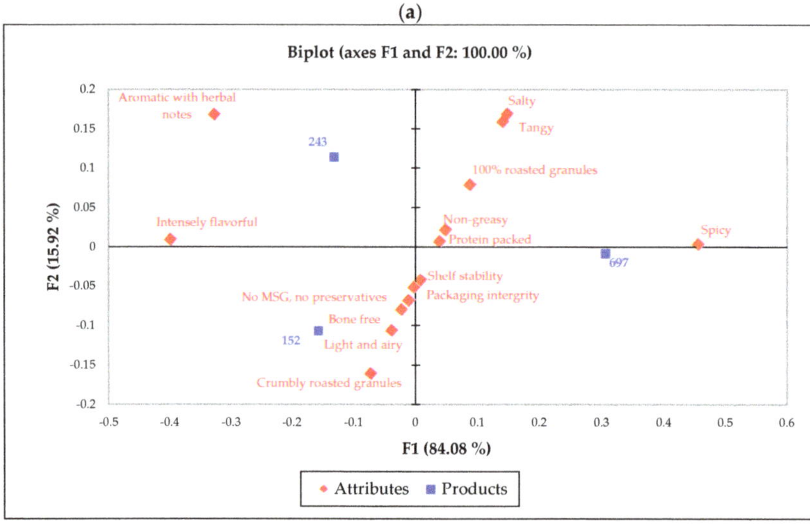

(b)

Figure 8. A biplot of the classes in the CLUSCATA analysis, displaying variance explained on the first and second axes: (**a**) sensory perception differences in Class 1 of the RPFP sample; (**b**) the differences in attributes among the RPFP samples in Class 2.

In contrast, Cluster 2 contained a slightly higher proportion of females (79%) and leaned toward older generations, with Generations Y (56%) and X (38%) predominating. This cluster also had a diverse educational background, with a significant number of participants below the undergraduate level (32%) and above master's degree (21%). The income distribution closely mirrored that in Cluster 1, although a slightly higher percentage reported earnings above THB 50,000 (approximately USD 1407, based on an exchange rate of THB 35.518 per USD on 7 February 2024). The online shopping experience was notably high, with the majority shopping online seldom (50%) or monthly (24%).

The detailed demographic analysis underscores the diversity within the consumer clusters, highlighting the importance of considering various consumer attributes in product development and marketing strategies to cater to the different sensory preferences revealed through the CATA analysis.

Table 8. Demographic distribution of consumer clusters in sensory evaluation of RPFP.

	%	
	Cluster 1	**Cluster 2**
Gender		
Female	77	79
Male	23	21
Generation		
X	23	38
Y	50	56
Z	27	6
Education		
Undergraduate	15	32
Bachelor's degree	69	35
Master's degree	4	12
Doctoral degree	12	21
Revenue (THB per month)		
<30,000	73	65
30,000–50,000	27	21
>50,000	0	9
Not specified	0	6
Frequency of consumption (per week)		
0 day	8	9
1 day	50	44
2 days	31	18
3 days	4	15
4 days	4	9
5 days	0	0
6 days	0	0
7 days	3	5
Online food shopping experience		
Experienced	77	94
Inexperienced	23	6
Frequency of online food shopping		
Once a month	12	24
Once a week	19	21
Seldom	46	50
Never	23	6
Preference		
243	27	41
152	69	56
No difference	4	3
243 overall liking score		
1–5	12	6
6–9	88	94
152 overall liking score		
1–5	4	9
6–9	96	91
697 overall liking score		
1–5	12	6
6–9	88	94

3.3.5. Consumer Purchase Intent

The assessment of consumer purchase intent in relation to RPFP offered valuable insights into market receptivity. The potential market demand for RPFP was determined by evaluating consumer intentions to purchase under a defined scenario. The participants were presented with a hypothetical offering: a 50 g package of the product, sealed within an aluminum pouch, priced at THB 59 (approximately USD 1.66, based on an exchange rate of THB 35.518 per USD on 7 February 2024), and available for purchase on an online platform, such as Shopee (shopee.co.th). This scenario was designed to mirror a realistic online shopping experience, aiming to gauge consumer purchase decisions in a digital marketplace context.

The collected data indicated a diverse range of purchase intentions across the different samples. For the herbal-flavored sample (243), a notable number of participants (67%) expressed a tentative "maybe buy" decision, suggesting a moderate level of interest that might be converted into actual purchases with appropriate marketing strategies or product adjustments. Conversely, the commercial benchmark sample (152) elicited a less enthusiastic response, with only 43% of respondents indicating a "maybe buy" intention, highlighting potential areas for improvement or differentiation to enhance its market appeal. The spicy-flavored sample (697) encountered some resistance, with 49% of the survey participants leaning toward a "probably won't buy" decision, indicating significant hesitance that could impact its market penetration and necessitate the reconsideration of its flavor profile or marketing approach.

These findings are systematically represented in a bar graph (Figure 9), which illustrates the distribution of purchase intentions across the evaluated samples. In this visual representation, consumer responses are categorized into definitive levels of purchase interest, ranging from "definitely would purchase" to "definitely would not purchase", thereby providing a clear overview of consumer predisposition toward each product variant.

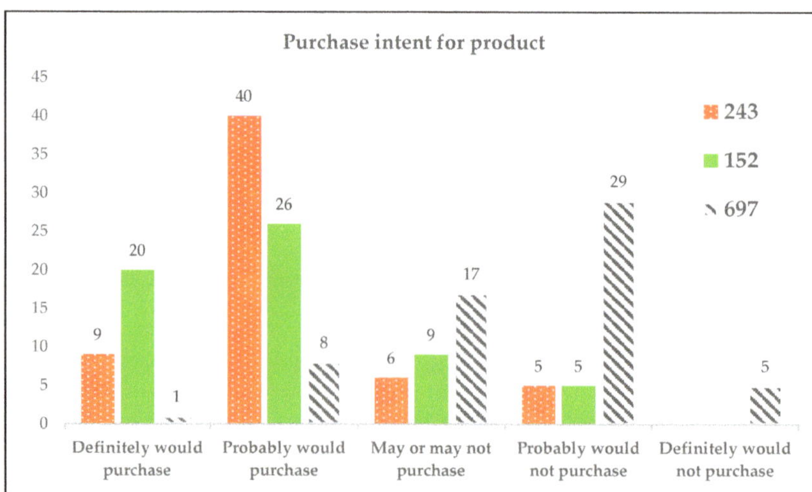

Figure 9. Distribution of purchase intentions for RPFP (samples 243, 152, and 697) among consumers, with these classified as "definitely would purchase", "probably would purchase", "may or may not purchase", "probably would not purchase", and "definitely would not purchase".

3.3.6. Consumer Preference

The consumer preference study aimed to elucidate the preference patterns for the RPFP samples, specifically samples 243 (herbal flavor) and 152 (benchmark), across various age demographics. This segment of the study was structured to identify any significant trends in consumer preferences that could inform product positioning and marketing strategies.

The participants were segmented into three age groups, namely, younger than 25 years (Generation Z), 26–43 years (Generation Y), and 44–58 years (Generation X), to assess how preferences might vary across different generations. The findings revealed that sample 152 (benchmark) was consistently favored across all age groups due to its mellow flavor profile, accessible texture, and subdued spicy and citrus notes, which collectively appealed to a broad spectrum of participants. For sample 243 (herbal flavor), while its robust flavor profile and authentic orange fish aroma, underscored by visible fish pieces, were appreciated for adding credibility to its origin, certain aspects, such as an overly intense flavor profile, pronounced sour notes, and a persistent fish aroma, were perceived negatively by some participants. Conversely, sample 152 was commended for its balanced flavor and fine, crumbly texture, although it was critiqued for potentially overemphasizing sweetness and saltiness and lacking the distinctive orange fish aroma that might have differentiated it further from competing products.

These insights are visualized in a bar graph (Figure 10), which provides a detailed comparison of consumer preferences for the RPFP samples across the designated age groups. This graphical representation facilitates a nuanced understanding of how each sample was received by different generational cohorts, highlighting the specific attributes that contributed to the overall preference for sample 152 over sample 243.

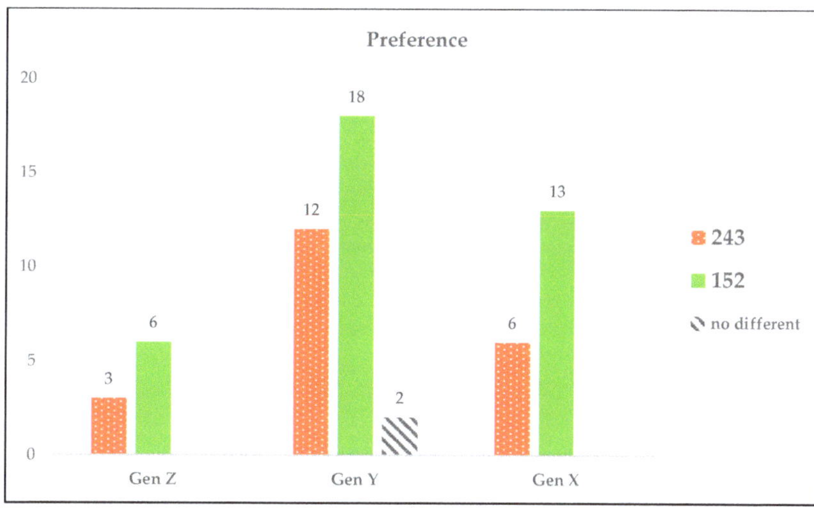

Figure 10. Comparative generational preferences for RPFP, illustrating the preferences among Generations Z (<25 years), Y (26–43 years), and X (44–58 years) for samples 152 (benchmark) and 243 (herbal flavor), including indications of no significant preference.

4. Discussion

Our study underscores the pivotal role of microbiological safety, sensory qualities, packaging efficiency, and alignment with consumer preferences in the online marketplace. In particular, the significant variations in nutritional composition between our developed RPFP variants and the commercial benchmark highlight the necessity of balancing traditional food product qualities with consumer health considerations. The protein content of the herbal flavor (28.97%) being lower than that of the commercial benchmark (40.17%) echoes findings from similar studies, indicating a potential gap in nutritional enhancement that could bridge traditional methods and commercial standards [41,42]. Additionally, the trade-off between flavor enhancement and nutritional cost, exemplified by the higher fat content in the spicy flavor (19.51%), demands careful consideration in product development for health-conscious consumers [43,44].

A noteworthy observation in our study is the discrepancy between the percentage of pickled fish added to each formula, as reported in Table 1, and the protein content of the finished products, as indicated in Table 3. This discrepancy may be attributed to variations in the protein content of the pickled fish, influenced by factors such as the fish type, size, age, and pickling process. Additionally, processing losses during production, including heat treatment and dehydration, could account for differences in the expected and observed protein content. These insights underline the complexity of achieving consistent nutritional profiles in food products and underscore the importance of considering these factors in product development and analysis.

Central to our findings is the discovery that the herbal flavor's superior dietary fiber content (14.23%) and the spicy flavor's heightened spiciness intensity underscore the intricate relationship between nutritional content and sensory attributes. This relationship is critical in online retail, where physical product evaluation is not possible, and purchasing decisions heavily rely on product descriptions and visual representations. Maintaining consistent sensory qualities can foster consumer trust and satisfaction, which are crucial for online sales success [45,46]. The emphasis on dietary fiber and spiciness opens avenues for targeted marketing strategies, addressing specific consumer health interests and taste preferences [47,48].

Our comprehensive evaluation of RPFP variants demonstrated their adherence to stringent microbiological safety standards, confirming their suitability for online retail. The meticulous analysis of microbial stability, including the assessment of pH levels and water activity over the storage period, underscores the effectiveness of our preservation techniques. These findings are crucial, as microbiological safety not only affects consumer health but also influences product reputation and trustworthiness in the online marketplace. Ensuring the microbiological integrity of RPFP variants through optimized processing and preservation methods reflects our commitment to delivering safe, high-quality food products to consumers.

The selection of packaging materials played a pivotal role in maintaining the sensory and nutritional integrity of the RPFP variants throughout the supply chain to the consumer. Our study highlights the transition from traditional polypropylene cups to laminated aluminum foil stand-up pouches, driven by the need for enhanced barrier properties, physical protection, and consumer convenience. These packaging solutions significantly contributed to extending the shelf life of the product by offering superior protection against oxygen, moisture, and microbial contamination, enhancing the overall product appeal and reinforcing brand image in the competitive online marketplace.

Our approach, evaluating RPFP variants through both nutritional and sensory lenses for online retail success, introduces a novel perspective in the field. This comprehensive assessment, including microbial safety and storage stability, offers a robust framework to ensure product quality in the digital marketplace. By addressing the literature gap in traditional food product enhancement for e-commerce, our study provides valuable insights into adapting traditional products to meet modern consumer demands [49,50].

However, our study's limitations, including the constrained sample size and focused market analysis, necessitate caution in generalizing these findings. Future research could broaden the scope of traditional products and market environments explored, validating our insights across a wider context. Additionally, delving into the long-term effects of storage on sensory and nutritional quality will enrich our understanding of product shelf-life and consumer satisfaction over time, further contributing to the field's knowledge base [51].

By navigating the complexities of online food retail with a multifaceted approach that prioritizes product integrity, sensory and nutritional expectations, and consumer preferences, RPFP stands well positioned for success in the competitive digital marketplace. This strategy not only addresses current market needs but also anticipates future consumer trends, establishing RPFP as a forward-thinking, consumer-centric offering.

5. Conclusions

Our study illuminated the critical importance of microbiological safety, sensory qualities, packaging efficiency, and consumer preferences in the online marketplace, particularly highlighting the intricate balance required between traditional food product attributes and modern consumer health expectations. The comprehensive evaluation conducted on roasted pickled fish powder (RPFP) variants demonstrated that variations in nutritional composition, specifically in protein and dietary fiber content, between the herbal and spicy flavors in comparison with a commercial benchmark significantly influence consumer trust and satisfaction in digital retail environments. This analysis underscores the potential for targeted marketing strategies that leverage these attributes to cater to specific consumer health interests and taste preferences.

Moreover, our investigation into the storage stability of RPFP variants revealed their robustness in maintaining microbiological safety and sensory attributes over time, underscoring the efficacy of optimized packaging solutions, particularly the transition to laminated aluminum foil stand-up pouches, in preserving product quality and extending shelf life. This focus on packaging efficiency not only enhances consumer appeal but also aligns with sustainability considerations crucial for online sales success.

Our novel approach in examining both the nutritional and sensory dimensions of RPFP variants for online retail success offers valuable insights for enhancing traditional food products in the e-commerce landscape. This dual focus contributes to closing the gap in the current literature, providing a robust framework for ensuring product quality and integrity in the digital marketplace.

However, it is crucial to acknowledge the limitations of our study, including the constrained sample size and the focused nature of our market analysis. These constraints may limit the generalizability of our findings, suggesting a need for caution in extrapolating our results to broader contexts. Future research should aim to expand the scope of traditional products and market environments explored, further validating our insights. Additionally, investigating the long-term effects of storage on sensory and nutritional quality will deepen our understanding of the factors that influence product shelf-life and consumer satisfaction, offering valuable directions for ongoing research in this domain.

In conclusion, by navigating the complexities of online food retail with a comprehensive approach that integrates product integrity, sensory and nutritional expectations, and consumer preferences, RPFP is strategically positioned for success in the competitive digital marketplace. Nevertheless, recognizing and addressing the limitations of this study are essential for advancing our knowledge and effectively meeting the dynamic demands of the digital age.

Author Contributions: Conceptualization, A.A. and P.B.; methodology, investigation, resources, data curation, writing—original draft preparation, writing—review and editing, supervision, project administration, and funding acquisition, A.A. All authors have read and agreed to the published version of the manuscript.

Funding: This research was supported by the Ministry of Foreign Affairs and International Cooperation of the Kingdom of Cambodia, through the Lancang-Mekong Cooperation (LMC) Special Fund, and administered by the Ministry of Women Affairs of the Kingdom of Cambodia. The project, titled "Women's Economic Empowerment in the Digital Age in Thailand", is registered with the Office of Innovation and Social Mission at Kasetsart University, reference number 00982/66.

Institutional Review Board Statement: The Kasetsart University Research Ethics Committee approved this study, which was carried out in compliance with the international guidelines for human research protection: the Declaration of Helsinki, the Belmont Report, CIOMS Guidelines, International Conference on Harmonization in Good Clinical Practice (ICH-GCP), and 45CFR 46.101(b) (COE No. COE67/013; date of approval: 9 February 2024).

Informed Consent Statement: Informed consent was obtained from all subjects involved in the study.

Data Availability Statement: The original contributions presented in the study are included in the article, further inquiries can be directed to the corresponding author.

Acknowledgments: The authors extend their gratitude to the Freshwater Fish Processing Community Enterprise (Aunt Uan Village No. 3) for their invaluable assistance in sample preparation. Special thanks are also due to the Khlongsamwa District Office for their support and facilitation of access to the community, making this research possible.

Conflicts of Interest: The authors declare no conflicts of interest. The funders had no role in the design of the study or in the decision to publish the results.

References

1. Ahuja, K.; Chandra, V.; Lord, V.; Peens, C. *Ordering in: The Rapid Evolution of Food Delivery*; McKinsey & Company: Chicago, IL, USA, 2021; p. 13.
2. Sloan, E. Top 10 Food Trends of 2021. In *Food Technology*; Institute of Food Technologists: Chicago, IL, USA, 2021.
3. Van Soest, T. Digital Transformation in the Food Industry: Trends, Examples, and Benefits. The Future of Commerce. 2023. Available online: https://www.the-future-of-commerce.com/2021/04/23/digital-transformation-in-food-industry/ (accessed on 12 February 2024).
4. Shuman, M. *Community Food Enterprise: Local Success in a Global Marketplace*; Article and Report Excerpts; Wallace Center at Winrock International: Arlington, VA, USA, 2010.
5. Deloitte China. *The Fresh Food Business: Spurring the 'Local Community' Trend Forward*; Deloitte China: Beijing, China, 2020; p. 13.
6. Chandra, V.; Gill, P.; Kohli, S.; Venkataraman, K.; Yoshimura, J.; Mathur, V. *The Next Horizon for Grocery E-Commerce: Beyond the Pandemic Bump*; Aull, B., Kohli, S., Eds.; McKinsey & Company: Chicago, IL, USA, 2022; p. 89.
7. Simmons, V.; Spielvogel, J.; Timelin, B.; Gi, M.T.P. *The Next S-Curve of Growth: Online Grocery to 2030*; McKinsey & Company: Chicago, IL, USA, 2022; p. 96.
8. Backley, J.H.; Herzog, L.J.; Foley, M.M. *Accelerating New Food Product Design and Development*, 2nd ed.; John Wiley & Sons Ltd.: Hoboken, NJ, USA; Institute of Food Technologists: Chicago, IL, USA, 2017; p. 400.
9. Ghule, P. Food Product Development Process: Steps & Flowchart. The Food Technologist. 2020. Available online: https://thefoodtechnologist.com/technology/food-product-development/ (accessed on 12 February 2024).
10. Moskowitz, H.R.; Porretta, S.; Silcher, M. *Concept Research in Food Product Design and Development*; Blackwell Publishing: Hoboken, NJ, USA, 2005.
11. Akhmadi, H.; Pratolo, S. Online Marketing of Food Products through Marketplace Platform: A Study of Community Based Online Marketplace of BEDUKMUTU. *E3S Web Conf.* **2021**, *232*, 02015. [CrossRef]
12. Akhmadi, H.; Fuadia Qurrotu Aini, Z.; Saffana, A.; Runanto, D. Online Purchase Behaviour of Food Product during COVID-19 Pandemic: A Study on Consumer of Bedukmutu Marketplace. *E3S Web Conf.* **2021**, *316*, 01013. [CrossRef]
13. Kwon, D.Y.; Nyakudya, E.; Jeong, Y.S. Fermentation: Food Products. In *Encyclopedia of Agriculture and Food Systems*; Van Alfen, N.K., Ed.; Academic Press: Oxford, UK, 2014; pp. 113–123.
14. Saithong, P.; Panthavee, W.; Boonyaratanakornkit, M.; Sikkhamondhol, C. Use of a starter culture of lactic acid bacteria in plaa-som, a Thai fermented fish. *J. Biosci. Bioeng.* **2010**, *110*, 553–557. [CrossRef] [PubMed]
15. Jittrepotch, N.; Rojsuntornkitti, K.; Kongbangkerd, T. Physico-chemical and sensory properties of Plaa-som, a Thai fermented fish product prepared by using low sodium chloride substitutes. *Int. Food Res. J.* **2015**, *2*, 721–730.
16. Ly, D.; Mayrhofer, S.; Domig, K.J. Significance of traditional fermented foods in the lower Mekong subregion: A focus on lactic acid bacteria. *Food Biosci.* **2018**, *26*, 113–125. [CrossRef]
17. Rodzi, N.A.R.M.; Lee, L.K. Traditional fermented foods as vehicle of non-dairy probiotics: Perspectives in South East Asia countries. *Food Res. Int.* **2021**, *150*, 110814. [CrossRef] [PubMed]
18. Narzary, Y.; Das, S.; Goyal, A.K.; Lam, S.S.; Sarma, H.; Sharma, D. Fermented fish products in South and Southeast Asian cuisine: Indigenous technology processes, nutrient composition, and cultural significance. *J. Ethn. Foods* **2021**, *8*, 33. [CrossRef]
19. Chan, S.X.Y.; Fitri, N.; Mio Asni, N.S.; Sayuti, N.H.; Azlan, U.K.; Qadi, W.S.M.; Dawoud, E.A.D.; Kamal, N.; Sarian, M.N.; Mohd Lazaldin, M.A.; et al. A Comprehensive Review with Future Insights on the Processing and Safety of Fermented Fish and the Associated Changes. *Foods* **2023**, *12*, 558. [CrossRef]
20. Onsurathum, S.; Pinlaor, P.; Haonon, O.; Chaidee, A.; Charoensuk, L.; Intuyod, K.; Boonmars, T.; Laummaunwai, P.; Pinlaor, S. Effects of fermentation time and low temperature during the production process of Thai pickled fish (pla-som) on the viability and infectivity of *Opisthorchis viverrini* metacercariae. *Int. J. Food Microbiol.* **2016**, *218*, 1–5. [CrossRef]
21. Begum, M.; Nowsad, A.A.; Al-Shahriar; Hasan, M.N.; Akter, M. Changes in Sensory Attributes and Consumers' Preference for Crispy Fish Pickles Prepared from Pangas (*Pangasianodon hypophthalmus*) at Different Storage Conditions. *Adv. Biol. Res.* **2022**, 3, 50–64.
22. Punyauppa-path, S.; Kiatprasert, P.; Punyauppa-path, P.; Rattanachaikunsopon, P.; Khunnamwong, P.; Limtong, S.; Srisuk, N. Distribution of *Kazachstania* Yeast in Thai Traditional Fermented Fish (Plaa-Som) in Northeastern Thailand. *J. Fungi* **2022**, *8*, 1029. [CrossRef] [PubMed]

23. Department of Medical Sciences. *Microbiological Quality Criteria for Food and Food Contact Containers*; Department of Medical Sciences: Nonthaburi, Thailand, 2017.
24. Aussanasuwannakul, A.; Teangpook, C.; Treesuwan, W.; Puntaburt, K.; Butsuwan, P. Effect of the Addition of Soybean Residue (Okara) on the Physicochemical, Tribological, Instrumental, and Sensory Texture Properties of Extruded Snacks. *Foods* **2022**, *11*, 2967. [CrossRef] [PubMed]
25. Sullivan, D.; Carpenter, D. *Methods of Analysis for Nutritional Labeling*; AOAC International: Rockville, MD, USA, 1993; Chapter 6; p. 106.
26. U.S. Food and Drug Administration. Yeasts, Molds and Mycotoxins. In *Bacteriological Analytical Manual (BAM)*; U.S. Food and Drug Administration: Silver Spring, MD, USA, 2001; Chapter 18.
27. U.S. Food and Drug Administration. Enumeration of *Escherichia coli* and Coliform Bacteria. In *Bacteriological Analytical Manual (BAM)*; U.S. Food and Drug Administration: Silver Spring, MD, USA, 2020; Chapter 4.
28. U.S. Food and Drug Administration. *Staphylococcus aureus*. In *Bacteriological Analytical Manual (BAM)*; U.S. Food and Drug Administration: Silver Spring, MD, USA, 2016; Chapter 12.
29. U.S. Food and Drug Administration. *Clostridium perfringens*. In *Bacteriological Analytical Manual (BAM)*; U.S. Food and Drug Administration: Silver Spring, MD, USA, 2001; Chapter 16.
30. U.S. Food and Drug Administration. *Bacillus cereus*. In *Bacteriological Analytical Manual (BAM)*; U.S. Food and Drug Administration: Silver Spring, MD, USA, 2020; Chapter 14.
31. *ISO 6579-1:2017/Amd.1:2020*; Microbiology of the Food Chain—Horizontal Method for the Detection, Enumeration and Serotyping of Salmonella—Part 1: Detection of Salmonella spp. ISO: Geneva, Switzerland, 2020.
32. Monteiro, M.L.G.; Mársico, E.T.; Lázaro, C.A.; da Silva Canto, A.C.V.C.; da Costa Lima, B.R.C.; da Cruz, A.G.; Conte-Júnior, C.A. Effect of transglutaminase on quality characteristics of a value-added product tilapia wastes. *J. Food Sci. Technol.* **2015**, *52*, 2598–2609. [CrossRef] [PubMed]
33. Lissitsa, S.; Kol, O. Generation X vs. Generation Y—A decade of online shopping. *J. Retail. Consum. Serv.* **2016**, *31*, 304–312. [CrossRef]
34. Parment, A. Generation Y vs. Baby Boomers: Shopping behavior, buyer involvement and implications for retailing. *J. Retail. Consum. Serv.* **2013**, *20*, 189–199. [CrossRef]
35. Lawless, H.T.; Heymann, H. *Sensory Evaluation of Food: Principles and Practices*; Springer: Berlin/Heidelberg, Germany, 2010; Volume 2.
36. Di Cairano, M.; Condelli, N.; Galgano, F.; Caruso, M.C. Experimental gluten-free biscuits with underexploited flours versus commercial products: Preference pattern and sensory characterisation by Check All That Apply Questionnaire. *Int. J. Food Sci. Technol.* **2022**, *57*, 1936–1944. [CrossRef]
37. Nurazizah, I.; Nur'utami, D.A. Application of check-all-that-apply (CATA) in sensory profile assessment of arabica dark roast and black pepper mixed coffee. *Future Food J. Food Agric. Soc.* **2021**, *9*, 1–12.
38. Patel, A.A.; Lopez, N.V.; Lawless, H.T.; Njike, V.; Beleche, M.; Katz, D.L. Reducing calories, fat, saturated fat, and sodium in restaurant menu items: Effects on consumer acceptance. *Obesity* **2016**, *24*, 2497–2508. [CrossRef]
39. Murillo, S.; Ardoin, R.; Prinyawiwatkul, W. Consumers' Acceptance, Emotions, and Responsiveness to Informational Cues for Air-Fried Catfish (*Ictalurus punctatus*) Skin Chips. *Foods* **2023**, *12*, 1536. [CrossRef]
40. *Lumivero XLSTAT*; Statistical and Data Analysis Solution; QSR International: Burlington, MA, USA, 2024.
41. Vasilaki, A.; Panagiotopoulou, E.; Koupantsis, T.; Katsanidis, E.; Mourtzinos, I. Recent insights in flavor-enhancers: Definition, mechanism of action, taste-enhancing ingredients, analytical techniques and the potential of utilization. *Crit. Rev. Food Sci. Nutr.* **2022**, *62*, 9036–9052. [CrossRef]
42. Tharrey, M.; Drogué, S.; Privet, L.; Perignon, M.; Dubois, C.; Darmon, N. Industrially processed v. home-prepared dishes: What economic benefit for the consumer? *Public. Health Nutr.* **2020**, *23*, 1982–1990. [CrossRef] [PubMed]
43. Xiao, Y.; Ji, X.; Xia, X. (Eds.) *Enhancement of Nutritional Profile/Biological Activity of Plant-Based Foods by Fermentation*; Frontiers Media SA: Lausanne, Switzerland, 2023.
44. Fanzo, J.; McLaren, R.; Bellows, A.; Carducci, B. Challenges and opportunities for increasing the effectiveness of food reformulation and fortification to improve dietary and nutrition outcomes. *Food Policy* **2023**, *119*, 102515. [CrossRef]
45. Ares, G.; Deliza, R. Studying the influence of package shape and colour on consumer expectations of milk desserts using word association and conjoint analysis. *Food Qual. Prefer.* **2010**, *21*, 930–937. [CrossRef]
46. Delarue, J.; Sieffermann, J.-M. Sensory mapping using Flash profile. Comparison with a conventional descriptive method for the evaluation of the flavour of fruit dairy products. *Food Qual. Prefer.* **2004**, *15*, 383–392. [CrossRef]
47. Mishra, B.P.; Mishra, J.; Paital, B.; Rath, P.K.; Jena, M.K.; Reddy, B.V.V.; Pati, P.K.; Panda, S.K.; Sahoo, D.K. Properties and physiological effects of dietary fiber-enriched meat products: A review. *Front. Nutr.* **2023**, *10*, 1275341. [CrossRef] [PubMed]
48. Arora, L.; Aggarwal, R.; Dhaliwal, I.; Gupta, O.P.; Kaushik, P. Assessment of sensory and nutritional attributes of foxtail millet-based food products. *Front. Nutr.* **2023**, *10*, 1146545. [CrossRef]
49. Barska, A.; Wojciechowska-Solis, J. E-Consumers and Local Food Products: A Perspective for Developing Online Shopping for Local Goods in Poland. *Sustainability* **2020**, *12*, 4958. [CrossRef]

50. Food and Agriculture Organization of the United Nations. Changing Consumer Preferences and Food Consumption Patterns. Available online: https://www.fao.org/3/cb8667en/online/src/html/changing-consumer-preferences-and-food-consumption-patterns.html (accessed on 12 February 2024).
51. Martínez, S.; Armesto, J.; Gómez-Limia, L.; Carballo, J. Impact of processing and storage on the nutritional and sensory properties and bioactive components of *Brassica* spp. A review. *Food Chem.* **2020**, *313*, 126065. [CrossRef]

foods

MDPI

Article

Transcriptome Reveals Regulation of Quorum Sensing of *Hafnia alvei* H4 on the Coculture System of *Hafnia alvei* H4 and *Pseudomonas fluorescens* ATCC13525

Yanan Wang [1,2], Xue Li [1,2], Gongliang Zhang [1,2], Jingran Bi [1,2] and Hongman Hou [1,2,*]

1 School of Food Science and Technology, Dalian Polytechnic University, No. 1 Qinggongyuan, Ganjingzi District, Dalian 116034, China; yananwang0226@163.com (Y.W.); lixuexiaomuzhu@163.com (X.L.); zgl_mp@163.com (G.Z.); bijr@dlpu.edu.cn (J.B.)

2 Liaoning Key Lab for Aquatic Processing Quality and Safety, No. 1 Qinggongyuan, Ganjingzi District, Dalian 116034, China

* Correspondence: houhongman@dlpu.edu.cn; Tel.: +86-411-8632-2020

Abstract: In the food industry, foodborne spoilage bacteria often live in mixed species and attach to each other, leading to changes in spoilage characteristics. Quorum sensing (QS) has been reported to be a regulating mechanism for food spoiling by certain kinds of bacteria. Here, the contents of biofilm, extracellular polysaccharides, and biogenic amines in the coculture system of *Hafnia alvei* H4 and *Pseudomonas fluorescens* ATCC13525 were significantly reduced when the QS element of *H. alvei* H4 was deleted, confirming that QS of *H. alvei* H4 is involved in the dual-species interactions. Then, transcriptomics was used to explore the regulatory mechanism at the mRNA molecular level. The deletion of the QS element decreased the transcript levels of genes related to chemotaxis, flagellar assembly, and the two-component system pathway of *H. alvei* H4 in the coculture system. Furthermore, a total of 732 DEGs of *P. fluorescens* ATCC13525 were regulated in the dual species, which were primarily concerned with biofilm formation, ATP-binding cassette transporters, and amino acid metabolism. Taken together, the absence of the QS element of *H. alvei* H4 weakened the mutual cooperation of the two bacteria in the coculture system, making it a good target for managing infection with *H. alvei* and *P. fluorescens*.

Keywords: *Hafnia alvei* H4; *Pseudomonas fluorescens*; cocultivation; QS regulation; transcriptome

Citation: Wang, Y.; Li, X.; Zhang, G.; Bi, J.; Hou, H. Transcriptome Reveals Regulation of Quorum Sensing of *Hafnia alvei* H4 on the Coculture System of *Hafnia alvei* H4 and *Pseudomonas fluorescens* ATCC13525. *Foods* **2024**, *13*, 336. https://doi.org/10.3390/foods13020336

Academic Editor: Boce Zhang

Received: 27 November 2023
Revised: 16 January 2024
Accepted: 19 January 2024
Published: 21 January 2024

1. Introduction

Aquatic foods are highly perishable due to their high protein content [1], in which microorganisms are necessary for food spoilage because of their ability to degrade nutrients [2]. The development and metabolic processes of microorganisms produce sulfides, amines, alcohols, ketones, aldehydes, and organic acids, which have disagreeable and undesirable odors [3]. But only a limited number of microorganisms, known as special spoilage microorganisms (SSOs), can defeat other microbes, grow significantly, and generate the appropriate metabolites [4,5].

Hafnia alvei was identified as the SSO of vacuum-packed spoiled meat [6] and spoiled seafood [7]. Furthermore, research has demonstrated that *H. alvei* can mediate the expression of various spoilage enzymes [8]. Amino acid decarboxylases, for instance, convert amino acid precursors into corresponding biogenic amines in *H. alvei*, which accelerates its spoilage [9,10]. *Pseudomonas fluorescens* is a common parasite in fish [11], poultry [12], dairy [13], and other fresh products. *P. fluorescens* is well-known for being able to generate dense biofilms [14,15], which contributes to the risk of food contamination with spoilage [16,17]. It has been discovered that the two most common Gram-negative bacteria found in damaged milk and chill-stored proteinaceous raw foods are *P. fluorescens* and *H. alvei* [18,19]. This suggests that these two organisms pose a threat to the food industry and food safety, and further research is urgently needed to control contamination from these bacteria to slow down the deterioration of the quality of fresh food. Naturally spoiled food contains numerous bacterial species and strains. The microbial interactions that occur during meat spoilage have a significant impact on the microbiota composition. Spoilage characteristics can be influenced by various relationships, including antagonism, mutualism, and commensalism, among the microbiota members [20]. Therefore, the study of the interaction between bacteria under coculture conditions is more in line with the real natural environment of food.

QS is a density-dependent mechanism that regulates gene expression in bacteria by secreting and detecting small molecules called autoinducers. The N-acyl-1-homoserine lactone (AHL) synthase and AHL reporter are involved in the QS system, which plays an important role in regulating biofilm formation and the expression of spoilage genes. This demonstrates the relationship between the QS system, biofilm formation, and the potential for spoilage [21]. A previous study has indicated that in *H. alvei* H4, there is a LuxI/R-type QS system present. Additionally, there was no AHL synthesis in the *luxI* mutant [7], indicating that quorum sensing is disrupted. However, the biological properties and importance of these systems have mainly focused on mono-species cultures with reasonably well-established conditions [22], where most microorganisms live in complex communities [23]. QS signals can be used as a strategy for bacteria to compete with other microbes coresiding in the same environment [24]. The overall role of QS as a regulator of interspecies interactions with mixed species is relatively poorly understood, especially in the context of QS-regulated effectors. *H. alvei* induces other bacterial species in the same environment to display food-quality-related behaviors [6], but the mechanism is not understood. Therefore, it is particularly necessary to explore the mechanism of interaction and QS between *H. alvei* H4 and other spoilage bacteria.

In this study, the effect of QS of *H. alvei* H4 WT on biofilm development, extracellular polysaccharide, and biogenic amine production in both *H. alvei* H4 and *P. fluorescens* ATCC13525 was assessed. Subsequently, dual-seq analyses were performed on the cells to uncover the regulatory mechanism of QS in regulating the spoilage ability of the two bacteria.

2. Materials and Methods

2.1. Bacterial Strains and Growth Conditions

The wild-type *H. alvei* H4 (*H. alvei* H4 WT) and a *luxI* mutant (*H. alvei* H4 Δ*luxI*), which had been previously constructed in our lab, were used in this study [7]; *P. fluorescens* ATCC13525 was purchased from the BeNa Culture Collection Centre. Every strain was regularly cultured in Luria Bertani (LB) medium (10 g/L tryptone, 10 g/L sodium chloride, and 5 g/L yeast extract) at 30 °C, and 20 μg/mL chloramphenicol was added for Δ*luxI* incubation.

2.2. Crystal Violet Biofilm Assay

The biofilms generated by *P. fluorescens* ATCC13525 and *H. alvei* H4 WT/Δ*luxI* single or dual species were compared using a crystal violet (CV) assay, in accordance with the method provided by Hou et al. [25]. The overnight-cultured *P. fluorescens* ATCC13525 and *H. alvei* H4 cultures were diluted 100-fold into 96-well plates (about 10^5 CFU/mL). After 12, 24, 36, and 48 h of incubation at 30 °C under static conditions, the culture medium was carefully collected and cleaned with sterile PBS (pH 7.0). Subsequently, 250 μL of anhydrous methanol was added and incubated for 15 min to preserve the cells. After staining with CV (0.1% *w*/*v*) for 15 min, the biofilm cells that clung to the lower surface and sides of each well were removed using deionized water (250 μL per well). The CV was resolubilized with 200 μL of 33% glacial acetic acid and then shaken at a speed of 300 rpm for a duration of 15 min. The optical density of the solution was subsequently measured at OD_{590} nm using a Spectra M2 spectrophotometer.

2.3. Extracellular Polysaccharides Assay

The procedure described by Harimawan et al. [26] was followed to extract extracellular polysaccharides of biofilm. Briefly, at each time interval (12, 24, 36, and 48 h), 10 mL of cultures was carefully aspirated and then centrifuged at 4 °C (8000× *g*, 15 min). Then, the precipitate was resuspended in saline, and it was placed in a water bath set at 80 °C for 30 min. After a 30 min centrifugation at 4 °C (12,000× *g*), the supernatant was passed through a 0.22 μm membrane. The content of exopolysaccharides in the supernatant was measured using the phenol sulfuric acid method.

2.4. Biogenic Amine Measurement Using HPLC

The amount of putrescine or cadaverine was measured using the previously described method [27]. The single colonies of three strains were selected and cultured in the amino acid medium at 30 °C and 150 rpm of shaking until the medium became turbid, and then, it was transferred again. At each time interval (12, 24, 48, and 72 h), 1 mL of cultures was carefully aspirated, 9 mL of 10% trichloroacetic acid (TCA) was added, and it was stored for 2 h at 4 °C. Subsequently, centrifuged at 4 °C (3000× *g*, 10 min) for 10 min, 200 μL of the liquid above the solid residue was pipetted, and 80 μL NaOH (2M), 120 μL of saturated $NaHCO_3$, and 800 μL of dansyl chloride (10 mg/mL, prepared in acetone) were added in order to complete the derivatization process. Following that, the mixture was placed in a warm bath set at 45 °C for 40 min. Subsequently, 50 μL of ammonia was added, and the mixture was allowed to sit at room temperature for 30 min. Finally, 550 μL of chromatography grade acetonitrile was added and centrifuged at 4 °C for 5 min at 3000× *g*. The resulting supernatant was filtered three times through a 0.22 μm filter membrane and stored at −80 °C. The same procedure was followed to detect biogenic amines produced by *H. alvei* H4 and Δ*luxI* in the presence of *P. fluorescens* ATCC1355.

The amount of BA was measured using an Agilent 1260 HPLC unit (Agilent Technologies Inc., Santa Clara, CA, USA) with an Agilent Zorbax SB-C 18 column (4.6 × 150 mm). There was a 1.0 mL/min flow rate. The column was eluted using an optimized gradient comprising ultrapure water (Solvent A) and acetonitrile (Solvent B) in a binary solution, following an optimized gradient: 0–10 min, 55% B; 10–15 min, 55–65% B; 15–20 min, 65–80% B; 20–25 min, 80% B; 25–30 min, 80–90% B; 30–33 min, 90% B; 33–35 min, 90–55% B.

2.5. Transcriptome Analysis and Quantitative Reverse Transcription (RT-qPCR)

The method of transcriptome analysis was referenced from Xue et al. [28]. *H. alvei* H4 and *P. fluorescens* ATCC13525 in dual species were used to extract RNA using the RNA prep pure Cell/Bacteria Kit (Tiangen Biotech, Beijing, China). Before conducting the experiments, the mRNA was isolated from total RNA by employing specific probes to eliminate rRNA contaminants. The fragmentation process was then carried out by subjecting the mRNA to high temperatures in the presence of divalent cations in the First Strand Synthesis Reaction Buffer (5X). For the generation of the first strand cDNA, a random hexamer primer was utilized along with M-MuLV Reverse Transcriptase (RNase H-). Subsequently, in the DNA polymerase I system, the dUTP was used as a substitute for dTTP in order to synthesize the second strand of cDNA. Finally, to ensure the selection of cDNA fragments in the desired length range of 370~420 bp, the library fragments underwent purification using the AMPure XP system from Beckman Coulter in the USA. PCR was then performed with Phusion High-Fidelity DNA polymerase, Universal PCR primers and Index (X) Primer. At last, the Agilent Bioanalyzer 2100 System was used to evaluate the library quality after the PCR products had been purified using the AMPure XP System.

RT-qPCR experiments with SYBR Green were performed to validate genes of interest and specific genes associated with the spoilage factors or viability of *H. alvei* H4 and *P. fluorescens* ATCC13525. Threshold cycles for the endogenous control (16S rRNA) were used to standardize the data. The cycle number at which the intensity of the fluorescence reached a predefined threshold was identified as the threshold cycle or Ct. Each gene was quantitated using the $-2^{-\Delta\Delta Ct}$ method.

2.6. Statistical Analysis

Every experiment was repeated in triplicate. The statistical software IBM SPSS (v19.0, IBM SPSS Inc., Armonk, NY, USA) was used to perform a one-way analysis of variance (ANOVA) on the data, which were reported as mean ± standard deviation (SD). A portion of the data was examined on the online platform of NovoMagic Cloud Platform https://magic.novogene.com (accessed on 16 August 2023).

2.7. Accession Numbers

The raw and processed transcriptome data of *H. alvei* H4 and *P. fluorescens* ATCC13525 in dual RNA seq were deposited at the Gene Expression Omnibus (GEO) database under the accession number GSE245811.

3. Results and Discussion

3.1. Characteristics under Coculture Conditions

3.1.1. Biofilm Formation

In the food industry, bacteria can adhere to solid surfaces to form biofilms, which are extremely difficult to detect and eradicate [29]. In order to investigate whether biofilm formation is influenced by interspecies interactions between *H. alvei* H4 and *P. fluorescens* ATCC1355, crystal violet staining was used to monitor the biofilm formation of *H. alvei* H4 and *P. fluorescens* ATCC1355 in a mono-inoculated or co-inoculated system in LB medium. First, there was a synergistic interaction between *H. alvei* H4 and *P. fluorescens* ATCC1355, where an increase in optical density was observed for the dual-species biofilms compared with biofilms formed by a single-culture strain (Figure 1A). Consistent with previous reports, mixed species always showed a high potential for biofilm formation, both under natural conditions and in the laboratory environment [30]. The hypothesis is that the synergistic effect in dual-species biofilms could be attributed to QS of *H. alvei* H4 WT. In particular, reductions in dual-species biofilm formation were observed when the *luxI* gene was deleted. These findings suggest that QS of *H. alvei* H4 WT plays a favorable role in the development of biofilms involving two species (Figure 1A). It may be that the destruction of QS led to the weakening of other metabolic functions in *H. alvei* H4 WT, which, in turn, affected the activities of *H. alvei* H4 WT and the cocultured *P. fluorescens* ATCC13525.

3.1.2. Exopolysaccharide Assay

Polysaccharides are also major components of EPS in the *P. fluorescens* biofilm matrix [14]. The amount of polysaccharides in mono- and dual-species biofilms was quantified, as presented in Figure 1B. Similar results to the optical density biofilm assay were obtained for the exopolysaccharide secretion. All strains demonstrated a strong exopolysaccharide-forming ability at 30 °C, especially the wild-type coculture system, whose exopolysaccharide biomass was significantly higher than that of the single species, indicating that the interactions between *H. alvei* H4 WT and *P. fluorescens* ATCC13525 promoted the capacity of the biofilm. However, the EPS secretion viability of *P. fluorescens* ATCC13525 cocultured with *H. alvei* Δ*luxI* was significantly decreased, suggesting the critical role of QS in the regulation of the interaction between *H. alvei* H4 WT and *P. fluorescens* ACTT13525.

3.1.3. Biogenic Amine Production

The formation of BAs is a result of amino acid decarboxylation, which is associated with the activities of microorganisms [31]. Putrescine (PUT) and cadaverine (CAD) are two common BAs found in spoiled seafood, with putrescine being synthesized through ornithine decarboxylation or agmatine deamination and cadaverine originating from lysine decarboxylation [32]. The content of PUT and CAD in various culture systems was measured to investigate the impact of QS on these biogenic amines (Figure 1C,D). It was demonstrated that at each period, the amount of PUT and CAD produced by *H. alvei* H4 WT was higher than *P. fluorescens* ACTT13525, which was almost undetectable. The amount of *H. alvei* H4 Δ*luxI* biogenic amines was substantially reduced ($p < 0.05$) compared with the

WT strain, and when they combined with *P. fluorescens* ACTT13525, the wild-type coculture system was still at an advantage in terms of yield. The higher levels of biogenic amines found in the dual species could be attributed to the influence of QS on the production of amino acid decarboxylase.

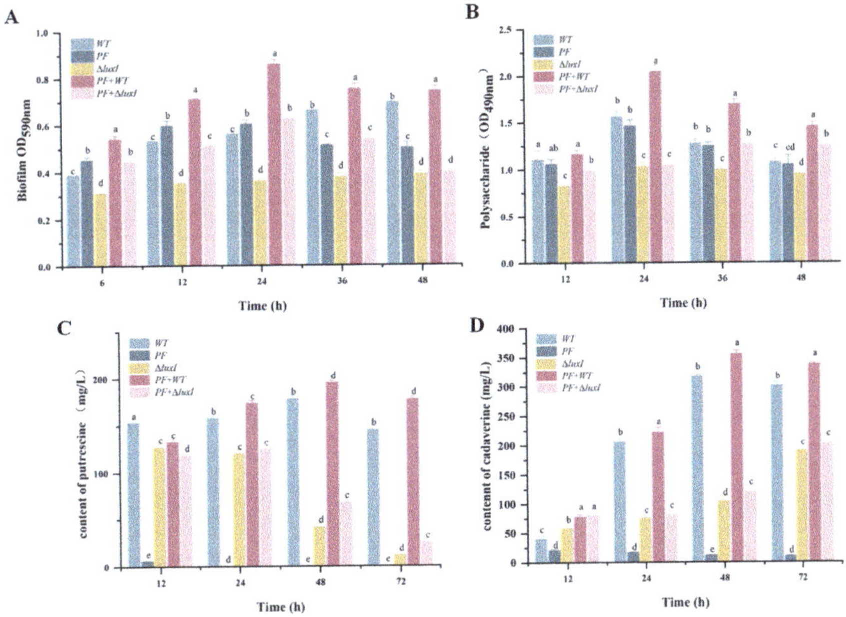

Figure 1. Changes in spoilage phenotypes in monoculture and coculture systems at different times. (**A**) Biofilm biomass; (**B**) polysaccharide; (**C**) putrescine; (**D**) cadaverine. WT: *Hafnia alvei* H4 wild type; PF: *Pseudomonas fluorescens* ATCC13525; Δ*luxI*: *Hafnia alvei* H4 *luxI* mutant strain; PF + WT: *P. fluorescens* ATCC13525 cocultured with *H. alvei* H4; PF + Δ*luxI*: *P. fluorescens* ATCC13525 cocultured with *H. alvei* H4 Δ*luxI* a, b, c, d and e represent significant differences between groups.

3.2. Transcriptome Analysis under Coculture Conditions

3.2.1. Transcriptional Responses in *H. alvei* H4 and *P. fluorescens* ATCC13525

The cocultures formed by duo species (*H. alvei* H4 WT + *P. fluorescens* ACTT13525) and duo species (*H. alvei* H4 Δ*luxI* + *P. fluorescens* ACTT13525) underwent RNA-seq. Sequences read from two different transcriptomes were mapped to the reference genomes of two bacteria (*H. alvei* and *P. fluorescens*), respectively, using bioinformatic methods (Figure 2A). Among these wild-type coculture system biological replicates, an average of 55.38% clean reads were mapped to the *H. alvei* reference genome (data set WP-H), and 43.93% clean reads were mapped to the *P. fluorescens* ATCC13525 reference genome (data set WP-P). Among these deletion-type coculture systematic biological replicates, an average of 50.52% clean reads were mapped to the *H. alvei* reference genome (data set IP-H), and 48.73% clean reads were mapped to the *P. fluorescens* ATCC13525 reference genome (data set IP-P).

In the differential gene expression analysis of RNA-seq data, *H. alvei* H4 and *P. fluorescens* ATCC13525 in the wild-type coculture system were used as controls. In *H. alvei* H4 Δ*luxI*, 1428 of the identified genes were significantly differentially expressed (DEGs), of which 675 were upregulated, and 753 were downregulated by comparing IP-H and WP-H. Similarly, a total of 792 DEGs were detected in *P. fluorescens* ACTT13525 in IP-P, including 418 upregulated and 374 downregulated DEGs. Volcano plots were used to visualize the distinct transcriptional profiles of DEGs in the two comparisons (Figure 2). To obtain a more profound understanding of the potential functional differences resulting from DEGs, a comprehensive functional grouping analysis based on the GO and KEGG database mapping was carried out (Figure 3). The enrichment genes in the three GO categories are summarized in Figure 3; all DEGs were categorized into three distinct groups: biological processes, cellular components, and molecular functions. The most enriched terms in these three categories in *H. alvei* H4 were the peptide metabolic process, organelles, and the structural constituents of ribosomes, respectively (Figure 3A). In *P. fluorescens* ATCC13525, after being cocultured with *H. alvei* H4 Δ*luxI*, the oxidation–reduction process, ribosome, and oxidoreductase activity were the most enriched terms in these three categories, respectively (Figure 3C). The KEGG database was used to further analyze the biological functions related to DEGs, and the results reveal 20 significantly changed metabolic pathways (Figure 3).

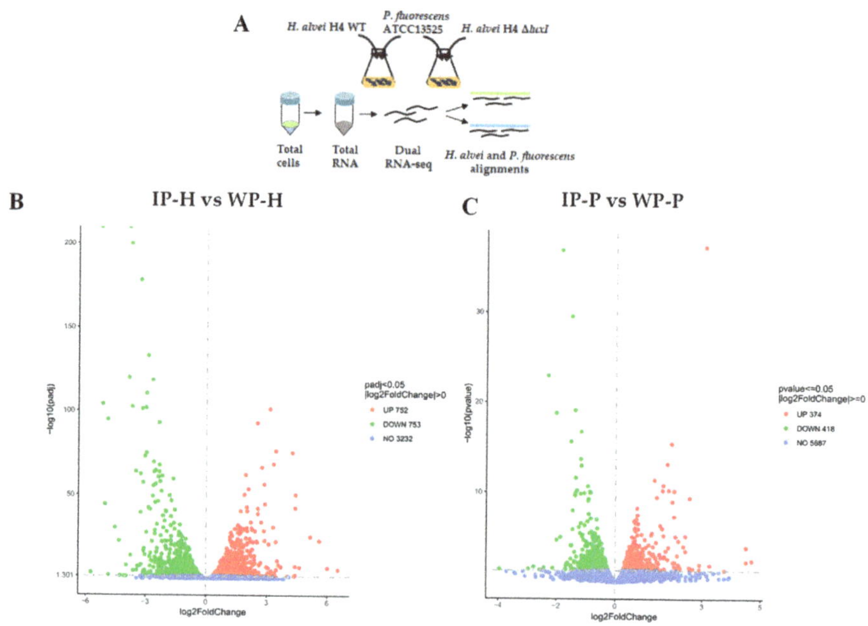

Figure 2. Experimental setup and total analysis of differentially expressed genes in *Hafnia alvei* H4 and *Pseudomonas fluorescens* ATCC13525. (**A**) Experimental setup for culture preparation and RNA extraction. (**B**) Volcano plot displaying differential regulation of *H. alvei* H4 genes in IP-H vs. WP-H; (**C**) Volcano plot displaying differential regulation of *P. fluorescens* ATCC13525 genes in IP-P vs. WP-P. Green and red dots represent down- and upregulated genes, respectively, and blue dots represent genes whose expressions were not altered significantly.

Many differential genes are involved in (but not limited to) flagellar assembly, bacterial chemotaxis, transcriptional regulators, the signal transduction system, and amino acids metabolism (Tables 1 and 2). The following is a detailed discussion of the KEGG enrichment pathways that are engaged in the reaction network.

3.2.2. Analysis of Genes Associated with Biofilm Formation

Flagellar Assembly and Bacterial Chemotaxis

Bacterial attachment to the surface is an essential initial step in biofilm formation, which is facilitated by flagella. It has been suggested that the flagellum is responsible for motility in Gram-negative bacteria [33]. The primary components of flagellar are the matrix, hook, and filaments. As shown in Table 1, gene downregulation caused by QS disruption was mainly focused on flagellar structural genes (*flgBCDE*, *flgFGHIJKL*, *fliCfliDfliS*, fliJK, *fliMN*, and *flhA*), two transcriptional regulatory protein genes (*flgM* and *fliA*), and several genes encoding chemotaxis proteins (*motA* and *cheW*) in *H. alvei* H4 Δ*luxI* (Table 1). The flagellar basal-body protein and lateral flagellar protein are encoded by the *flgC* and *flgL* genes. FlgD and FliN are important components of the flagellar hook and the flagellar cytoplasmic C-ring (the mechanical core of the flagella), respectively, while MotA is a motility protein present on the surface of the bacteria that provides the power for normal flagellar motility [34–36]. The downregulation of the above genes may reduce the fixation ability and driving ability of the substrate, ultimately leading to a decrease in bacterial adhesion.

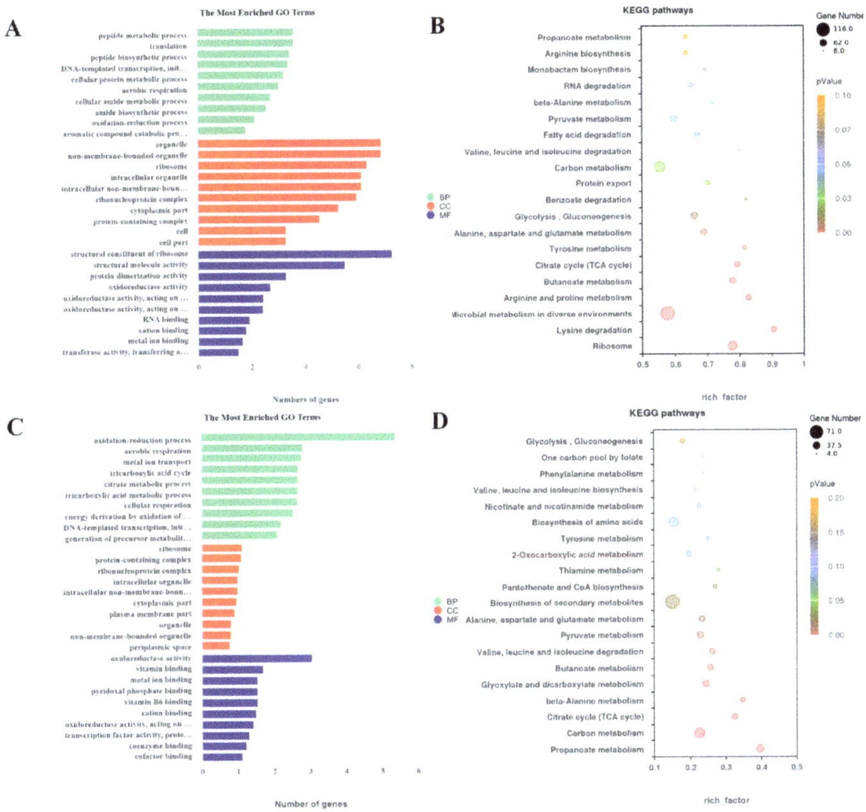

Figure 3. Statistical enrichment of DEGs in GO and KEGG pathways (MF: molecular function, BP: biological process, CC: cellular component). (**A**) GO terms of DEGs in *Hafnia alvei* H4 of IP-H vs. WP-H; (**B**) Scatter diagram of the top 20 KEGG enrichment pathways for DEGs in *H. alvei* H4 of IP-H vs. WP-H; (**C**) GO terms of DEGs in *Pseudomonas fluorescens* ATCC13525 of IP-P vs. WP-P; (**D**) Scatter diagram of the top 20 KEGG enrichment pathways for DEGs in *P. fluorescens* ATCC13525 of IP-P vs. WP-P.

Table 1. Validation of selected DEGs related to flagellar assembly, bacterial chemotaxis, BA metabolism, and two-component system in *H. alvei* H4 (IP-H vs. WP-H).

Gene Name	log2 Radio	Gene Description
flagellar assembly		
fliL	−4.32515228	flagellar basal body-associated protein FliL
fliN	−5.766718961	flagellar motor switch protein FliN
fliM	−2.837097495	flagellar motor switch protein FliM
fliJ	−2.408896556	flagellar export protein FliJ
ERL64_RS17280	−1.339639572	flagellar hook-length control protein FliK
fliS	−2.016760592	flagellar export chaperone FliS

Table 1. *Cont.*

Gene Name	log2 Radio	Gene Description
motA	−2.425518042	flagellar motor stator protein MotA
flgB	−3.035717278	flagellar basal body rod protein FlgB
flgC	−3.475398667	flagellar basal body rod protein FlgC
flgD	−3.258609823	flagellar hook assembly protein FlgD
flgE	−3.074676572	flagellar hook protein FlgE
ERL64_RS17360	−3.436893106	flagellar basal body rod protein FlgF
flgG	−3.336673997	flagellar basal body rod protein FlgG
flgK	−1.888983693	flagellar hook-associated protein FlgK
flgL	−2.226392467	flagellar hook-associated protein FlgL
ERL64_RS17345	−1.831507544	flagellar basal body P-ring protein FlgI
bacterial chemotaxis		
ERL64_RS14790	−2.265739966	methyl-accepting chemotaxis protein
Tsr	−2.265395717	methyl-accepting chemotaxis protein
ERL64_RS14800	−2.424604914	chemotaxis response regulator protein
cheR	−2.098189509	protein-glutamate O-methyltransferase CheR
ERL64_RS14800	−2.424604914	chemotaxis response regulator protein-glutamate methylesterase
two-component system		
envZ	−0.966750293	two-component system sensor histidine kinase EnvZ
ompR	−0.459688448	two-component system response regulator OmpR
phoR	−1.146705289	phosphate regulon sensor histidine kinase PhoR
phoQ	0.940798155	two-component system sensor histidine kinase PhoQ
phoP	1.035332663	two-component system response regulator PhoP
rpoN	−1.164441481	RNA polymerase factor sigma-54
BA metabolism		
ldcc	−3.212727378	lysine decarboxylase
speF	−0.860138669	ornithine decarboxylase SpeF
speA	−0.82353036	biosynthetic arginine decarboxylase
speB	−0.669157809	agmatinase
puuA	3.06324221	glutamine synthetase family protein
puuB	2.140552542	FAD-binding oxidoreductase
puuC	2.239399155	aldehyde dehydrogenase PuuC
puuD	2.663951099	gamma-glutamyl-gamma-aminobutyrate hydrolase
patD	1.429802723	amino butyraldehyde dehydrogenase
ERL64_RS21280	3.401803121	amma-amino butyraldehyde dehydrogenase
ERL64_RS02325	3.06324221	glutamine synthetase family protein
speG	0.419542916	spermidine N1-acetyltransferase
potF	−0.987904791	spermidine/putrescine ABC transporter substrate-binding protein PotF
potI	−2.362303979	putrescine ABC transporter permease PotI
potH	−2.568324997	putrescine ABC transporter permease PotH
potG	−0.987904791	putrescine ABC transporter ATP-binding subunit PotG

Table 2. Selected DEGs related to biofilm formation, amino acid metabolism, and ABC transport in *P. fluorescens* ATCC13525 (IP-P vs. WP-P).

Gene Name	log2 Radio	Gene Description
biofilm formation		
fliL	−0.925171985	flagellar basal body-associated protein FliL
flgM	−0.699499627	flagellar biosynthesis anti-sigma factor FlgM
fliA	−0.523949137	RNA polymerase sigma factor FliA
CPH89_RS16845	−1.029022096	OmpA family protein
cheW	−0.506899146	chemotaxis protein CheW
CPH89_RS23015	−0.547698862	methyl-accepting chemotaxis protein
tssC	−1.370116898	type VI secretion system contractile sheath large subunit
tssG	2.543096874	type VI secretion system baseplate subunit TssG

Table 2. *Cont.*

Gene Name	log2 Radio	Gene Description
tssH	0.939045148	type VI secretion system ATPase TssH
CPH89_RS09505	1.808730303	type VI secretion system tip protein VgrG
amino acid metabolism		
CPH89_RS22860	−1.373666561	methionine gamma-lyase
CPH89_RS03900	−1.088510925	L-serine ammonia-lyase
CPH89_RS10590	−0.592860749	homoserine kinase
mmsB	−1.524151588	3-hydroxyisobutyrate dehydrogenase
CPH89_RS06815	0.574998597	catalase
katB	1.644147633	catalase KatB
CPH89_RS08005	2.317725314	D-amino acid dehydrogenase
CPH89_RS00185	0.475801482	aspartate/tyrosine/aromatic aminotransferase
CPH89_RS21450	1.152619749	aspartate aminotransferase family protein
CPH89_RS16850	0.716263227	argininosuccinate synthase
CPH89_RS21350	0.422007256	glutamine synthetase family protein
gltB	0.609812179	glutamate synthase large subunit
CPH89_RS00200	0.920911912	aldehyde dehydrogenase family protein
CPH89_RS27825	1.897595222	CoA-acylating methylmalonate-semialdehyde dehydrogenase
ABC transport		
RbsB	−0.923885631	sugar ABC transporter substrate-binding protein
CPH89_RS28860	−1.034779256	ABC transporter ATP-binding protein
CPH89_RS10545	−0.845831876	MetQ/NlpA family ABC transporter substrate-binding protein
CPH89_RS24050	−2.801510174	ABC transporter permease
CPH89_RS19375	−1.480685108	ABC transporter ATP-binding protein
CPH89_RS16095	−1.459658475	amino acid ABC transporter permease
CPH89_RS23690	2.876670285	carbohydrate ABC transporter permease
urtD	2.872150064	urea ABC transporter ATP-binding protein UrtD
CPH89_RS08665	1.314199994	sulfonate ABC transporter substrate-binding protein
ssuB	0.972531529	aliphatic sulfonates ABC transporter ATP-binding protein
CPH89_RS15185	0.63533128	glycine betaine ABC transporter substrate-binding protein

In line with these observations, the downregulated genes of *P. fluorescens* ATCC13525 in response to QS of *H. alvei* H4 WT were two genes encoding transcription regulators and RNA polymerase sigma factor, including *flgM*, encoding the anti-σ-28 factor, which plays a crucial role in regulating the flagellar number and flagellar assembly rate, and *fliA* (as a σ factor regulating bacterial flagella gene expression) [37]. These results indicate that the downregulation of *flgM* and *fliA* in the IP-P group was due to the absence of the QS element in *H. alvei* H4 WT and further indicated that QS of *H. alvei* H4 WT exerted a positive effect on the flagellar synthesis transcriptional regulators of *P. fluorescens* ATCC13525, rather than directly on the basic structure of the flagellum.

Bacterial chemotaxis is a crucial mechanism in various biological processes, such as the formation of biofilms and the regulation of quorum sensing, etc. It relies on the mobility of bacteria, wherein their movement is triggered by chemical stimuli [38]. The entire process of bacterial chemotaxis is governed by a complex interplay among different proteins, including methyl-accepting chemotactic proteins (MCPs) and cytoplasmic proteins (Ches). MCPs are responsible for sensing signals, while cytoplasmic proteins process sensory signals and then transmit control signals to flagellar motors for chemotactic movement [39]. MCPs are able to cross the cell membrane, detect chemical changes in the surrounding environment, and then send out signals by themselves to induce bacteria to adapt to growth. The downregulation of MCPs leads to the decreased ability of bacteria to sense signals in the environment [40]. In addition, a previous study showed that the phosphorylated form of cytoplasmic proteins cheY binds to FliM and FliN, switching the rotation direction of the rotation of flagellar motors [41]. In this study, the expression of *tsr* (encoding MCPs) was downregulated in the IP-H group, and the genes encoding FliM and FliN were also significantly downregulated (Table 1). Therefore, the loss of the QS element can lead to the inhibition of the expression of chemotactic-related genes in *H. alvei* H4 WT during coculture, further affecting the content of biofilms. In *P. fluorescens* ATCC13525, the chemotactic adaptation protein CheW, which interacts with MCPs and CheA and is necessary for the MCP-mediated control of CheA activity, was likewise downregulated [42]. This suggests that QS in *H. alvei* H4 WT influences the chemotactic expression of *P. fluorescens* ATCC13525 in a coculture system. The expression of these chemotactic genes in *H. alvei WT* and *P. fluorescens* is also regulated by the AHL-mediated QS system, which is consistent with the case in *Aeromonas* [43].

Signal Transduction System

Bacteria have multiple signal transduction systems, and two-component systems (TCSs) are significant signaling pathways that control chemotaxis, biofilm formation, and responses to environmental stimuli [44]. The primary constituents of the TCS are histidine protein kinases (HKs) and response regulators (RRs). The EnvZ/OmpR system is a common two-component system, which creates a viable way for bacteria to survive in adverse conditions [45]. Briefly, OmpR is activated by EnvZ and binds to the target gene promoter, activating the expression of corresponding genes such as *ompF* and *ompC*. Thereby, it senses external signal stimuli and modulates the motility of microbial cells, thus affecting biofilm formation [46]. As shown in Table 2, the downregulation of these signal transduction genes for *envZ* (encoding HKs) and *ompR* (encoding RRs) further proves that QS of *H. alvei* H4 WT may influence biofilm formation by modulating its own signaling in the coculture, which is similar to the result that DEGs enriched in TCSs may sense external signal stimuli and modulate the motility of microbial cells, thus influencing the formation of biofilms [46]. We found that in response to the destruction of QS, most of the DEGs related to signal transduction were involved in the two-component system of *H. alvei* H4 WT.

However, the genes for ABC transporters in *P. fluorescens* ATCC13525 were among the most strongly downregulated genes when cocultured with *H. alvei* H4 Δ*luxI*. ABC transporters play a crucial role in facilitating the transportation of essential nutrients and the removal of harmful substances across the membrane in microorganisms. This effectively enhances the microorganisms' ability to resist stressful conditions [47]. Additionally, these transporters are responsible for carrying a diverse range of substrates, such as ions and macromolecules, as well as materials required for biofilm formation [48]. *RbsB* is a ribose-binding protein, a subset of which is also the primary chemoreceptor for chemotaxis. The research of Zhang et al. [49] suggests that *RbsB* may play an important role in QS systems, and Armbruster et al. [50] also demonstrate that the deletion of *RbsB* significantly affects the absorption of AI-2 and biofilm formation. In this study, the expression of *RbsB* in the IP-P group was downregulated (Table 2), which may explain the decrease in biofilm formation during the coculture of *P. fluorescens* ATCC13525 and *H. alvei* H4 Δ*luxI* (Figure 1A).

Bacteria can cooperate with each other to form dense biofilms on the surfaces of various food production equipment, which are difficult to remove and cause food spoilage. The above results demonstrate that disrupting QS of *H. alvei* H4 WT can downregulate genes related to biofilm formation in cocultures, thereby reducing biofilm formation (Figure 1), and is an effective way to inhibit food spoilage.

3.2.3. Amino Acid Metabolism

BAs are organic compounds with biological activity that result from the decarboxylation of precursor amino acids caused by microorganisms' amino acid decarboxylases [51]. Therefore, the contents of precursor amino acids are highly likely to affect the production of biogenic amines by microorganisms. According to our previous results, CAD and PUT are the main BAs found in *H. alvei* H4 WT. CAD originates from lysine decarboxylation. *ldcc* is a key gene (encoding lysine decarboxylase) in the decarboxylation of lysine for the synthesis of CAD. According to Table 1, *ldcc* was downregulated in the IP-H group compared to the WP-H group for *H. alvei* H4, while it was not activated in *P. fluorescens* ATCC13525 of the IP-P group compared to the WP-P group, which may inhibit the decarboxylation of lysine to produce CAD. The synthesis of PUT involves a relatively complex multi-step process [52], which is synthesized via ornithine decarboxylation or agmatine deamination. In the IP-H group, there was a notable decrease in the expression of three genes involved in the PUT synthesis pathway: *speA*, *speB*, and *speF*. Specifically, *speA* encodes arginine decarboxylase, *speB* encodes agmatinase, and *speF* encodes ornithine decarboxylase (Table 1), whereas no change occurred in *P. fluorescens* ATCC 13525 of the IP-P group compared to the WP-P group. The downregulation of these genes suggests a hindered process in the synthesis of PUT. On the contrary, the expression of *puuABCD* in the putrescine degradation pathway was upregulated in *H. alvei* H4 Δ*luxI* (Table 1), which accelerated the consumption rate of PUT. In addition, several pot genes encoding PUT transporter protein were downregulated in Table 1. The decrease in putrescine content was attributed to a faster rate of putrescine consumption due to increased expression of *puuABCD* and a lower rate of putrescine synthesis due to the downregulation of *speABF*. The content of cadaverine in the later stage was higher than that of putrescine (Figure 2), possibly because the anabolic pathway of cadaverine was relatively simple.

In addition, a high increase in amino acid metabolism may enable bacterial cells to adapt the structure and function of biofilm in response to environmental changes. [46]. It has also been found that amino acid biosynthesis is essential for biofilm formation [53]. In the transcription results for *P. fluorescens* ACTT13525 in the IP-P group, *CPH89_RS06815* and *katB*, which are responsible for encoding peroxidase HPII involved in tryptophan metabolism, were upregulated (Table 2). It is suspected that the metabolism of tryptophan is accelerated, leading to a decrease in biofilm content in the deletion coculture system. Meanwhile, the metabolisms of other amino acids, including threonine and D-amino-acid were dysregulated when cocultured with *H. alvei* H4 Δ*luxI*. The genes *CPH89_RS00185*, encoding aspartate/tyrosine/aromatic aminotransferase, and *CPH89_RS08005*, encoding D-amino acid dehydrogenase, were upregulated. In contrast, *CPH89_RS21450*, encoding an aspartate aminotransferase family protein involved in the biosynthesis of arginine, an essential source of protein and nucleic acid biosynthesis, is downregulated in the IP-P group. This suggests that when the QS system of *H. alvei* H4 WT in the coculture system was frustrated, an imbalanced state and disturbance of the amino acid metabolism in the cells were formed. Combined with the above considerations, QS of *H. alvei* H4 WT in the coculture affected the amino acid metabolism of *P. fluorescens* ACTT13525 at the transcriptional level.

When food is contaminated by microorganisms, the microorganisms will metabolize amino acids in proteins to produce biogenic amines. Appropriate measures need to be taken to control the content of biogenic amines in food. And the absence of QS in a coculture inhibits the production of biogenic amines in mixed bacteria by regulating amino acid metabolism, which provides ideas to inhibit the production of biogenic amines in real food processing environments.

3.3. Real-Time Quantitative PCR Validation (RT-qPCR)

In order to confirm the validity of the expression of several genes associated with biofilm formation and spoilage obtained by RNA-seq, a total of 13 genes, including *fliC*, *motA*, *tsr*, *puuB*, *puuC*, *speB*, *speF*, and *ldcc* from *H. alvei* H4, and 5 genes belonging to various metabolic pathways (*cheW*, *katB*, *CPH89-RS06815*, *CPH89-RS10590*, and *RbsB*) from *P. fluorescens* ACTT13525 were selected for transcription polymerase chain reaction analysis (Figure 4). In *H. alvei* H4 Δ*luxI*, the gene belonging to the flagellar component (*fliC*) and bacterial chemotaxis-related genes, including methyl-accepting chemotaxis protein (*tsr*) and flagellar motor protein (*motA*), were significantly downregulated in the dual species. And in *P. fluorescens* ACTT13525 cocultured with *H. alvei* H4 Δ*luxI*, the *cheW* (related to bacterial chemotaxis) and *RbsB* (related to ABC transporters) were also downregulated. These results indicate that flagella and chemotactic-related genes are, indeed, downregulated after QS of *H. alvei* H4 WT is destroyed, thus affecting the formation of biofilm in the coculture system. In addition, three genes (*speA*, *speB*, and *speF*) within the PUT synthesis pathway also exhibited downregulation, while the expressions of *puuB* and *puuC* involved in the putrescine degradation pathway of *H. alvei* H4 were upregulated in the IP-H group compared to the WP-H group. This proves that when QS of *H. alvei* H4 WT is disrupted, the gene expression of the PUT synthesis pathway is inhibited, while the gene expression of the degradation pathway is promoted. Furthermore, the expression levels of *CPH89_RS06815* and *katB* involved in the tryptophan metabolism of *P. fluorescens* ACTT13525 were upregulated, while the gene *CPH89_RS10590* (related to amino acid metabolism) was downregulated, which suggests that QS of *H. alvei* induces metabolic disorders in *P. fluorescens*. Moreover, the results of the study indicate that the patterns of gene expression obtained from both dual RNA-seq and RT-qPCR were generally consistent, which shows that the RNA-Seq data are credible and reproducible. Additionally, a strong correlation was observed between the two data sets for each species (*H. alvei* H4 and *P. fluorescens* ATCC13525), with Pearson's correlation coefficients of 0.9086 (Figure 4C) and 0.9861 (Figure 4D), respectively, ($p < 0.02$). This further supports the validity of the RNA-Seq data.

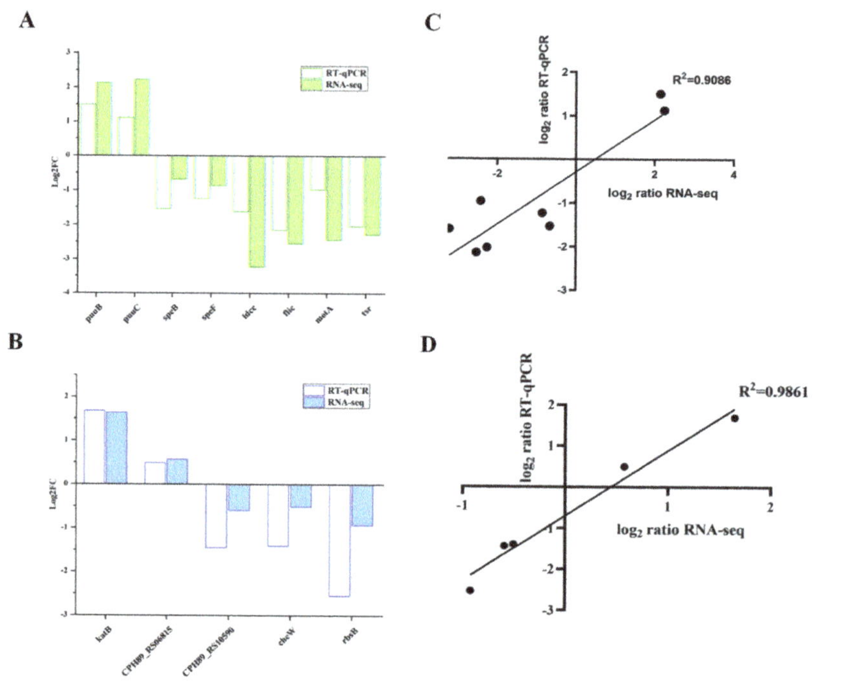

Figure 4. Validation of the selected DEG expression of *Hafnia alvei* H4 and *Pseudomonas fluorescens* ATCC13525 in dual species. (**A**) Quantitative reverse transcription PCR (RT-qPCR) of differentially expressed genes of *H. alvei* H4. (**B**) RT-qPCR of differentially expressed genes of *P. fluorescens* ATCC13525. (**C**) The regression line of RNA-Seq data (X-axis) plotted against the RT-qPCR data (Y-axis) of *H. alvei* H4. (**D**) The regression line of RNA-Seq data (X-axis) plotted against the RT-qPCR data (Y-axis) of *P. fluorescens* ATCC13525.

4. Conclusions

Overall, all the results show that QS of *H. alvei* not only regulated its own genes but also promoted microbial communication and spoilage by influencing gene expression in cocultured strains. That is, the effects of coexistence are "two-way", where the presence of *H. alvei* affects the metabolism and behavior of *P. fluorescens* and vice versa. It was further established that QS of *H. alvei* H4 WT plays an important role in a coculture system. Based on phenotyping experiments (Figure 1), we mainly analyzed flagellar assembly, bacterial chemotaxis and the two-component system, the ABC transporter system and amino acid metabolism pathways. The schematic is shown in Figure 5. All of them are directly or indirectly related to biofilm and biogenic amine formation [54]. Specifically, the disruption of QS in *H. alvei* H4 WT inhibits the flagellar composition of *H. alvei* and the expression of the flagellar transcriptional regulators of *P. fluorescens*. At the same time, the expression of chemotactic genes in the two strains is also inhibited, thereby reducing biofilm formation. In addition, the regulation of the signaling systems is also a contributing factor to the decrease in biofilms. This includes the regulation of QS of *H. alvei* H4 WT through its two-component system, as well as the regulation of the ABC transporter of *P. fluorescens*. The reduction in the accumulation of biogenic amines is primarily caused by the downregulation of certain amino acid decarboxylase enzymes. In addition, QS of *H. alvei* H4 WT will result in an imbalance of amino acids and metabolic disorders in *P. fluorescens*, consequently impacting phenotypic changes in a coculture. This means that *H. alvei* H4 and *P. fluorescens* ATCC13525 undergo transcriptional changes in their interactions, which depend on the production of QS of *H. alvei* H4 WT. Disrupting QS of *H. alvei* H4 WT will inevitably reduce the capacity for spoilage caused by both *H. alvei* H4 WT and *P. fluorescens* ATCC13525. It indicated that QS of *H. alvei* H4 WT might offer a new target to reduce the food spoilage development caused by *H. alvei* and *P. fluorescens*. It is possible to regulate the expression of spoilage indicators such as biofilm and biogenic amines in the entire coculture system by interfering with QS. More rigorous studies are necessary to further explore the role of *H. alvei* quorum sensing in coculture regulation at the protein level.

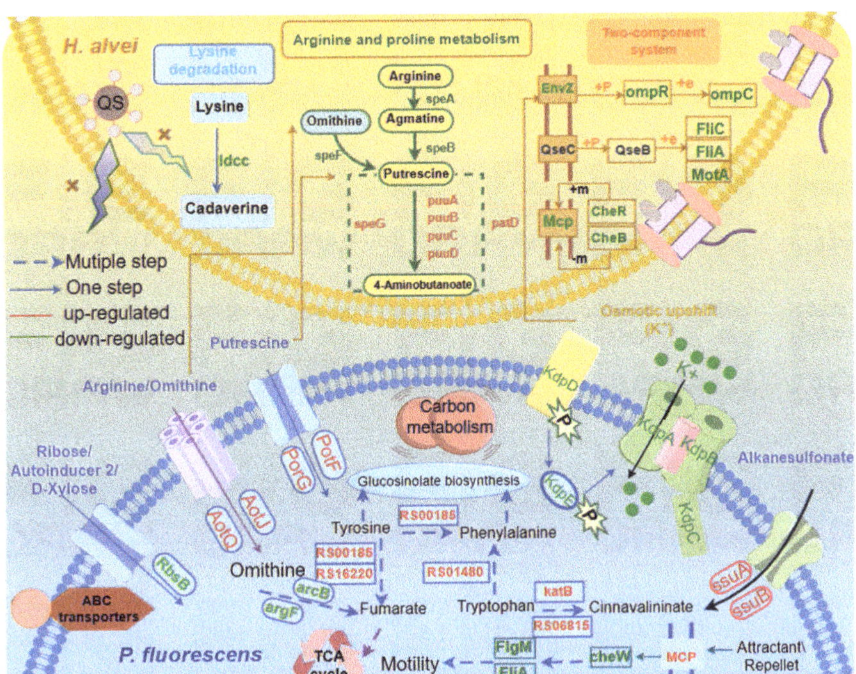

Figure 5. Schematic representation of certain biological pathways in *H. alvei* H4 and *P. fluorescens* affected by QS of *H. alvei* in a coculture system. This model represents the transcriptional response of genes with significant differences in transcription levels (flagellar assembly, bacterial chemotaxis, lysine degradation, arginine and proline metabolism, and two-component system) of *H. alvei* H4 and (ABC transporters, bacterial chemotaxis, phenylalanine, tyrosine, and tryptophan biosynthesis) of *P. fluorescens* ATCC13525 after the *luxI* gene knockout from *H. alvei* H4 in dual species. The yellow gradient background area represents *H. alvei* H4, and the blue gradient background area represents *P. fluorescens* ATCC13525. Most DEGs were integrated and are indicated in red (upregulated), and green (downregulated), respectively.

Author Contributions: Conceptualization, Y.W.; Methodology Y.W., X.L. and G.Z.; Investigation, Y.W., X.L. and J.B.; Data curation, H.H.; Writing—original draft preparation, Y.W.; Writing—review and editing, H.H.; Supervision, H.H.; Project administration, H.H.; Funding acquisition, H.H. All authors have read and agreed to the published version of the manuscript.

Funding: This work was supported by the National Key Research and Development Program project (No. 2022YFD2100500); the APC was funded by H.H.

Institutional Review Board Statement: Not applicable.

Informed Consent Statement: Not applicable.

Data Availability Statement: The data presented in this study are available on request from the corresponding author. The data are not publicly available due to the need for the first author the apply for a Master's degree.

Conflicts of Interest: The authors declare no conflicts of interest. The funders had no role in the design of the study; in the collection, analyses, or interpretation of data; in the writing of the manuscript; or in the decision to publish the results.

References

1. Özogul, F.; Polat, A.; Özogul, Y. The effects of modified atmosphere packaging and vacuum packaging on chemical, sensory and microbiological changes of sardines (*Sardina pilchardus*). *Food Chem.* **2004**, *85*, 49–57. [CrossRef]
2. Wang, Y.; Feng, L.; Lu, H.; Zhu, J.; Kumar, V.; Liu, X. Transcriptomic analysis of the food spoilers *Pseudomonas fluorescens* reveals the antibiofilm of carvacrol by interference with intracellular signaling processes. *Food Control* **2021**, *127*, 108115. [CrossRef]
3. Robson, A.A.; Kelly, M.S.; Latchford, J.W. Effect of temperature on the spoilage rate of whole, unprocessed crabs: *Carcinus maenas*, *Necora puber* and *Cancer pagurus*. *Food Microbiol.* **2007**, *24*, 419–424. [CrossRef] [PubMed]
4. Xu, Y.F.; Lin, H.; Sui, J.X.; Cao, L.M. Production and Characterization of Egg Yolk Antibodies (Igy) against Two Specific Spoilage Organisms (SSO) in Aquatic Products. *Adv. Mater. Res.* **2011**, *343–344*, 519–529. [CrossRef]
5. Han, F.; Li, M.; Lin, H.; Wang, J.; Cao, L.; Khan, M.N. The novel Shewanella putrefaciens-infecting bacteriophage Spp001: Genome sequence and lytic enzymes. *J. Ind. Microbiol. Biotechnol.* **2014**, *41*, 1017–1026. [CrossRef]
6. Bruhn, J.B.; Christensen, A.B.; Flodgaard, L.R.; Nielsen, K.F.; Larsen, T.O.; Givskov, M.; Gram, L. Presence of Acylated Homoserine Lactones (AHLs) and AHL-Producing Bacteria in Meat and Potential Role of AHL in Spoilage of Meat. *Appl. Environ. Microbiol.* **2004**, *70*, 4293–4302. [CrossRef]
7. Zhu, Y.L.; Hou, H.M.; Zhang, G.L.; Wang, Y.F.; Hao, H.S. AHLs Regulate Biofilm Formation and Swimming Motility of Hafnia alvei H4. *Front. Microbiol.* **2019**, *10*, 1330. [CrossRef]
8. Shen, Y.; Cui, F.C.; Wang, D.F.; Li, T.T.; Li, J.R. Quorum Quenching Enzyme (PF-1240) Capable to Degrade AHLs as a Candidate for Inhibiting Quorum Sensing in Food Spoilage Bacterium Hafnia alvei. *Foods* **2021**, *10*, 2700. [CrossRef]
9. Hwang, C.C.; Lee, Y.C.; Huang, Y.R.; Lin, C.M.; Shiau, C.Y.; Hwang, D.F.; Tsai, Y.H. Biogenic amines content, histamine-forming bacteria and adulteration of bonito in tuna candy products. *Food Control* **2010**, *21*, 845–850. [CrossRef]
10. Kanki, M.; Yoda, T.; Ishibashi, M.; Tsukamoto, T. Photobacterium phosphoreum caused a histamine fish poisoning incident. *Int. J. Food Microbiol.* **2004**, *92*, 79–87. [CrossRef]
11. Ge, Y.; Zhu, J.; Ye, X.; Yang, Y. Spoilage potential characterization of Shewanella and Pseudomonas isolated from spoiled large yellow croaker (*Pseudosciaena crocea*). *Lett. Appl. Microbiol.* **2017**, *64*, 86–93. [CrossRef] [PubMed]
12. Wang, H.; Cai, L.; Li, Y.; Xu, X.; Zhou, G. Biofilm formation by meat-borne Pseudomonas fluorescens on stainless steel and its resistance to disinfectants. *Food Control* **2018**, *91*, 397–403. [CrossRef]
13. Zarei, M.; Yousefvand, A.; Maktabi, S.; Pourmahdi Borujeni, M.; Mohammadpour, H. Identification, phylogenetic characterisation and proteolytic activity quantification of high biofilm-forming *Pseudomonas fluorescens* group bacterial strains isolated from cold raw milk. *Int. Dairy J.* **2020**, *109*, 104787. [CrossRef]
14. Mann, E.E.; Wozniak, D.J. *Pseudomonas* biofilm matrix composition and niche biology. *FEMS Microbiol. Rev.* **2012**, *36*, 893–916. [CrossRef] [PubMed]
15. Delille, A.; Quilès, F.; Humbert, F. In Situ Monitoring of the Nascent *Pseudomonas fluorescens* Biofilm Response to Variations in the Dissolved Organic Carbon Level in Low-Nutrient Water by Attenuated Total Reflectance-Fourier Transform Infrared Spectroscopy. *Appl. Environ. Microbiol.* **2007**, *73*, 5782–5788. [CrossRef]
16. Puga, C.H.; Orgaz, B.; SanJose, C. Listeria monocytogenes Impact on Mature or Old *Pseudomonas fluorescens* Biofilms during Growth at 4 and 20 °C. *Front. Microbiol.* **2016**, *7*, 00134. [CrossRef]
17. Li, T.; Yang, B.; Li, X.; Li, J.; Zhao, G.; Kan, J. Quorum sensing system and influence on food spoilage in *Pseudomonas fluorescens* from turbot. *J. Food Sci. Technol.* **2018**, *55*, 3016–3025. [CrossRef]
18. Chen, T.R.; Wei, Q.K.; Chen, Y.J. *Pseudomonas* spp and *Hafnia alvei* growth in UHT milk at cold storage. *Food Control* **2011**, *22*, 697–701. [CrossRef]

19. Liu, M.; Gray, J.M.; Griffiths, M.W. Occurrence of Proteolytic Activity and N-Acyl-Homoserine Lactone Signals in the Spoilage of Aerobically Chill-Stored Proteinaceous Raw Foods. *J. Food Prot.* **2006**, *69*, 2729–2737. [CrossRef]
20. Nychas, G.-J.E.; Skandamis, P.N.; Tassou, C.C.; Koutsoumanis, K.P. Meat spoilage during distribution. *Meat Sci.* **2008**, *78*, 77–89. [CrossRef]
21. Fu, L.; Wang, C.; Liu, N.; Ma, A.; Wang, Y. *Quorum sensing* system-regulated genes affect the spoilage potential of *Shewanella baltica. Food Res. Int.* **2018**, *107*, 1–9. [CrossRef] [PubMed]
22. Whiteley, M.; Diggle, S.P.; Greenberg, E.P. Progress in and promise of bacterial quorum sensing research. *Nature* **2017**, *551*, 313–320. [CrossRef] [PubMed]
23. Kaspar, J.R.; Lee, K.; Richard, B.; Walker, A.R.; Burne, R.A. Direct interactions with commensal streptococci modify intercellular communication behaviors of *Streptococcus mutans. ISME J.* **2020**, *15*, 473–488. [CrossRef] [PubMed]
24. Sikdar, R.; Elias, M. Quorum quenching enzymes and their effects on virulence, biofilm, and microbiomes: A review of recent advances. *Expert Rev. Anti-Infect. Ther.* **2020**, *18*, 1221–1233. [CrossRef]
25. Hou, H.M.; Zhu, Y.L.; Wang, J.Y.; Jiang, F.; Qu, W.Y.; Zhang, G.L.; Hao, H.-S. Characteristics of N-Acylhomoserine Lactones Produced by Hafnia alvei H4 Isolated from Spoiled Instant Sea Cucumber. *Sensors* **2017**, *17*, 772. [CrossRef]
26. Harimawan, A.; Ting, Y.-P. Investigation of extracellular polymeric substances (EPS) properties of *P. aeruginosa* and *B. subtilis* and their role in bacterial adhesion. *Colloids Surf. B: Biointerfaces* **2016**, *146*, 459–467. [CrossRef]
27. Sang, X.; Ma, X.; Hao, H.; Bi, J.; Zhang, G.; Hou, H. Evaluation of biogenic amines and microbial composition in the Chinese traditional fermented food grasshopper sub shrimp paste. *Lwt* **2020**, *134*, 109979. [CrossRef]
28. Xue, Z.P.; Cu, X.; Xu, K.; Peng, J.H.; Liu, H.R.; Zhao, R.T.; Wang, Z.; Wang, T.; Xu, Z.S. The effect of glutathione biosynthesis of *Streptococcus thermophilus* ST-1 on cocultured *Lactobacillus delbrueckii* ssp. *bulgaricus ATCC11842. J. Dairy Sci.* **2023**, *106*, 884–896. [CrossRef]
29. Winkelströter, L.K.; Teixeira, F.B.d.R.; Silva, E.P.; Alves, V.F.; De Martinis, E.C.P. Unraveling Microbial Biofilms of Importance for Food Microbiology. *Microb. Ecol.* **2013**, *68*, 35–46. [CrossRef]
30. Jahid, I.K.; Ha, S.D. The Paradox of Mixed-Species Biofilms in the Context of Food Safety. *Compr. Rev. Food Sci. Food Saf.* **2014**, *13*, 990–1011. [CrossRef]
31. Zhang, C.; Zhu, F.; Jatt, A.N.; Liu, H.; Niu, L.; Zhang, L.; Liu, Y. Characterization of co-culture of *Aeromonas* and *Pseudomonas* bacterial biofilm and spoilage potential on refrigerated grass carp (*Ctenopharyngodon idellus*). *Lett. Appl. Microbiol.* **2020**, *71*, 337–344. [PubMed]
32. Linares, D.M.; Martín, M.; Ladero, V.; Alvarez, M.A.; Fernández, M. Biogenic Amines in Dairy Products. *Crit. Rev. Food Sci. Nutr.* **2011**, *51*, 691–703. [CrossRef] [PubMed]
33. Rossi, P.; Xing, Q.; Bini, E.; Portaliou, A.G.; Clay, M.C.; Warren, E.M.; Khanra, N.K.; Economou, A.; Kalodimos, C.G. Chaperone Recycling in Late-Stage Flagellar Assembly. *J. Mol. Biol.* **2023**, *435*, 167954. [CrossRef] [PubMed]
34. Saijo-Hamano, Y.; Uchida, N.; Namba, K.; Oosawa, K. In Vitro Characterization of FlgB, FlgC, FlgF, FlgG, and FliE, Flagellar Basal Body Proteins of Salmonella. *J. Mol. Biol.* **2004**, *339*, 423–435. [CrossRef] [PubMed]
35. Altarriba, M.; Merino, S.; Gavín, R.; Canals, R.o.; Rabaan, A.; Shaw, J.G.; Tomás, J.M. A polar flagella operon (*flg*) of *Aeromonas hydrophila* contains genes required for lateral flagella expression. *Microb. Pathog.* **2003**, *34*, 249–259. [CrossRef]
36. Deme, J.C.; Johnson, S.; Vickery, O.; Aron, A.; Monkhouse, H.; Griffiths, T.; James, R.H.; Berks, B.C.; Coulton, J.W.; Stansfeld, P.J.; et al. Structures of the stator complex that drives rotation of the bacterial flagellum. *Nat. Microbiol.* **2020**, *5*, 1553–1564. [CrossRef] [PubMed]
37. Wilkinson, D.A.; Chacko, S.J.; Vénien-Bryan, C.; Wadhams, G.H.; Armitage, J.P. Regulation of Flagellum Number by FliA and FlgM and Role in Biofilm Formation by *Rhodobacter sphaeroides*. *J. Bacteriol.* **2011**, *193*, 4010–4014. [CrossRef]
38. Hazelbauer, G.L. Bacterial Chemotaxis: The Early Years of Molecular Studies. *Annu. Rev. Microbiol.* **2012**, *66*, 285–303. [CrossRef]
39. Karmakar, R. State of the art of bacterial chemotaxis. *J. Basic Microbiol.* **2021**, *61*, 366–379. [CrossRef]
40. Porter, S.L.; Wadhams, G.H.; Armitage, J.P. Signal processing in complex chemotaxis pathways. *Nat. Rev. Microbiol.* **2011**, *9*, 153–165. [CrossRef]
41. Minamino, T.; Kinoshita, M.; Namba, K. Directional Switching Mechanism of the Bacterial Flagellar Motor. *Comput. Struct. Biotechnol. J.* **2019**, *17*, 1075–1081. [CrossRef]
42. Kentner, D.; Sourjik, V. Spatial organization of the bacterial chemotaxis system. *Curr. Opin. Microbiol.* **2006**, *9*, 619–624. [CrossRef] [PubMed]
43. Li, Y.; Wang, Y.; Han, S.; Qin, M.; Wu, X.; Niu, W.; Gao, C.; Wang, H. N-acyl-homoserine lactones-mediated quorum sensing promotes intestinal colonization of *Aeromonas veronii* through facilitating cheA gene expression. *Aquaculture* **2024**, *579*, 740189. [CrossRef]
44. Sieuwerts, S.; Molenaar, D.; van Hijum, S.A.F.T.; Beerthuyzen, M.; Stevens, M.J.A.; Janssen, P.W.M.; Ingham, C.J.; de Bok, F.A.M.; de Vos, W.M.; van Hylckama Vlieg, J.E.T. Mixed-Culture Transcriptome Analysis Reveals the Molecular Basis of Mixed-Culture Growth in Streptococcus thermophilus and Lactobacillus bulgaricus. *Appl. Environ. Microbiol.* **2010**, *76*, 7775–7784. [CrossRef] [PubMed]
45. Wang, L.C.; Morgan, L.K.; Godakumbura, P.; Kenney, L.J.; Anand, G.S. The inner membrane histidine kinase EnvZ senses osmolality via helix-coil transitions in the cytoplasm. *EMBO J.* **2012**, *31*, 2648–2659. [CrossRef] [PubMed]

46. Yan, J.; Yang, Z.; Xie, J. Comparative Transcriptome Analysis of *Shewanella putrefaciens* WS13 Biofilms Under Cold Stress. *Front. Cell. Infect. Microbiol.* **2022**, *12*, 851521. [CrossRef]
47. Zhang, B.Z.; Jiang, C.S.; Cao, H.; Zeng, W.; Ren, J.P.; Hu, Y.F.; Li, W.T.; He, Q.G. Transcriptome analysis of heat resistance regulated by quorum sensing system in *Glaesserella parasuis*. *Front. Microbiol.* **2022**, *13*, 968460. [CrossRef]
48. Xie, Z.; Jian, H.; Jin, Z.; Xiao, X.; Kelly, R.M. Enhancing the Adaptability of the Deep-Sea Bacterium *Shewanella piezotolerans* WP3 to High Pressure and Low Temperature by Experimental Evolution under H_2O_2 Stress. *Appl. Environ. Microbiol.* **2018**, *84*, e02342-17. [CrossRef]
49. Zhang, P.Y.; Xu, P.P.; Xia, Z.J.; Wang, J.; Xiong, J.; Li, Y.Z. Combined treatment with the antibiotics kanamycin and streptomycin promotes the conjugation of *Escherichia coli*. *FEMS Microbiol. Lett.* **2013**, *348*, 149–156. [CrossRef]
50. Armbruster, C.E.; Pang, B.; Murrah, K.; Juneau, R.A.; Perez, A.C.; Weimer, K.E.D.; Swords, W.E. RbsB (NTHI_0632) mediates quorum signal uptake in nontypeable *Haemophilus influenzae* strain 86-028NP. *Mol. Microbiol.* **2011**, *82*, 836–850. [CrossRef]
51. Li, J.; Zhou, L.; Feng, W.; Cheng, H.; Muhammad, A.I.; Ye, X.; Zhi, Z. Comparison of Biogenic Amines in Chinese Commercial Soy Sauces. *Molecules* **2019**, *24*, 1522. [CrossRef] [PubMed]
52. Sun, Z.; Zhang, Y.; Lin, X.; Zhang, S.; Chen, Y.; Ji, C. Inhibition Mechanism of *Lactiplantibacillus plantarum* on the Growth and Biogenic Amine Production in *Morganella morganii*. *Foods* **2023**, *12*, 3625. [CrossRef] [PubMed]
53. Jijakli, K.; Jensen, P.A.; Whiteson, K.L. Metabolic Modeling of *Streptococcus mutans* Reveals Complex Nutrient Requirements of an Oral Pathogen. *mSystems* **2019**, *4*, 419507. [CrossRef] [PubMed]
54. Zhang, X.; Zheng, L.; Lu, Z.; Zhou, L.; Meng, F.; Shi, C.; Bie, X. Biochemical and molecular regulatory mechanism of the pgpH gene on biofilm formation in *Listeria monocytogenes*. *J. Appl. Microbiol.* **2023**, *134*, lxac086. [CrossRef]

foods

MDPI

Article

Line Laser Scanning Combined with Machine Learning for Fish Head Cutting Position Identification

Xu Zhang [1], Ze Gong [1], Xinyu Liang [2], Weichen Sun [1], Junxiao Ma [1] and Huihui Wang [1,3,*]

1 School of Mechanical Engineering & Automation, Dalian Polytechnic University, Dalian 116034, China; zhangxu_dlut@163.com (X.Z.); 18252648000@163.com (Z.G.); sswwcc1999@126.com (W.S.); dlpujxxy@126.com (J.M.)
2 School of Food Science & Technology, Dalian Polytechnic University, Dalian 116034, China; y15714110553@126.com
3 National Engineering Research Center of Seafood, Dalian 116034, China
* Correspondence: whh419@126.com

Abstract: Fish head cutting is one of the most important processes during fish pre-processing. At present, the identification of cutting positions mainly depends on manual experience, which cannot meet the requirements of large-scale production lines. In this paper, a fast and contactless identification method of cutting position was carried out by using a constructed line laser data acquisition system. The fish surface data were collected by a linear laser scanning sensor, and Principal Component Analysis (PCA) was used to reduce the dimensions of the dorsal and abdominal boundary data. Based on the dimension data, Least Squares Support Vector Machines (LS-SVMs), Particle Swarm Optimization-Back Propagation (PSO-BP) networks, and Long and Short Term Memory (LSTM) neural networks were applied for fish head cutting position identification model establishment. According to the results, the LSTM model was considered to be the best prediction model with a determination coefficient (R^2) value, root mean square error (RMSE), mean absolute error (MAE), and residual predictive deviation (RPD) of 0.9480, 0.2957, 0.1933, and 3.1426, respectively. This study demonstrated the reliability of combining line laser scanning techniques with machine learning using LSTM to identify the fish head cutting position accurately and quickly. It can provide a theoretical reference for the development of intelligent processing and intelligent cutting equipment for fish.

Keywords: fish; head cutting position; linear laser scanning; identification model

Citation: Zhang, X.; Gong, Z.; Liang, X.; Sun, W.; Ma, J.; Wang, H. Line Laser Scanning Combined with Machine Learning for Fish Head Cutting Position Identification. *Foods* **2023**, *12*, 4518. https://doi.org/10.3390/foods12244518

Academic Editor: Ioan Cristian Trelea

Received: 29 November 2023
Revised: 9 December 2023
Accepted: 14 December 2023
Published: 18 December 2023

1. Introduction

The main process in fish processing includes scaling, gutting, cleaning, and head/tail cutting, where head removal is an important part of cutting planning and directly affects the processing quality and meat yield [1]. The main types of head cutting processes are manual and mechanical [2]. The manual method is time-consuming and laborious, with low processing efficiency and high skill requirements for the processors [3], which cannot be adapted to the needs of short-term and high-volume production of bulk fish. Due to the biodiversity of fish, even the size of the heads of the same specification and the same batch of raw material varies greatly, and existing mechanical cuts are processed according to a pre-set cutting position [2], which is unsatisfactory in terms of reducing meat yield and non-compliance with handling. To be specific, if the processing volume is set too large, it will lead to a lower meat yield and waste; if the processing volume is set too small, the cutting tool will be easily damaged by cutting at the gill cover, resulting in cutting failure and even causing failure of the entire equipment operation. Therefore, how to achieve automatic identification of the fish head position so as to control the accurate cutting of the fish head is a problem that needs to be urgently solved for flexible, intelligent, and efficient processing of bulk fish.

Dynamic 3D reconstruction is one of the important technologies for fast and accurate contour feature identification in advanced manufacturing, and food and agricultural product processing [4,5]. The acquisition of the measured object contour data is mainly realized by means of computer vision, lasers, ultrasound, NMR, and X-rays. Among them, single/binocular vision usually establishes the inference rules by capturing image information such as shape, texture, and color, and has been widely used in food size and volume measurements [6]. Ultrasound, NMR, and X-rays are mostly used to probe the interior of objects, such as the application of NMR technology in fresh meat quality evaluation, X-ray for dark field imaging in medicine, etc. Laser sensors are suitable for high-volume installation and have been effectively applied in shape monitoring in animal husbandry and in planting and pruning of agricultural crops, which are not susceptible to the influence of the external lighting environment [7–10]. Due to the technical characteristics of industrial fish head removal processing, the high measurement accuracy required, the complex and variable processing environment, and the large installation capacity of the equipment, this study selected laser scanning technology for fish surface data information collection. In addition, to achieve accurate cutting control of the fish head, it is necessary to build a reliable mathematical model for predictions. The traditional single-factor linear regression prediction method is simple in principle and mature in technology, but it lacks a self-learning capability and is difficult to achieve accurate descriptions of complex non-linear models. In recent years, machine learning has performed superiorly in the field of intelligent control, gradually replacing traditional prediction methods [11,12]. Least Squares Support Vector Machines (LS-SVMs) overcome the shortcomings of artificial neural networks and SVM to achieve global optimality with the principle of structural risk minimization, which is more widely used in the field of food testing, and has been used in the identification of salmon moisture content, the prediction of acetic acid content in beer, and the prediction of changes in freshness indicators during the refrigeration of trout fillets with high accuracy [13–16]. Particle Swarm Optimization-Back Propagation (PSO-BP) neural networks addresses the problem where traditional Back-Propagation (BP) neural networks easily to fall into local optima and has great advantages in weight and bias initialization, learning rate adjustment, and convergence speed, and is widely used in agriculture and has achieved good results in tea water content prediction, coliform amount prediction, grain yield prediction, and so on [17–19]. A Long and Short Term Memory (LSTM) neural network is a supervised neural network that addresses the long-term dependency problem in Recurrent Neural Networks (RNNs); as a non-linear model, LSTM can be used as a complex non-linear unit to construct larger deep neural networks, mostly for the quantitative analysis of important elements of samples, prediction of important components, and so on [20,21]. The above studies provide a lot of references and support for the present study on the identification of fish head cutting position using fish body surface feature information and the establishment of reliable models.

The aim of this study was to explore the feasibility of achieving automatic fish head position identification by using a constructed fish surface contour laser scanning system and to propose a rapid fish head identification method based on 3D contour information so as to provide fast and accurate cutting path planning for automated and intelligent head removal processing. The main research contents of this paper are as follows: (1) constructing a laser scanning system for fish body data information to realize the automatic acquisition of contour information on the outer surface of the fish body; (2) proposing a data validity discrimination and filtering method suitable for feature extraction of contour information of the fish body radial section; (3) taking the dimensionality reduced feature values as inputs and the fish head cutting position as outputs, MPR-, LS-SVM-, and LSTM-based fish head ideal cutting position identification models were established to achieve accurate identification of the fish head cutting position.

2. Materials and Methods

2.1. Materials

This study was carried out on crucian carp. The experimental samples were purchased from Dalian Wholesale Fish Market, China. A total of 204 crucian carp were randomly selected and placed in a thermostat with ice for rapid transportation back to the laboratory. The parameters of the fish body shape were defined as shown in Figure 1; the distance between the mouth and the trailing edge of the gill cover was defined as the head length (Figure 1a), the maximum distance from the front of the mouth to the end of the tail fin was defined as the total length (Figure 1b), and the maximum distance between the dorsal and ventral parts of the fish was defined as the maximum width (Figure 1c), setting the cut line passing through this location as the ideal cut line for the head; the height of the highest point of the fish to the level of the conveyor belt when the fish was placed horizontally was defined as the maximum thickness (Figure 1d). The statistics of the manual measurements of the 204 samples are shown in Table 1.

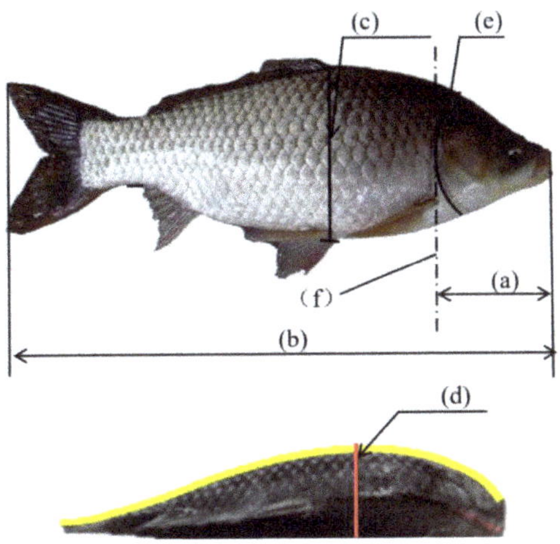

Figure 1. Relationship between head cutting position and fish body parameters: (a) fish head length; (b) total length; (c) maximum width; (d) maximum thickness; (e) ideal cutting line; (f) gill.

Table 1. Index statistics of fish sample data.

Sample Size	Statistical Indicator	a/mm	b/mm	c/mm	d/mm	Weight/g
204	Maximum value	63.1	239.4	103.7	51.54	569
	Minimum value	50.1	200.8	94.6	45.3	473
	Mean	223.6	54.6	100.9	47.49	532.1
	Standard deviation	2.5	8.9	2.2	1.85	21.6

2.2. Data Acquisition and Pre-Processing

2.2.1. Data Acquisition System

A fish data information laser scanning system was used to collect 3D contour information of the fish body. As shown in Figure 2b, the system mainly consisted of a line laser scanning sensor (LLT-2600 scanContral2D/3D, Micro-Epsilon, Ortenburg, German), a drive mechanism, a dark box, and a data processing unit. The laser scanning sensor scanned 640 points at a time, with a scanning frequency of 300 Hz, and outputted the measurement results via Ethernet (Modbus TCP protocol). The drive mechanism consisted of a conveyor belt and a drive unit, which can transport the material horizontally and linearly under the drive of a servo motor, and the motor speed was set at 6.6 rpm. The dark box was equipped with two groups of strip light sources, which were placed on both sides of the dark box, as shown in Figure 2b; the data processing unit was controlled by a computer, which was used to realize pre-processing such as data segmentation and filtering of the collected raw fish body information (Figure 2c).

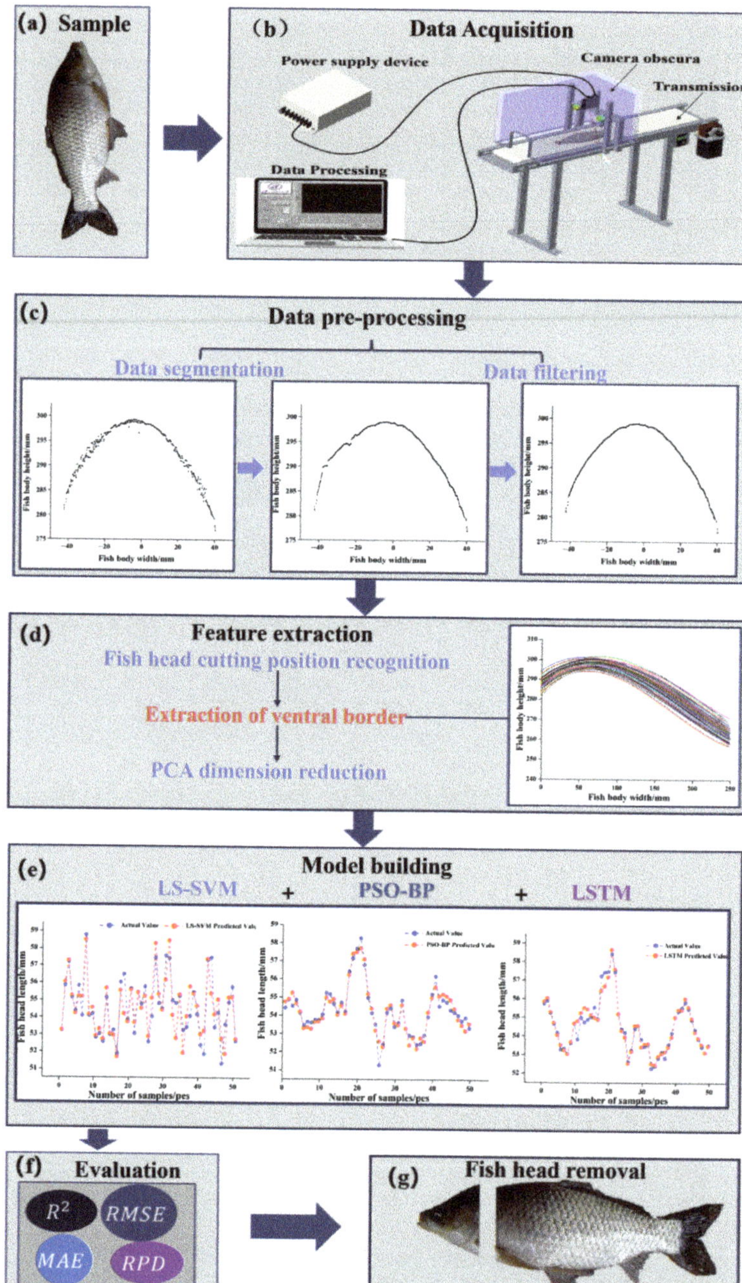

Figure 2. Fish head cutting position determination flow chart: (**a**) sample; (**b**) data acquisition; (**c**) data pre-processing; (**d**) feature extraction; the different colored lines represent the ventral–dorsal dividing line extracted from different samples; (**e**) model building; (**f**) evaluation; (**g**) fish head removal.

2.2.2. Data Acquisition Process

The laser sensor was calibrated to send a laser ray vertically downwards, with the line laser perpendicular to the conveyor belt transport direction. The laser scanning schematic is shown in Figure 3a,b; the intersection points between the vertical line of the laser source and the horizontal plane where the conveyor belt surface was located was defined as the O point, the height was in the Z direction, the laser direction was defined as the Y direction, and the conveyor belt conveying the reverse direction was defined as the X direction. During data acquisition, the head of the fish was orientated in the same direction as the movement direction, and when the sample triggered the timing procedure, the laser sensor started scanning the fish to obtain the 3D point cloud information of the surface contour of the fish. Due to the huge amount of real-time data obtained by laser scanning, the sampling frequency was set to 0.42 Hz without affecting the accuracy of the calculation, and the obtained contour information was actually a number of point cloud data containing the contour information of the fish body cross-section (Figure 3c), which was stored in the form of an array.

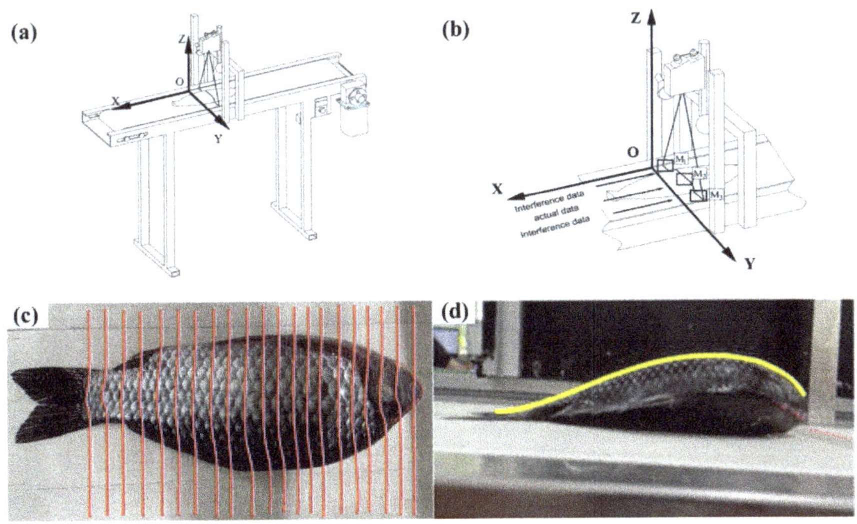

Figure 3. Data acquisition process: (**a**) coordinate diagram; (**b**) scanning process diagram; (**c**) radial section contour diagram; (**d**) Y projection contour diagram.

2.2.3. Data Validity Discernment

As shown in Figure 3b, since the width of the line laser is larger than the fish it scans, the interference data formed on the surface of the conveyor belt on both sides of the fish body will inevitably be collected simultaneously during the process of acquiring the fish body contour data. In this study, the threshold segmentation method was used to remove the interfering data while retaining useful information on the surface of the fish body for subsequent processing. The calculation process of the threshold segmentation method is as follows.

$$[M] = \begin{cases} [M_1, M_3], & \Delta h_1 < T, \Delta h_2 < T \\ [M_2], & \Delta h_1 > T, \Delta h_2 > T \end{cases} \tag{1}$$

$$\Delta h_1 = |M_1 - M_2|, \tag{2}$$

$$\Delta h_2 = |M_2 - M_3| \tag{3}$$

where [M] represents the array after threshold segmentation. $[M_2]$ represents the array of radial sections of the fish body. $[M_1]$ and $[M_3]$ represent the radial cross-sectional array of the conveyor belt on both sides of the fish body. T represents the segmentation threshold. Δh_1 represents the absolute value of the difference between adjacent elements' height value of the right endpoint of $[M_1]$ and height value of the left endpoint of $[M_2]$. Δh_2 represents the absolute value of the difference between adjacent elements' height value of the right endpoint of $[M_2]$ and height value of the left endpoint of $[M_3]$.

2.2.4. Data Filtering

The absorption and reflection of light by the fish itself, the influence of external lighting, and the vibration of the conveyance mechanism during movement can cause high-frequency fluctuations in the data and form noise [22]. Since the Kalman filter has a good suppression effect on the random fluctuation noise generated by the data, and the median filter can reduce the fluctuation range of the data, the Kalman filter and the median filter were adopted to denoise the data. In this study, the covariance of the system noise of the Kalman filter was set to 0.0001, the covariance of the measurement noise was set to 0.1, the covariance of the system noise was set to 1, and the left and right rank of the median filter were set to 2 and −1, respectively.

2.3. *Fish Head Cut Position Identification*

2.3.1. Feature Extraction

As shown in Figure 3d, the fish head cut position is on the contour line composed of the highest points of the radial section data of the fish body, namely the boundary between the abdomen and the back of the fish body, which is defined as the ventral–dorsal demarcation line of the fish body in this study. The sampled data on the ventral–dorsal demarcation line of the fish body were taken as input and the real value of the fish head cut position was taken as output to construct the fish head cut position identification model and to achieve the prediction of the fish head cut position. The volume of data on the ventral–dorsal demarcation line is large, and there is a strong correlation between some data points, making it a large amount of redundant information and affecting the calculation accuracy. Principal Component Analysis (PCA) is an unsupervised machine learning algorithm that can transform multiple variables into a few composite variables, eliminating redundant information and reducing computational effort [23]. In this study, PCA was chosen to reduce the dimensionality of the ventral–dorsal divide, using a few principal components instead of the entire ventral–dorsal divide data. As shown in Equations (4)–(6), the collected ventral–dorsal dividers were transformed by the Z-score method to standardize them, as a way to eliminate the difference in magnitude and the difference in order of magnitude between different indicators. The correlation coefficient matrix between the independent variables was solved by using a standardized data matrix, and the characteristic roots were obtained according to the characteristic equation of the correlation coefficient matrix, as shown in Equation (8), with the cumulative contribution of the variance at 95% to determine the extracted principal components for subsequent studies. As shown in Equations (9) and (10), the indicator coefficient matrix of each principal component was obtained according to the component matrix and multiplied with the standardized data matrix to obtain the principal component values.

$$\overline{z}_j = \frac{\sum_{i=1}^{n} z_{ij}}{s_j}, s_j = \sqrt{\frac{\sum_{i=1}^{n} \left(z_{ij} - \overline{z}_j\right)^2}{n-1}} \tag{4}$$

$$\tilde{z}_{ij} = \frac{z_{ij} - \overline{z}_j}{s_j} \tag{5}$$

$$Z_{ij} = \frac{z_{ij} - \overline{z}_j}{s_j}, D = \frac{{Z_{ij}}^T Z_{ij}}{n-1} \tag{6}$$

$$\left|D - \lambda I_p\right| = 0 \tag{7}$$

$$\frac{\sum_{j=1}^{m} \lambda_i}{\sum_{j=1}^{p} \lambda_i} \geq 0.95 \tag{8}$$

$$U_i = P_i / \sqrt{\lambda_i} \tag{9}$$

$$F_i = \sum_{i=1}^{n} U_i \times ZX_i \tag{10}$$

where $\overline{z}_j$ represents the mean value of the ventral–dorsal divide, s_j represents the standard deviation of the ventral–dorsal divide, $\widetilde{z}_{ij}$ represents the standardized data, Z_{ij} represents the standardized matrix combined by column, D represents the correlation coefficient matrix, λ represents the eigenvalues of this matrix, P_I represents the component matrix, U_I represents the index coefficient matrix, and F_i represents the principal component values of the ventral–dorsal divide of the fish.

2.3.2. Establishment of Fish Head Cut Position Identification Models

The principal component values of 204 fish body samples after PCA dimensionality reduction were used as model inputs and the length of the fish head was the output to construct the LS-SVM, PSO-BP, and LSTM models for the identification of the cutting position of the fish head. Among them, a total of 154 samples were randomly selected as the training set, and the remaining 50 samples were used for the testing set.

LS-SVM Model

The LS-SVM maps linearly indistinguishable data in space to a high-dimensional feature space by means of a constructed kernel function that makes the data divisible in the feature space [24,25]. Given that the Gaussian radial basis function (RBF) is capable of non-linear mapping and has fewer parameters, the RBF kernel function was chosen to construct the LS-SVM model in this study. The penalty parameter sig2 and the radial basis kernel parameter gam are two important parameters of the RBF kernel function, which are closely related to the accuracy and generalization ability of the model [26]. In this study, the particle swarm algorithm (PSO) was used to find the best values for the above two parameters, and the maximum number of iterations was taken as 100, with the search range of sig2 being 0.1 to 100 and the search range of gam being 0.01 to 100.

PSO-BP Model

BP neural network is an error backpropagation algorithm, and its learning process can be summarized as signal forward propagation, error backpropagation, weights, and threshold update. The network includes input, hidden, and output layers. Using the gradient descent method, the weights and thresholds between different network layers are adjusted inversely by comparing the model output values with the expected values to reduce the error along the gradient direction [19]. The approximate solution that satisfies the error accuracy is sought through several iterations, and its structure is shown in Figure 4.

In a traditional BP network, it is easy to fall into local optimal solutions in the training process, which makes the final model accuracy too low [17]. To avoid this problem, this study used the PSO algorithm with global search ability to optimize the network weight of the BP neural network. Figure 5 shows the structure of the PSO-BP neural network built in this study, where the inputs are three independent variables, the number of hidden layers was ten layers, and the parameters of the neural network were set as follows: the number of iterations was 1000, the training objective was 0.00001, and the learning rate was set to 0.09. The PSO parameters were set as follows: learning factor $c_1 = c_2 = 2$, inertia weights $\omega_{max} = 0.9$ and $\omega_{min} = 0.3$ with a maximum iteration of 200.

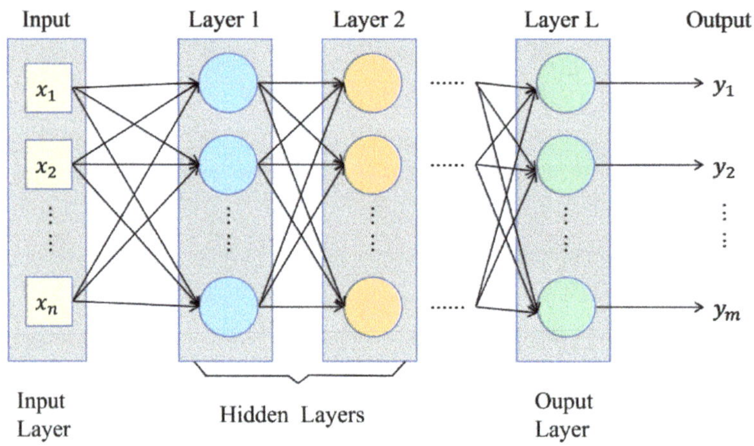

Figure 4. BP neural network unit structure.

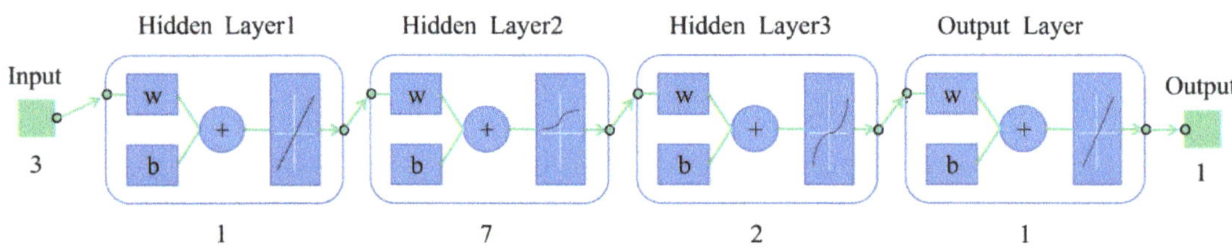

Figure 5. PSO-BP unit structure.

LSTM Model

The LSTM model is a special type of Recurrent Neural Network (RNN), consisting of memory blocks that add input and output channels to the hyperbolic tangent function (tanh), which can correlate the feature data of each fish with each other and analyze their non-linear relationships [27]. The cell structure of the LSTM model is shown in Figure 6, where f_t, i_t, and o_t denote the forgetting layer, input layer, and output layer, respectively; x_t is the input of the current cell; C_{t-1} and h_{t-1} are the outputs of the last network cell; W and V are the weight matrices; b is the bias term; and σ is the sigmoid function layer. The LSTM network used in this study has an input layer with three inputs, a hidden layer activated by the tanh function, one forgetting layer, one fully connected layer, and one output layer. The gradient threshold size used was 1, the initial learning rate was 0.005, and the maximum number of iterations was 200.

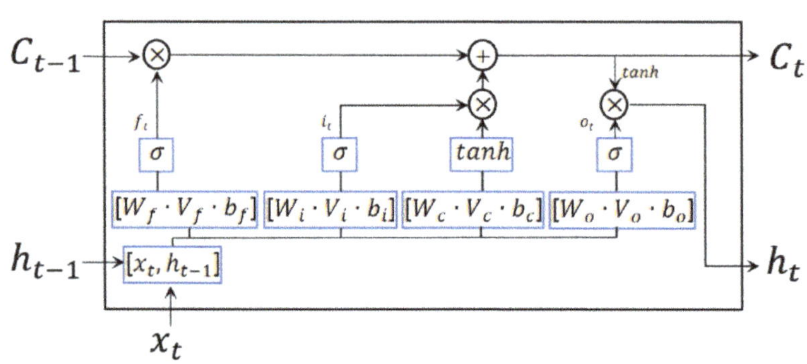

Figure 6. LSTM unit structure.

2.4. Model Evaluation Metrics

In order to verify the identification effect of the model, the coefficient of determination (R^2), the root mean square error (RMSE), the mean absolute error (MAE), and the relative analysis error (RPD) were used as evaluation indicators of the prediction model. R^2 is an important parameter in model evaluation, and usually $R^2 > 0.82$ indicates that the method can be used for practical applications. $R^2 > 0.9$ indicates that the model has excellent fish head cutting position identification. Smaller RMSE values represent better model prediction performance [28]. MAE can better reflect the actual situation of prediction value error, and a smaller value represents a smaller prediction error [28,29]. RPD can intuitively reflect the prediction ability of the model; when RPD > 1.4, the model reliability is poor. When the RPD > 2.5, the model prediction is accurate and reliable [30,31]. The formulas of the above evaluation indexes are as follows:

$$R^2 = 1 - \frac{\sum_{i=1}^{n}\left(y_i - \hat{y}_i\right)^2}{\sum_{i=1}^{n}(y_i - \bar{y}_i)^2} \tag{11}$$

$$RMSE = \sqrt{\frac{\sum_{i=0}^{n}\left(y_i - \widehat{y}_i\right)^2}{n}} \tag{12}$$

$$MAE = \frac{1}{n}\sum_{i=1}^{n}\left|y_i - \widehat{y}_i\right| \tag{13}$$

$$RPD = \frac{SD}{RMSEP} = \sqrt{\frac{1}{1 - {R_p}^2}} \tag{14}$$

3. Results

3.1. Data Pre-Processing

3.1.1. Data Segmentation

As shown in Figure 7a, the initial point cloud data obtained after laser scanning included the conveyor belt contour and the fish surface contour point cloud data. The maximum horizontal height of the conveyor belt contour was 250.32 mm, and this part was removed according to the method described in Section 2.3.2 to achieve partial dimensionality reduction of the data and eliminate the interference information irrelevant to the fish contour (Figure 7b).

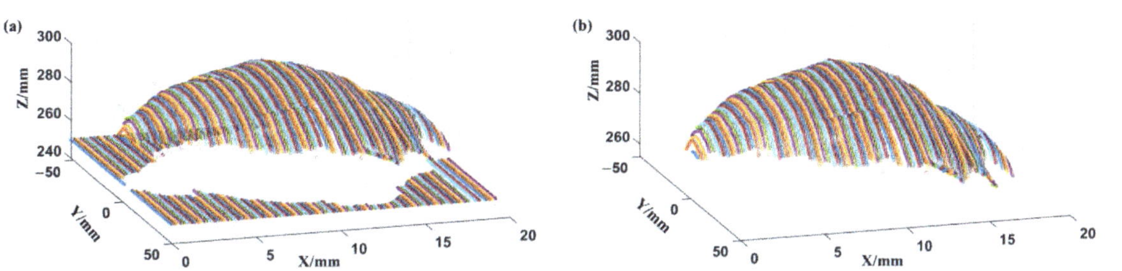

Figure 7. Upper surface data collection of fish: (**a**) laser point cloud data collection; (**b**) fish surface data after segmentation. The different colored lines represent the results of a single line laser scan and consist of multiple points that form a radial profile.

3.1.2. Data Filtering

The results of the original data filtering of the radial cross-section profile of the fish body are shown in Figure 8. As can be seen in Figure 8a, the original data had more obvious noise and data fluctuation, and the radial cross-section profile of the fish body was not accurate. After adding the Kalman filter, the large-scale noise generated by system vibration was basically removed, indicating that it has a good suppression effect on the noise generated by random fluctuations in this study (Figure 8b), and the radial profile curve was further refined. But, due to the existence of data fluctuations, the profile curve was still not smooth and complete. After adding the median filter, the range of fluctuations of the original data was reduced (Figure 8c), and the radial profile curve gradually became continuous and complete. When the data were subjected to Kalman and median filters in turn, the contour curve became smooth and the high-frequency noise was obviously improved (Figure 8d), which is closer to the real contour curve. Therefore, in this study, the Kalman filter and the median filter for the pre-processing of the radial contour of the fish body were used together.

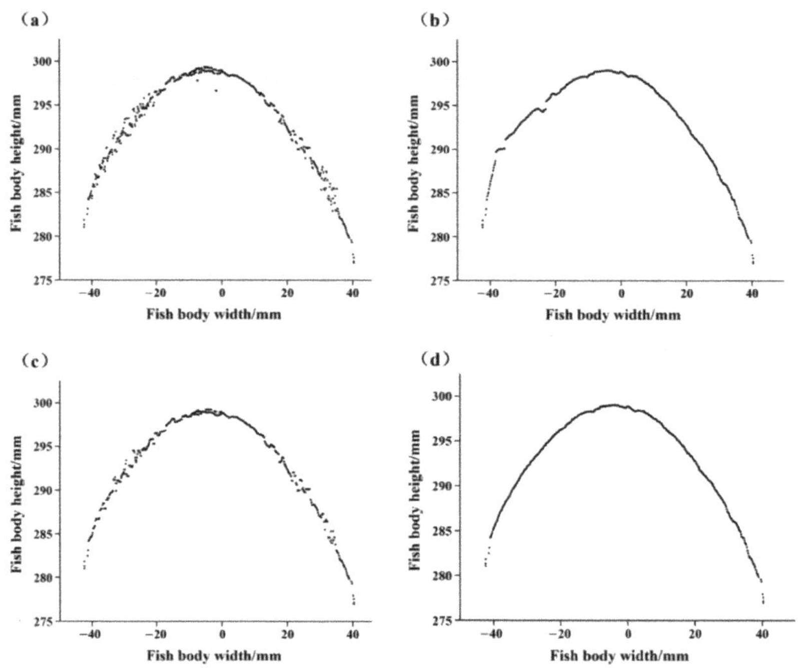

Figure 8. Data filtering: (**a**) original data; (**b**) Kalman-filtered data; (**c**) median-filtered data; (**d**) Kalman- and median-filtered data.

3.2. Fish Head Cut Position Identification

3.2.1. Extraction of Ventral–Dorsal Demarcation Line

For the fish body, the thickness of the ventral–dorsal part of the fish decreased slowly, and the thickness of the tail part of the fish decreased rapidly. The maximum point of the thickness of the fish was at the ventral–dorsal part, and the position of the head cut was before the maximum point of the thickness. The ventral–dorsal demarcation line of a fish is a cloud of points consisting of the maximum values of all radial cross-section heights of the fish body. The ventral–dorsal demarcation lines of 204 fish bodies are shown in Figure 9. It illustrates that although the size of the samples was not uniform, the corresponding ventral–dorsal demarcation lines had the same trend of changes, which increased rapidly and reached the maximum value before about one-third of the total length, and then began to decrease. As shown in the figure (Figure 9), the ventral–dorsal demarcation line could reflect the change law of the radial thickness of the fish body, which can be used as the basis for fish head cutting. Taking the ventral–dorsal demarcation line as input, three supervised machine learning methods (LS-SVM, PSO-BP, and LSTM) were used to train and predict the model, respectively, to achieve the identification of the fish head cutting position.

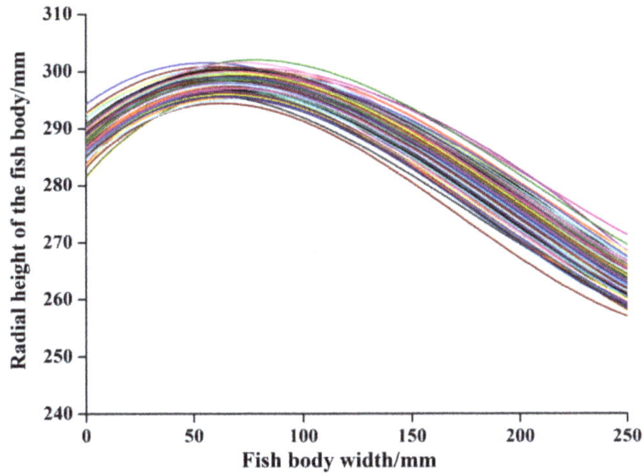

Figure 9. Abdominal and dorsal dividing lines. The different colored lines represent the ventral–dorsal dividing line extracted from different samples.

3.2.2. Data Dimensionality Reduction of Abdominal and Dorsal Dividing Lines

In order to reduce the volume of data for the identification of fish head cut features and to achieve rapid identification of fish head cut locations, PCA was used to reduce the dimensionality of the abdominal and dorsal dividing lines. Factor analysis was performed on the measured ventral–dorsal demarcation lines of the 204 samples, and the results of the obtained feature values are shown in Table 2. The cumulative variance contribution of the first three principal components reached 95.080%, indicating that the first three principal components can significantly reflect 95.080% of the information of the original data. In addition, it can also be seen from the gravel plot shown in Figure 10 that the trend of the first, second, and third eigenvalues was more obvious, and from the fourth eigenvalue onwards, the trend of the eigenvalues tended to be stable, so the first three principal components were taken for subsequent modeling.

Table 2. Total variance of interpretation.

Component	Initial Eigenvalue			Extraction of Squares and Loading			Rotate the Square and Load		
	Total	Variance of %	Cumulative %	Total	Variance of %	Cumulative %	Total	Variance of %	Cumulative %
1	43.031	71.718	71.718	43.031	71.718	71.718	34.740	57.901	57.901
2	11.990	19.983	91.702	11.990	19.983	91.702	16.894	28.157	86.058
3	1.999	3.3319	95.033	1.999	3.331	95.033	5.385	8.974	95.033
4	0.842	1.403	96.436						
5	0.386	0.643	97.079						
6	0.340	0.567	97.646						
7	0.261	0.434	98.080						
8	0.139	0.232	98.312						
9	0.112	0.187	98.496						
10	0.097	0.161	98.661						
		......							
204	-1.04×10^{-15}	-1.74×10^{-15}	100.00						

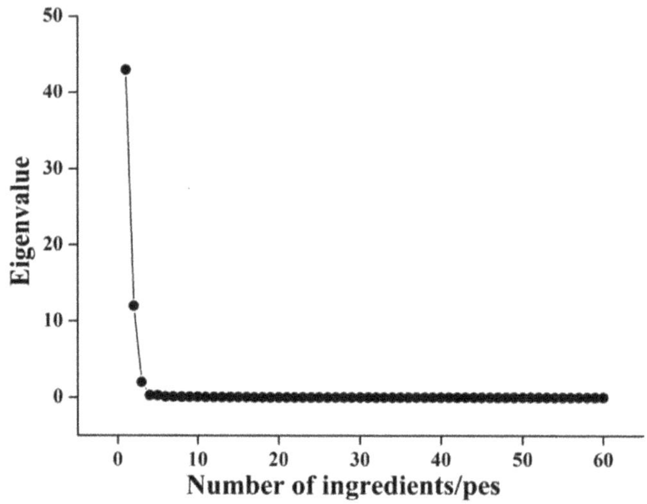

Figure 10. Gravel map extracted based on covariance matrix.

3.2.3. Fish Head Cutting Position Identification Model

Taking the three principal component values obtained as the model input and the actual cut position of the fish head as the model output, all 204 samples were randomly divided into two groups, one group of 154 samples as the training set and the other group of 50 samples as the test set, to construct the LS-SVM, PSO-BP, and LSTM fish head cutting position recognition models.

LS-SVM Model

In order to obtain an LS-SVM model with high prediction accuracy and stability, the two parameters of the RBF kernel function in the model, penalty coefficient sig2 and kernel parameter gam, need to be optimized. The results of parameter search using PSO showed that the LS-SVM model had the best recognition effect when the penalty parameter sig2 = 0.01 and the optimal kernel parameter gam = 15.3284; the fitness curve is shown in Figure 11, from which, after 29 iterations, the fitness curve started to smooth out and the validation parameters reached the optimal solution. The recognition results of the LS-SVM model for the fish head cutting position are shown in Figure 12. The R_c^2, RMSE, and MAE of the training set were 0.9125, 0.2622, and 0.1857, respectively, and the R_p^2, RMSE, MAE, and RPD of the test set were 0.9094, 0.8548, 0.6123 and 2.4041, respectively. The results showed that its R_p^2 was more than 0.9, which indicates that the LS-SVM model has good generalization ability and performance, and RMSE and MAE were low, which represents fewer outliers and errors in the results predicted by the model, but its RPD values were less than 2.5, indicating that the model is not stable enough and has limited identification capability.

PSO-BP Model

The connection weights of each layer of the BP neural network were encoded into particles, and the PSO algorithm was used to search for the optimal network weights within the set number of iterations, in which the population size of the particle swarm was set to be 20 to prevent overconsumption of computational resources. In order to avoid the particle speed growth being too fast or too slow, the value of the speed was set to $[-1, 1]$, ensuring that the solution is within a reasonable range of the particle position to avoid overstepping the boundaries, and the position was set to $[-2, 2]$. The training results of the PSO-BP neural network for fish head cutting position identification are shown in Figure 13. After the computation of the PSO-BP model, the model's recognition results of the fish head cutting position are shown in Figure 14. The R_c^2, RMSE, and MAE of the training set were 0.9168, 0.5203, and 0.3277, respectively, and the R_p^2, RMSE, MAE, and RPD of the test set were 0.9295, 0.5126, 0.3143, and 2.513, respectively. Compared with the LS-SVM model, its larger R_p^2 value represents a better model performance, its smaller RMSE and MAE, represent a further reduction in the error of the PSO-bp model prediction, and its RPD was more than 2.5, which indicates that its recognition has strong reliability and is suitable for the recognition of the ideal cutting position of fish heads.

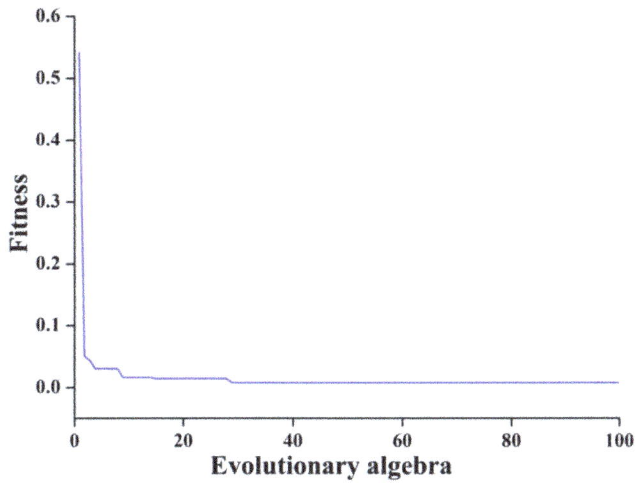

Figure 11. Fitness diagram.

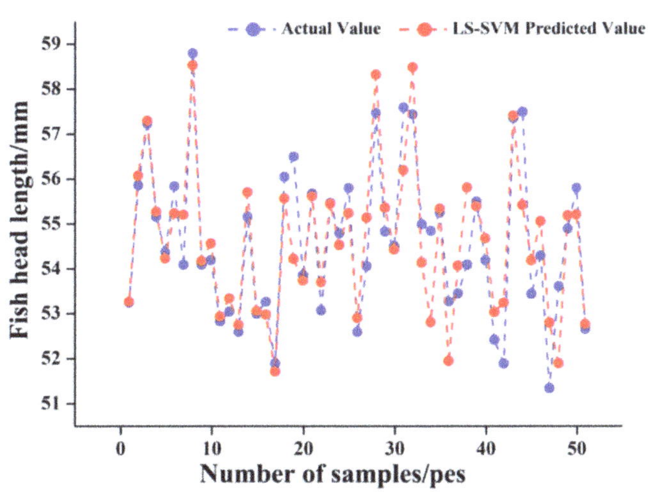

Figure 12. Comparison diagram of predicted and measured fish head length based on LS-SVM.

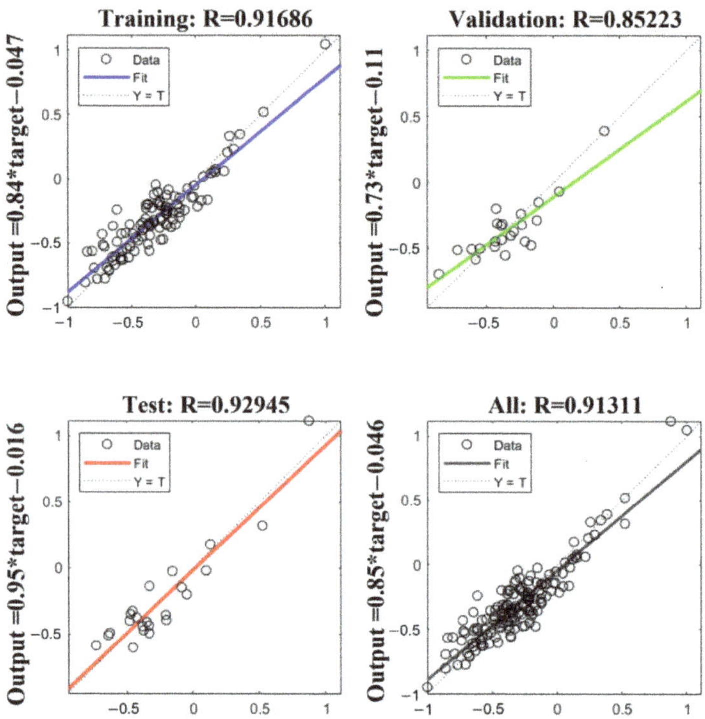

Figure 13. PSO-BP unit structure training results.

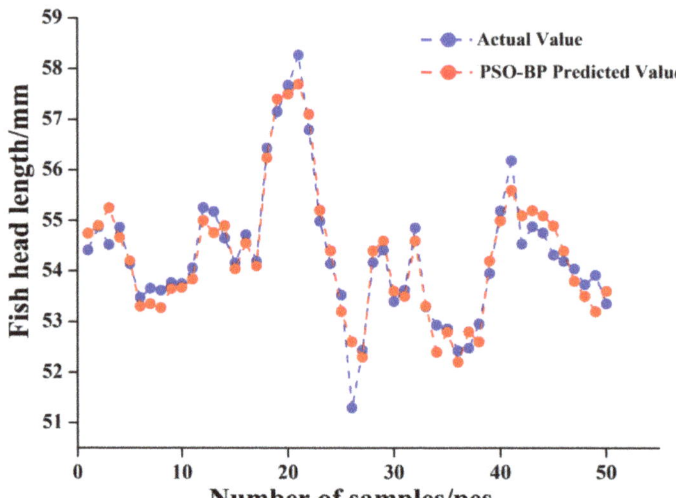

Figure 14. Comparison diagram of predicted and measured fish head length based on PSO-BP.

LSTM Model

During the LSTM training, the forgetting, input, and output layers were activated by sigmoid functions, and the entire output data range was transformed in the [0,1] interval to keep the data normalized. In the model-building process, if a neuron parameter produces large volatility, the overall fit of the model will be biased towards that neuron; therefore, during the training process, to reduce the impact of overfitting on the prediction model, dropout regularization was added, and the dropout rate was taken as 0.2. As can be seen from Figure 15, during 200 iterations, there was a brief upward trend in the loss function of the training and test sets, followed by a rapid decline, which slowed down during the subsequent iterations, indicating that the LSTM output values fit the true values better and better. The results of the LSTM model for the identification of fish head cutting positions are shown in Figure 16. The R_c^2, RMSE, and MAE of the training set were calculated by the LSTM model to be 0.9705, 0.1964, and 0.1477, respectively, and the R_p^2, RMSE, MAE, and RPD of the test set were 0.9480, 0.2957, 0.1933, and 3.1426, respectively. Among the analyzed models, the LSTM model's R_p^2 reached up to 0.9480, with the best generalization ability and performance, and it had a lower RMSE and MAE, which represents minimum and stable prediction error, and a larger RPD value of more than 2.5, which indicates that its identification has strong reliability and is suitable for the recognition of the ideal cutting position of fish heads.

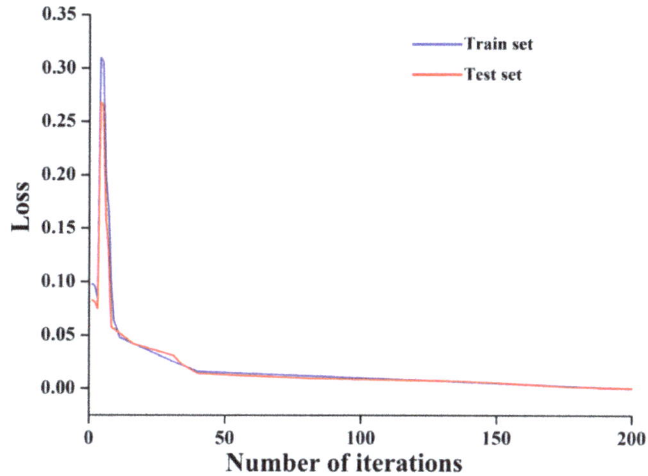

Figure 15. The loss function of training iterations and the target value of the LSTM model.

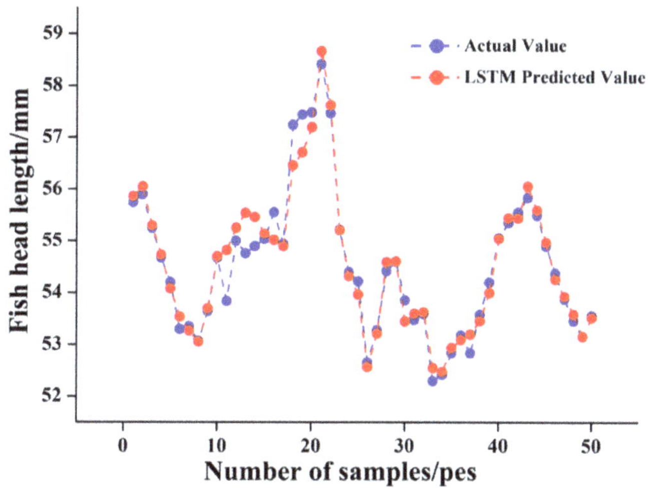

Figure 16. Comparison diagram of predicted and measured fish head length based on LSTM.

In the present study, the interference information and random fluctuation noise in the original data were processed by threshold segmentation and Kalman and median filtering, respectively, which can accurately restored the radial profile curve of the sample. The principal component value of the ventral–dorsal demarcation lines after PCA dimensionality reduction can be used as the feature information for identifying the cutting position of the fish head. In the established MPR, LS-SVM, and LSTM fish head cutting position identification models, the overall performance of the LSTM model was the best: the error between the predicted value and the actual value of the model was the smallest, and the reliability was also high. This shows that the line laser scanning technology combining with machine learning has the potential for fish head cutting position recognition. At the same time, the identification based on the information of the ventral–dorsal dividing line proved to be effective. The ventral–dorsal demarcation line is a line composed of the highest points in each radial profile section, which is identified by the morphological characteristics of the fish. It will not be limited to the posture and individual size, so the accuracy of the identification method based on the ventral–dorsal demarcation line will not be affected by the placement and size of the fish. The spindle-shaped and flat-hammer-shaped fishes with a similar morphology to that of the samples used in this study (crucian carp) all have a ventral–dorsal dividing line, and therefore, the method proposed in this study could have strong applicability to bulk fish with similar shapes and head-removing needs, such as grass carp, silver carp, etc., and even high value-added fish such as salmon and tuna. With a further increase in sample size, the accuracy and generalization of the identification model could also be continuously improved. In addition, due to the high precision and efficiency of line laser scanning, the application of this method is more conducive to the on-line detection of bulk fish. However, in actual large-scale fish processing, when two or more fish are stacked up and placed, it is hard to extract the ventral–dorsal demarcation line of an individual sample fish, so the method should be applied to the situation where raw materials are scanned one by one.

4. Conclusions

In this study, we used line laser scanning technology to nondestructively collect the surface information of the fish body, restore the outer contour information of the fish body through data processing, and extract the feature variables of the ventral–dorsal demarcation line through PCA to reduce the computation time and improve the identification ability. Using the obtained feature variables as the inputs, fish head identification models based on LS-SVM, PSO-BP, and LSTM were established. The results showed that the data processed by threshold segmentation, Kalman filtering, and median filtering can restore the radial contour of the fish body more accurately. In the model comparison experiments, the LSTM model outperformed the LS-SVM and PSO-BP model with the highest R^2 and RPD of 0.9480 and 3.1426, and lowest RMSE and MAE of 0.2957 and 0.1933, respectively, in the test set. These results validate the potential of combining line laser scanning techniques with machine learning for fish head cutting position identification. In future studies, experiments will be conducted on more types and sizes of fish to improve the reliability and practicality of the model.

Author Contributions: Conceptualization, X.Z. and H.W.; methodology, X.Z. and H.W.; software, Z.G. and J.M.; validation, X.Z., Z.G. and J.M.; formal analysis, X.Z.; investigation, X.L.; resources, X.L. and W.S.; data curation, J.M.; writing—original draft preparation, Z.G., W.S. and X.L.; writing—review and editing, H.W.; supervision, H.W.; project administration, H.W. All authors have read and agreed to the published version of the manuscript.

Funding: This work was supported by the Natural Science Foundation of China (32372434) and Scientific Research Fund of Liaoning Province Education Department (LJKZ0511 and LJKZ0542).

Data Availability Statement: Data are contained within the article.

Conflicts of Interest: The authors declare no conflict of interest.

References

1. Azarmdel, H.; Mohtasebi, S.S.; Jafari, A.; Muñoz, A.R. Developing an orientation and cutting point determination algorithm for a trout fish processing system using machine vision. *Comput. Electron. Agric.* **2019**, *162*, 613–629. [CrossRef]
2. Qing, C.; Branch, D. Method study on mechanized head cutting for marine small fish. *Fish. Mod.* **2012**, *39*, 38–42.
3. Wang, H.H.; Zhang, X.Y.; Li, P.P.; Sun, J.L.; Yan, P.T.; Zhang, X.; Liu, Y.Q. A New Approach for Unqualified Salted Sea Cucumber Identification: Integration of Image Texture and Machine Learning under the Pressure Contact. *J. Sens.* **2020**, *2020*, 8834614. [CrossRef]
4. Mustafa, A.; Volino, M.; Kim, H.; Guillemaut, J.-Y.; Hilton, A. Temporally Coherent General Dynamic Scene Reconstruction. *Int. J. Comput. Vis.* **2021**, *129*, 123–141. [CrossRef]
5. Wang, J.H.; Yang, Y.X.; Zhou, Y.G. Dynamic Three-Dimensional Surface Reconstruction Approach for Continuously Deformed Objects. *IEEE Photonics J.* **2021**, *13*, 1–15. [CrossRef]
6. Liu, Z.; Xiang, C.Q.; Chen, T. Automated Binocular Vision Measurement of Food Dimensions and Volume for Dietary Evaluation. *Comput. Sci. Eng.* **2018**. Available online: https://ieeexplore.ieee.org/abstract/document/8365083 (accessed on 13 December 2023).
7. Tsoulias, N.; Xanthopoulos, G.; Fountas, S.; Zude, M. In-situ detection of apple fruit using a 2D LiDAR laser scanner. In Proceedings of the IEEE International Workshop on Metrology for Agriculture and Forestry, Trento, Italy, 4–6 November 2020.
8. Antequera, T.; Caballero, D.; Grassi, S.; Uttaro, B.; Perez-Palacios, T. Evaluation of fresh meat quality by Hyperspectral Imaging (HSI), Nuclear Magnetic Resonance (NMR) and Magnetic Resonance Imaging (MRI): A review. *Meat Sci.* **2020**, *172*, 108340. [CrossRef] [PubMed]
9. Ando, M.; Sugiyama, H.; Maksimenko, A.; Rubenstein, E.; Roberson, J.; Shimao, D.; Hashimoto, E.; Mori, K. X-ray dark-field imaging and its application to medicine. *Radiat. Phys. Chem.* **2004**, *71*, 899–904. [CrossRef]
10. Weisgerber, J.N.; Medill, S.A.; McLoughlin, P.D. Parallel-Laser Photogrammetry to Estimate Body Size in Free-Ranging Mammals. *Wildl. Soc. Bull.* **2015**, *39*, 422–428. [CrossRef]
11. Chen, H.; Liu, Z.; Gu, J.; Ai, W.; Wen, J.; Cai, K. Quantitative analysis of soil nutrition based on FT-NIR spectroscopy integrated with BP neural deep learning. *Anal. Methods* **2018**, *10*, 5004–5013. [CrossRef]
12. Cobourn, W.G.; Dolcine, L.; French, M.; Hubbard, M.C. A comparison of nonlinear regression and neural network models for ground-level ozone forecasting. *Air Repair* **2000**, *50*, 1999–2009. [CrossRef] [PubMed]
13. Lin, W.M.; Tu, C.S.; Yang, R.F.; Tsai, M.T. Particle swarm optimisation aided least-square support vector machine for load forecast with spikes. *Iet Gener. Transm. Distrib.* **2016**, *10*, 1145–1153. [CrossRef]
14. Zhu, F.L.; He, Y.; Shao, Y.N. Application of Near-Infrared Hyperspectral Imaging to Predicting Water Content in Salmon Flesh. *Spectrosc. Spectr. Anal.* **2015**, *35*, 113–117.
15. Zhang, Y.; Jia, S.; Zhang, W. Predicting acetic acid content in the final beer using neural networks and support vector machine. *J. Inst. Brew.* **2012**, *118*, 361–367. [CrossRef]
16. Khoshnoudi-Nia, S.; Moosavi-Nasab, M. Nondestructive Determination of Microbial, Biochemical, and Chemical Changes in Rainbow Trout (*Oncorhynchus mykiss*) During Refrigerated Storage Using Hyperspectral Imaging Technique. *Food Anal. Methods* **2019**, *12*, 1635–1647. [CrossRef]
17. Wang, F.R.; Xie, B.M.; Lue, E.L.; Zeng, Z.X.; Mei, S.; Ma, C.Y.; Guo, J.M. Design of a Moisture Content Detection System for Yinghong No. 9 Tea Leaves Based on Machine Vision. *Appl. Sci.* **2023**, *13*, 1806. [CrossRef]
18. Deng, Y.; Xiao, H.J.; Xu, J.X.; Wang, H. Prediction model of PSO-BP neural network on coliform amount in special food. *Saudi J. Biol. Sci.* **2019**, *26*, 1154–1160. [CrossRef] [PubMed]
19. Zhang, L.; Liu, J.; Zhi, L. Research on Grain Yield Prediction Method Based on Improved PSO-BP. *J. Appl. Sci.* **2014**, *14*, 1990–1995. [CrossRef]
20. Örnek, M.N.; Örnek, H.K. Developing a deep neural network model for predicting carrots volume. *J. Food Meas. Charact.* **2021**, *15*, 3471–3479. [CrossRef]
21. Zhong, N.; Li, Y.P.; Li, X.Z.; Guo, C.X.; Wu, T. Accurate prediction of salmon storage time using improved Raman spectroscopy. *J. Food Eng.* **2021**, *293*, 110378. [CrossRef]
22. Gao, C.; Bao, S.; Zhou, C.; He, X.; Shiju, E.; Sun, J.; Gong, J. Research on the line point cloud processing method for railway wheel profile with a laser profile sensor. *Measurement* **2023**, *211*, 112640. [CrossRef]
23. Zabalza, J.; Ren, J.; Ren, J.; Liu, Z.; Marshall, S. Structured covariance principal component analysis for real-time onsite feature extraction and dimensionality reduction in hyperspectral imaging. *Appl. Opt.* **2014**, *53*, 4440–4449. [CrossRef] [PubMed]
24. Zhang, X.L.; Liu, F.; He, Y.; Li, X.L. Application of Hyperspectral Imaging and Chemometric Calibrations for Variety Discrimination of Maize Seeds. *Sensors* **2012**, *12*, 17234–17246. [CrossRef] [PubMed]
25. Tarhouni, M.; Laabidi, K.; Lahmari-Ksouri, M.; Zidi, S. System Identification based on Multi-kernel Least Squares Support Vector Machines (Multi-kernel LS-SVM). In Proceedings of the International Joint Conference on Computational Intelligence, Dhaka, Bangladesh, 14–15 December 2018.
26. Zhang, X.; Sun, J.L.; Li, P.P.; Zeng, F.Y.; Wang, H.H. Hyperspectral detection of salted sea cucumber adulteration using different spectral preprocessing techniques and SVM method. *Lwt-Food Sci. Technol.* **2021**, *152*, 112295. [CrossRef]
27. Wang, J.; Si, H.P.; Gao, Z.; Shi, L. Winter Wheat Yield Prediction Using an LSTM Model from MODIS LAI Products. *Agriculture* **2022**, *12*, 1707. [CrossRef]

28. Qi, J.; Du, J.; Siniscalchi, S.M.; Ma, X.L.; Lee, C.H. Analyzing Upper Bounds on Mean Absolute Errors for Deep Neural Network-Based Vector-to-Vector Regression. *IEEE Trans. Signal Process.* **2020**, *68*, 3411–3422. [CrossRef]
29. Weijie, W.; Yanmin, L. Analysis of the Mean Absolute Error (MAE) and the Root Mean Square Error (RMSE) in Assessing Rounding Model. In *OP Conference Series: Materials Science and Engineering*; IOP Publishing: Bristol, UK, 2018; p. 012049.
30. Cheng, J.H.; Sun, D.W.; Pu, H.B.; Zhu, Z.W. Development of hyperspectral imaging coupled with chemometric analysis to monitor *K* value for evaluation of chemical spoilage in fish fillets. *Food Chem.* **2015**, *185*, 245–253. [CrossRef]
31. Khoshnoudi-Nia, S.; Moosavi-Nasab, M. Prediction of various freshness indicators in fish fillets by one multispectral imaging system. *Sci. Rep.* **2019**, *9*, 14704. [CrossRef]

foods

MDPI

Article

The Preparation and Properties of Amino-Carboxymethyl Chitosan-Based Antibacterial Hydrogel Loaded with ε-Polylysine

Yixi Li [1,2], Yulong Qiu [1,2], Hongman Hou [1,2], Gongliang Zhang [1,2], Hongshun Hao [1,2] and Jingran Bi [1,2,*]

1 School of Food Science and Technology, Dalian Polytechnic University, No. 1, Qinggongyuan, Ganjingzi District, Dalian 116034, China; lyxixi3@163.com (Y.L.); qiuyulong2022@163.com (Y.Q.); zgl_mp@163.com (G.Z.); houhongman@dlpu.edu.cn (H.H.); beike1952@163.com (H.H.)
2 Liaoning Key Lab for Aquatic Processing Quality and Safety, No. 1, Qinggongyuan, Ganjingzi District, Dalian 116034, China
* Correspondence: bijr@dlpu.edu.cn

Abstract: In this paper, amino-carboxymethyl chitosan (ACC) was prepared through amino carboxymethylation, which introduces -COOH and -NH_2 groups to the chitosan (CS) chains. Meanwhile, dialdehyde starch (DAS) was produced by oxidizing corn starch using sodium periodate. To attain the optimal loading and long-time release of ε-polylysine (ε-PL), the ACC/DAS hydrogels were synthesized through the Schiff base reaction between the amino group on ACC and the aldehyde group in DAS. The molecular structure, microcosmic properties, loading capacity, and bacteriostatic properties of the four types of hydrogels containing different mass concentrations of ACC were investigated. The results showed that the dynamic imine bond C=N existed in the ACC/DAS hydrogels, which proved that the hydrogels were formed by the cross-linking of the Schiff base reaction. With the increasing mass concentration of the ACC, the cross-sectional morphology of the hydrogel became smoother, the thermal stability increased, and the swelling behavior was gradually enhanced. The tight network structure improved the ε-PL loading efficiency, with the highest value of 99.2%. Moreover, the loading of ε-PL gave the hydrogel good antibacterial properties. These results indicate that ACC/DAS hydrogel is potential in food preservation.

Keywords: hydrogel; amino-carboxymethyl chitosan; dialdehyde starch; antimicrobial peptide; bacterial inhibition

Citation: Li, Y.; Qiu, Y.; Hou, H.; Zhang, G.; Hao, H.; Bi, J. The Preparation and Properties of Amino-Carboxymethyl Chitosan-Based Antibacterial Hydrogel Loaded with ε-Polylysine. *Foods* **2023**, *12*, 3807. https://doi.org/10.3390/foods12203807

Academic Editor: Filomena Nazzaro

Received: 21 September 2023
Revised: 12 October 2023
Accepted: 16 October 2023
Published: 17 October 2023

1. Introduction

Microbiologically derived natural bioactive ingredients have become popular in recent years for use in functional foods and food preservation due to their nutritional value and bioactivity. They are being used to meet people's desire for a healthy lifestyle [1]. However, the application of these bioactive compounds in the food industry is hampered by their unstable nature, which results in poor solubility in water, high susceptibility to enzymes, and low thermostability [2]. ε-polylysine (ε-PL) was initially obtained from the *Streptomyces albulus* sp. *Lysinopolymerus* by the researchers Shima and Sakai [3]. It is classified as a cationic homo-polyamide and is composed of approximately 25–30 lysine residues, which are connected by an amide linkage that bridges the 3-amino and carboxyl groups [4]. The potent antimicrobial activity, low toxicity, and high water solubility of ε-PL make it an ideal option for preventing or controlling bacterial growth in food products [5]. However, the potent antimicrobial activity of the cationic ε-PL can be reduced in alkaline foods through the precipitation of the resulting electrostatic complexes, which prevents ε-PL from interacting with microbial surfaces [6].

Hydrogels are three-dimensional networks composed of hydrophilic polymer chains that are interconnected through physical or chemical connections [7]. The distinctive physical characteristics exhibited by hydrogels render them advantageous as vehicles for pharmaceuticals and physiologically active compounds. The porosity of hydrogel can be effectively controlled by manipulating the degree of cross-linking. This mechanism enables molecules of various sizes to penetrate the hydrogel structure. The diffusion, swelling control, and chemical processes such as hydrolysis or enzymatic hydrogel networks can alter the bioactive release rate from a hydrogel [8]. The Schiff base reaction, characterized by its rapid and mild nature, has found extensive utilization in hydrogel applications due to its favorable attributes. Chitosan (CS) is hydrophobic, biodegradable, biocompatible, non-toxic, and antibacterial, and the rich amino groups are widely used in Schiff base reactions [9]. Guan et al. introduced glutaraldehyde to prepare chitosan sodium alginate polyelectrolyte hydrogels via the Schiff base reaction, which compounded a sparse, porous, and amorphous structure, thereby improving their swelling behavior [10]. However, it is important to note that natural CS demonstrates limited water solubility and relatively low antibacterial efficacy. The enhancement of CS's solubility and antibacterial efficacy can be achieved by synthesizing functional chitosan derivatives through chemical modification [11]. Amino-carboxymethyl chitosan (ACC) has superior biological and physicochemical characteristics compared to CS, thus making it a suitable option for numerous applications. By introducing more amino groups into the CS structure, the Schiff base reaction with aldehyde group compounds may be considerably enhanced while preserving the properties of the CS chain. Additionally, ACC has better biocompatibility, water retention capability, and greater viscosity than CS, thereby qualifying it to be used as a hydrogel matrix [12]. Dialdehyde starch (DAS) has excellent physical, chemical, and biochemical properties and is a starch modifier with a large number of reactive aldehyde groups. Luo et al. demonstrated that chitosan-dialdehyde starch membranes based on Schiff bases are biocompatible with common mammalian cells [13].

In order to fabricate a long-term release carrier for ε-PL, a bio-based hydrogel was constructed in this paper. Firstly, CS was modified through amino-carboxymethylation to increase the amino content of the polymer chain. Meanwhile, DAS was obtained from maize starch through the oxidation of $NaIO_4$. The amino group located on ACC underwent a reaction with the aldehyde groups presented on DAS, resulting in the formation of a hydrogel through a Schiff base reaction. The characterization of ACC/DAS hydrogels with varying ACC concentrations was conducted using several analytical techniques, including scanning electron microscopy (SEM), Fourier transform infrared spectroscopy (FTIR), X-ray diffraction (XRD), and thermogravimetry (TGA), and their swelling characteristics were also analyzed. The loading capacity of ACC/DAS with ε-PL was quantified, and the antibacterial characteristics of the ACC/DAS/ε-PL hydrogels were examined. The aim of this study was to explore the potential applications of ACC based hydrogels in food preservation.

2. Materials and Methods

2.1. Materials

CS (80~97% degree of deacetylation and molecular weight = 30,000 Da) was obtained from Sangon Biotech (Shanghai, China). Corn starch was obtained from Yuanye Biotechnology (Shanghai, China). Sodium periodate ($NaIO_4$, 98%) and ε-PL were obtained from Aladdin Biochemical Technology (Shanghai, China). Ethyl-3-(3-dimethylaminopropyl) carbodiimide hydrochloride (EDC, 98.5%) was obtained from MacLean Biochemical Technology (Shanghai, China). The chemicals ethylenediamine anhydrous (ED, 99%), chloroacetic acid, and dimethyl sulfoxide (DMSO, 99.5%) were obtained from the Damao Chemical Reagent Factory (Tianjin, China). *Staphylococcus aureus* (*S. aureus*, ATCC 6538) and *Escherichia coli* (*E. coli*, CICC 10670) were obtained from the China General Microbial Strain Repository Management Centre (Wuhan, China).

2.2. Synthesis of ACC and DAS

A solution of CS (5 g) was prepared by dissolving it in 50 mL of dimethyl sulfoxide (DMSO) at 30 °C for 60 min. Alkalized CS was prepared by adding sodium hydroxide (90% w/v) and mixing at 30 °C for 2 h. After the addition of 25 g chloroacetic acid, the mixture was stirred at 60 °C for 4 h, and the mixture was then filtered and rinsed with 75% volume ethanol. The purified material was subjected to a drying process at 50 °C for 12 h. Then, the solution underwent dissolution, followed by pH adjustment to 3.65 with the addition of hydrochloric acid with a concentration of 1 mol/L. This process resulted in the formation of a white precipitate. The precipitate was determined to be carboxymethyl chitosan (CC). The sample underwent two rinses using anhydrous ethanol, followed by a roasting process at 60 °C for 6 h. Following this, CC was dissolved in phosphate buffered saline (PBS) at a concentration of 2.5 wt% and a pH of 7.4. The solution was then agitated continuously for the duration of the night. ED was added at a molar ratio of 20:1 (ED:CC). Hydrochloric acid (HCl) was added to adjust the pH of the solution to 5.0. The compound EDC was subsequently introduced to the compound CC at a molar ratio of 2:1. After subjecting the solution to agitation at 37 °C for 4 h, it was thereafter subjected to dialysis for a period of 72 h in water. A dialysis membrane with a molecular weight cut off (MWCO) of 3.5 kDa was used to dialyze the solution. This continued until the dialysate no longer included any traces of ethylenediamine and EDC. The acquisition of ACC was achieved through the process of freezing the reaction product at a temperature of −80 °C.

The corn starch (2 g) was dissolved in 23 mL of hydrochloric acid solution with a final concentration of 0.6 mol/L. $NaIO_4$ was then added at a molar ratio of 2:1 ($NaIO_4$:starch) and the mixture was stirred at 40 °C for 2 h. Next, the resultant product underwent three washes with deionized water and was subjected to drying at 40 °C for a period of 12 h. The resulting product was identified as DAS.

2.3. Determination of Amino Content of ACC and Aldehyde Group Content of DAS

Initially, varying concentrations (1 mL) of β-alanine solution were combined with 1 mL of sodium bicarbonate (4 wt%) and 1 mL of 2,4,6-trinitrobenzenesulfonic acid (0.1 wt%). The mixture was then reacted at 40 °C for 2 h before being diluted fivefold. The absorbance at 415 nm was measured using a UV-vis spectrophotometer (UV-1750, Shimadzu, Japan), and a standard curve was subsequently developed with the following formula: y = 1.3111x + 0.2936 (R^2 = 0.9991), where y is the absorbance at 415 nm and x is the molar concentration of β-alanine. To determine the amino content of ACC, the ACC was substituted for β-alanine.

The aldehyde group content of DAS was determined using the alkali solubilization method by dissolving 0.2 g of DAS in 10 mL NaOH (0.25 mol/L), which was heated at 70 °C for 2 h and then cooled. Next, 15 mL of H_2SO_4 (0.125 mol/L) was added to the solution and shaken well until turning light-yellow. Then, 0.2 g of activated carbon was added and the solution was shaken for 2 min. The solution was filtered and five drops of phenolphthalein were added and titrated with 0.1 mol/L NaOH. The aldehyde group content was calculated using the Formula (1):

$$-CHO(\%) = \frac{(C_1V_1 + C_2V_2 - 2C_3V_3)}{m \times 1000} \times 100 \quad (1)$$

C_1 is the concentration of NaOH when dissolving DAS (0.25 mol/L), V_1 is the volume of NaOH (10 mL), C_2 is the concentration of NaOH used in the titration (0.1 mol/L), V_2 is the volume of NaOH solution during the titration, C_3 is the concentration of H_2SO_4 (0.125 mol/L), V_3 is the volume of H_2SO_4 (15 mL), m is the mass of DAS, and 160 is the average relative molecular mass of DAS glucose unit.

2.4. Preparation of ACC/DAS Hydrogels

Solutions of ACC with concentrations of 4.5 w/v%, 6 w/v%, 7.5 w/v%, and 9 w/v% were prepared by dissolving 4.5 g, 6 g, 7.5 g, and 9 g of ACC in 100 mL of PBS at a pH of 7.4, respectively. The DAS solution with a concentration of 6 wt% was prepared by dissolving 6 g of DAS in 100 mL of PBS. Next, 10 mL ACC solutions were placed in an acrylic plate with a 90 mm diameter and dried at 40 °C for 30 min until the ACC solutions were semi-solidified. Then, 10 mL of DAS solution was poured on top of the ACC layer and dried at 40 °C for 3 h to obtain four hydrogels with molar ratios of 1:1.6, 1:1.25, 1:1, and 1:0.6 of -NH_2:-CHO, named as the ACC_1/DAS, ACC_2/DAS, ACC_3/DAS, and ACC_4/DAS hydrogels.

2.5. Characterizations of ACC, DAS, and ACC/DAS Hydrogels

The ACC/DAS hydrogels were placed in liquid nitrogen and rapidly frozen before being sectioned. Scanning electron microscopy (SEM, JSM-7800F, JEOL, Tokyo, Japan) was used to assess the morphological features of the ACC/DAS hydrogels. The FTIR spectra of the ACC, DAS, and ACC/DAS hydrogels were obtained from 4000 to 400 cm^{-1} with a range resolution of 4 cm^{-1} using an FTIR spectrometer (Spectra II, PerkinElmer, Waltham, MA, USA). The XRD profiles of the ACC, DAS, and ACC/DAS hydrogels were collected at ambient temperature via a diffractometer (XRD-6100, Shimadzu, Kyoto, Japan) with a scanning range of 5° to 60° at a rate of about 5° per minute. The thermal stability of the ACC/DAS hydrogels was measured using a thermogravimetric analyzer (TGA550, TA, Newcastle, DE, USA) with a temperature range of 0–600 °C, a heating rate of 30 °C/min, and a nitrogen flow rate of 20 mL/min. For the determination of the mechanical properties, the hydrogels were cut into strips of 10 × 100 mm. The mechanical properties of the ACC/DAS hydrogels were tested using an automated tensile testing machine (INSTRON5965, Instron, Norwood, MA, USA), with an initial distance of 40 mm between the clamping plates and at a testing speed of 50 mm/s. The test was carried out at a speed of 1.5 m/s.

2.6. Swelling Degree of ACC/DAS Hydrogels

The ACC_1/DAS, ACC_2/DAS, ACC_3/DAS, and ACC_4/DAS hydrogels were immersed in deionized water for 10, 20, 30, 40, 50, 60, 70, 80, 90, 100, 110, and 120 min, respectively. Then, the hydrogels were taken out from the solution and promptly weighed through the elimination of surface moisture using filter paper [14]. The degree of swelling (SW) was determined using the following Formula (2):

$$Swell\ degree(\%) = \frac{W_S - W_D}{W_D} \times 100 \quad (2)$$

W_S is the weight of the hydrogel subsequent to the process of swelling and W_D is the weight of the hydrogel in its dry state prior to swelling.

2.7. Loading Efficiency of ACC/DAS Hydrogels

The ε-PL solution (2 mg/mL) was diluted with 1 mol/L PBS (pH 6.9) to yield ε-PL standard solutions with concentrations of 0.02, 0.05, 0.08, 0.1, 0.15, 0.2, and 0.25 mg/mL, respectively. Then, 0.2 mL of each standard solution was added to 0.8 mL of 0.5 mmol/L methyl orange. The samples were shaken for 30 min at 30 °C; they were centrifuged at 8000 rpm for 5 min, the supernatant was diluted 10 times and the absorbance was read at A_1 at 465 nm. The absorbance A_0 was measured using a mixture of PBS and methyl orange as the control, and the UV spectrophotometer was zeroed with PBS. The standard curve—represented by the equation y = 1.8277x + 0.0368 (R^2 = 0.9938)—was derived using the mass concentration of ε-PL as the independent variable and the difference in absorbance $\Delta A = (A_0 - A_1)$ as the dependent variable.

ACC/DAS/ε-PL hydrogels are made by dissolving 0.2 g of ε-PL in 10 mL of the prepared ACC solution, as described in 2.4. Four ACC/DAS hydrogels with different ACC contents were added to a centrifuge tube containing 10 mL of deionized water, respectively. After shaking for 60 s, the samples underwent centrifugation at a speed of 10,000 revolutions per minute for a duration of 3 min [15]. The mass concentration of ε-PL in the supernatant was determined by means of colorimetric analysis with methyl orange. A volume of 0.2 mL of the supernatant was combined with 0.8 mL of the methyl orange solution with a concentration of 0.5 mmol/L. The supernatant underwent dilution 10 times, and the absorbance at a wavelength of 465 nm was quantified using a UV spectrophotometer. The absorbance value at OD_{465} was added into the ε-PL standard curve to obtain the ε-PL concentration; the loading efficiency was computed using the following Formula (3):

$$Loading\ effciency(\%) = \frac{M_0 - C_s V}{M_0} \times 100 \tag{3}$$

where the symbol M_0 is used to represent the initial mass of ε-PL added. C_s is the weight concentration of ε-PL in the supernatant, and V is the volume of the supernatant (10 mL).

2.8. Antibacterial Activity of ACC/DAS Hydrogels

Staphylococcus aureus and *Escherichia coli* were selected as the experimental bacteria. They were cultured in liquid LB medium until logarithmic growth. The ACC/DAS/ε-PL hydrogels were cut into 2 × 3.5 cm rectangles and added into bacterial culture medium. The samples were subjected to incubation at a temperature of 37 °C, while being agitated at a speed of 50 revolutions per minute. A volume of 0.2 mL of the culture was extracted at time intervals of 6, 8, 10, 12, 24, 48, and 60 h, and ACC/DAS hydrogels without ε-PL addition were used as the control. The absorbance at a wavelength of 600 nm was subsequently determined using a UV-visible spectrophotometer [16]. Formula (4) for calculating the antimicrobial rate is as follows:

$$Antibacterial\ rate\ (\%) = \frac{OD_1 - OD_2}{OD_1} \times 100 \tag{4}$$

In the equation, OD_1 represents the bacterial optical density seen in the control group, whereas OD_2 represents the bacterial optical density seen in the group treated with the hydrogels.

2.9. Statistical Analysis

All the experiments were performed in at least three independent trials. The statistical analyses were conducted using SPSS 16.0. (SPSS Inc., Chicago, IL, USA).

3. Results

3.1. Characterizations of ACC and DAS

ACC is a class of amination derivatives following the carboxymethylation of CS, containing -COOH and $-NH_2$ functional groups. Through the determination using trinitrobenzene sulfonic acid, the amino content of ACC was 0.83 ± 0.02 mmol/g. In Figure 1a, the FTIR spectra of CS, CC, and ACC reveal distinct absorption bands at 1655, 1593, 1323, and 1381 cm^{-1}, which are attributed to amides I, II, III, and the $-CH_3$ vibration band of the CS chain, respectively. Both the CC and ACC spectra exhibited two prominent peaks at 1605 cm^{-1} and 1419 cm^{-1}. The results are consistent with the description of Fan et al. [17]. These peaks can be attributed to the asymmetrical and symmetrical stretching of the -COO groups [18]. The observed absorption peak at 1632 cm^{-1} in the ACC spectra indicated the stretching vibration associated with the amide group. This finding suggests that the grafting of ethylenediamine onto CC has been accomplished effectively. It is evident that ACC has a notably greater amino content in comparison to CS [19]. Figure 1b displays the X-ray diffractograms of CS, CC, and ACC. There were noticeable differences in the peak height, size, and position between the CS and ACC. The CS sample displayed two prominent peaks at 2θ = 13.1° and 23.50°, indicating its crystalline nature [20]. On the other hand, the CC sample revealed two distinct peaks at 2θ = 12.30° and 20.36°. These peaks can be explained by the abundance of -OH and $-NH_2$ groups within the CS molecule, which facilitate the creation of many hydrogen bonds [21]. The incorporation of carboxymethyl and amino groups into the structural unit of ACC resulted in a weakening of the hydrogen bonding interactions between the hydroxyl groups present on the chitosan chain. As a consequence, there was a decrease in the degree of crystallinity observed in chitosan, as evidenced by the broadening of the diffraction peaks for ACC [22]. In addition, the crystallinity of ACC was lower than that of CC due to the introduction of amino groups.

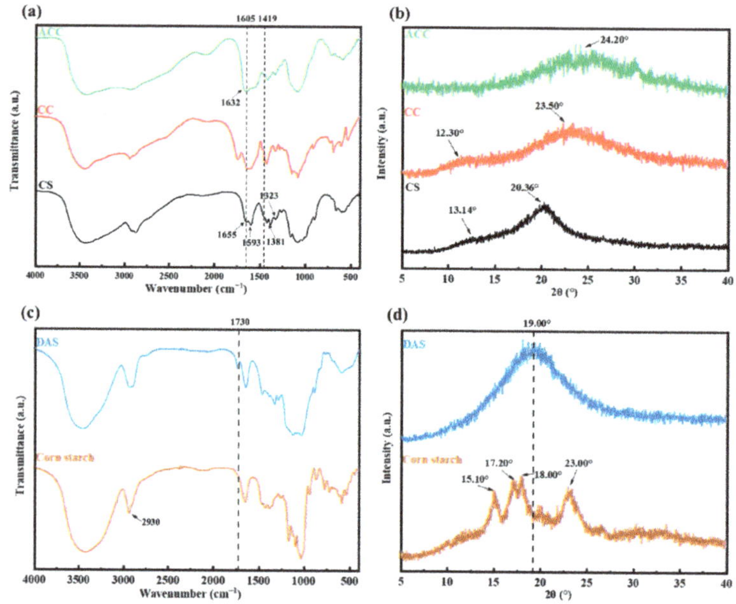

Figure 1. (**a**) FTIR spectra; (**b**) XRD patterns of CS, CC, and ACC; (**c**) FTIR spectra; (**d**) XRD patterns of corn starch and DAS.

DAS is a significant type of modified starch that is produced by oxidizing starch with the potent oxidizing agent periodate. Using the alkali solubilization method, the aldehyde group contents of DAS were determined as 89.8 ± 0.77%. Analysis of the structural modifications of starch subjected to oxidation through sodium periodate was conducted using FTIR, as depicted in Figure 1c. The peak at around 3310 cm^{-1} was associated with the vibrational stretching of -OH, while the peak at 2930 cm^{-1} was assigned to the symmetric and asymmetric stretching vibrations of the C-H group. In contrast to native starch, the FTIR analysis of DAS revealed a significant peak at 1730 cm^{-1}, which was related to the stretching vibration of C=O [23]. This would suggest that the oxidation of starchy maize by $NaIO_4$ was effective [24]. The composition of the starch granules included both crystalline and non-crystalline regions. The starch has a crystalline structure of type A, as evidenced by the presence of XRD peaks at approximately 15.10°, 17.20°, 18.00°, and 23.00° (Figure 1d) [25]. The oxidation and acidic hydration events of corn starch result in the conversion of the hydroxyl group to an aldehyde. This chemical transformation causes the hydrogen bond between the starch components to break down, ultimately disrupting the crystal structure. The disappearance of the diffraction peaks intrinsic to starch in the XRD pattern of DAS after acidolysis and oxidation provides evidence that the A-type crystallization of the starch was disturbed by sodium periodate [26].

3.2. Preparation and Characterizations of ACC/DAS Hydrogels

As depicted in Scheme 1, the CS surface exhibits a substantial presence of reactive groups—namely amino and carboxyl groups—which therefore undergo modifications through chemical reactions. In this thesis, a two-step process of carboxymethylation and amino-carboxymethylation was used to enhance the amino group on the surface of chitosan. Simultaneously, the hydroxyl groups within the corn starch molecules were aldehydized through the process of sodium periodate oxidation. Afterwards, the amino group on the ACC surface was cross-linked with the aldehyde group on the DAS to form imine bonds, and a three-dimensional network skeleton structure of long-chain polysaccharides was constructed. The ε-PL was efficiently immobilized on the hydrogel matrix through hydrogen bonding, thus preparing an ACC/DAS/ε-PL system with long-lasting bacterial inhibition, in order to build a hydrogel that can effectively load and continuously deliver natural active substances.

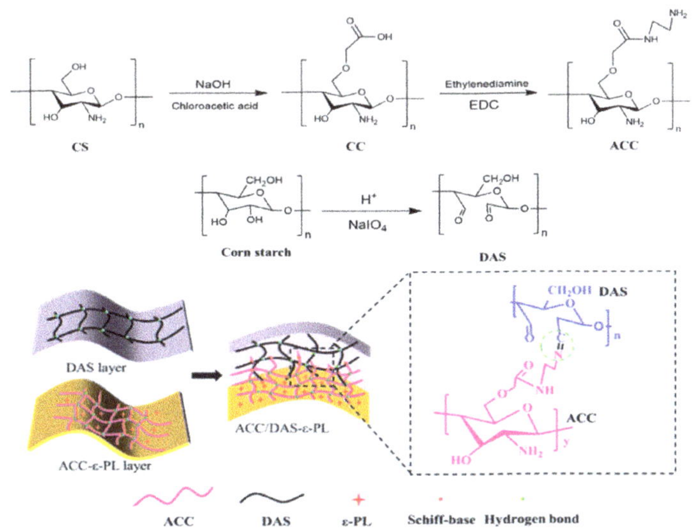

Scheme 1. Preparation approach of ACC/DAS hydrogels.

In Figure 2, the cross-section morphologies of the four ACC/DAS hydrogels were exhibited. When the molar ratio of the amino group on the ACC molecule to the aldehyde group on the DAS surface is 1:1.6, the cross-section of the hydrogel ACC_1/DAS is very rough and there is a large amount of visible aggregation. As the ACC content increases, the aggregation on the cross-section of the ACC_2/DAS hydrogel decreases and becomes more lamellar and denser. When the molar ratio of $-NH_2$ to -CHO reached 1:0.6, the cross-sectional microstructure of ACC_4/DAS was flat and smooth, with no aggregation. This may be due to the increased Schiff base reaction enhanced cross-link density, which built a tight and dense 3D network structure.

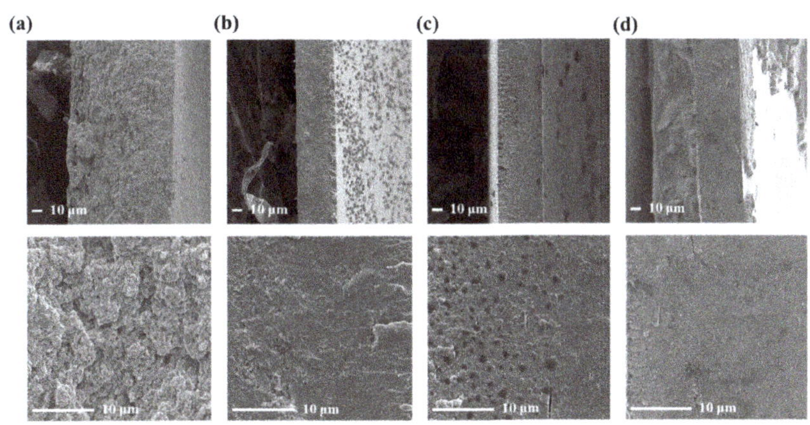

Figure 2. SEM image of (**a**) ACC_1/DAS; (**b**) ACC_2/DAS; (**c**) ACC_3/DAS; (**d**) ACC_4/DAS hydrogels.

In the FTIR spectrograms of the four ACC/DAS hydrogels with increasing amino contents in ACC (Figure 3a), the 1735 cm^{-1} absorption band of the aldehyde stretching vibration and the 1545 cm^{-1} absorption band of the amino stretching vibration gradually weakened due to the Schiff base reaction, while the 1630 cm^{-1} absorption band belonging to the imine group was correspondingly enhanced. This indicates that imine Schiff base crosslinks (-NCH-) are established between the ACC and DAS within the ACC/DAS hydrogel network [27].

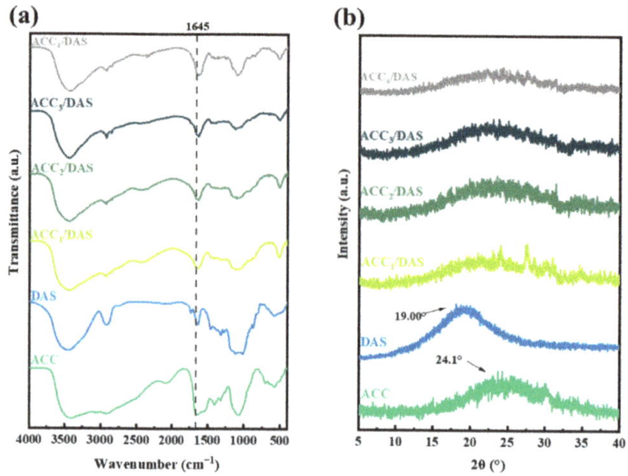

Figure 3. (**a**) FTIR spectra; (**b**) XRD patterns of ACC/DAS hydrogels.

The XRD patterns of the ACC/DAS hydrogels are shown in Figure 3b. When the ACC reacted with the DAS to form the hydrogel, the peaks of DAS at 19.0° and ACC at 24.1° disappeared. The XRD pattern of the ACC/DAS composite exhibits broad signals, indicating its amorphous structure. These results indicated that a strong Schiff base interaction between the DAS and ACC molecules occurred, and the long chains of the two molecules were arranged both separately and entangled with each other, thus disrupting the original arrangement [28].

The TGA plot in Figure 4 depicts the thermal decomposition pattern of the ACC/DAS hydrogels. There are two stages of mass loss for ACC/DAS hydrogels containing different concentrations of ACC. Between 80 and 160 °C, the hydrogel experienced weight loss caused by the evaporation of residual water molecules. This weight loss indicated a significant hydrogen bonding pattern in the hydrogel [29]. The second phase of degradation occurs rapidly between temperatures of 200 to 500 °C. In addition, the hydrogel undergoes a major mass loss phase, with the maximum weight loss rate at 203 °C [30]. At this point, it is believed that the thermal decomposition of the macromolecular backbone in the hydrogel substrate occurs, mainly due to the degradation of the glucosamine and acetylated groups [31]. Additionally, it is evident that the thermal stability of the hydrogel demonstrated a notable enhancement when the molar ratio of -NH_2:-CHO rose from 1:1.6 to 1:0.6. Conversely, a substantial decline in the thermal stability of the hydrogel is observed when the molar ratio of -NH_2:-CHO is 1:0.6. These results indicated that a sufficient amount of -NH_2 probably leads to a strong Schiff base reaction between the polymer components in the hydrogel, which builds a dense structure of the hydrogel and possesses thermal stability [32]. However, the strongest Schiff base reaction occurred in the ACC_4/DAS hydrogel, which produced the superfluous Schiff base fractures in the second phase and a decrease in the thermal stability [33].

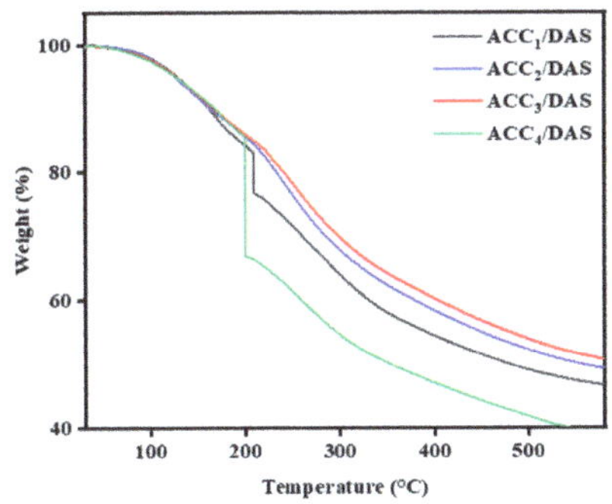

Figure 4. The thermogravimetric analysis of ACC/DAS hydrogels.

3.3. *Swelling Behavior of the Hydrogels*

The appearance of swelling in hydrogels is widely recognized as a crucial characteristic of hydrogels, and the hydrophilic nature of ACC/DAS hydrogels allows small molecules such as water to enter the three-dimensional network structure through swelling, providing a scaffolding structure for the loading of active substances. As shown in Figure 5, all four hydrogels swelled rapidly within 10 min, with the highest swelling degree of 67% for the ACC_3/DAS hydrogels. A decreasing trend in the swelling curve of ACC_1/DAS and ACC_2/DAS was clearly observed between 10 and 30 min, while the swelling degrees of ACC_3/DAS and ACC_4/DAS were still increasing due to the higher -NH_2 contents. When hydrogels are submerged in water, water molecules are forced to move into the polymer structure due to the presence of an osmolality gradient. The phenomenon of water diffusion leads to the expansion of hydrogels, hence causing an increase in the pore size within the hydrogel. ACC_3/DAS exhibited a longer swelling equilibrium time than the other hydrogels. Meanwhile, it is clear that the level of expansion of the hydrogel increases when the molar ratio of -NH_2:-CHO in the hydrogel is 1:1.6, 1:1.25, and 1:1, while the swelling degree of the ACC_4/DAS hydrogel is the lowest when the molar ratio of -NH_2:-CHO in the hydrogel is 1:0.6. This was due to the excessive crosslinking of the Schiff bases, which resulted in the polymer chains' tight cross-linked network, leading to a reduced pore size and poor flexibility, which in turn restricts the passage of water molecules, and the swelling properties of the hydrogels are limited by the spatial site resistance effect [34].

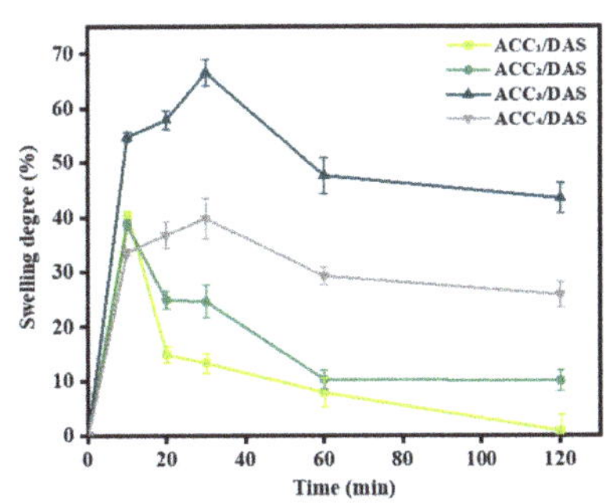

Figure 5. Degree of swelling plotted against time for ACC/DAS hydrogels.

The mechanical properties of the four ACC/DAS hydrogels with different ACC contents are shown in the Table 1, and it can be seen that the tensile strength and elastic modulus of the hydrogels improved significantly, while the breaking elongation gradually reduced with the increase in the amino acid content in the ACC, which may be due to the formation of the Schiff base bond through the reaction of -NH_2 and -CHO in the hydrogel.

These results indicated that the Schiff base reaction could contribute to the cross-link of the network, leading to the toughness of the ACC/DAS hydrogel being enhanced.

Table 1. The mechanical properties of ACC/DAS hydrogels.

Samples	Tensile Strength (MPa)	Breaking Elongation (%)	Elastic Modulus (MPa)
ACC_1/DAS	43.3 ± 7.4 [a]	40.5 ± 0.03 [a]	81.47 ± 5.4 [a]
ACC_2/DAS	43.7 ± 5.7 [a]	10.2 ± 0.04 [b]	586.2 ± 3.7 [b]
ACC_3/DAS	59.3 ± 7.6 [b]	7.6 ± 0.04 [b]	663.7 ± 1.9 [b]
ACC_4/DAS	63.2 ± 8.4 [c]	6.4 ± 0.01 [b]	741.1 ± 2.3 [b]

Different letters in different columns represent significant differences between samples ($p < 0.05$).

3.4. Loading Efficiency of the Hydrogels

As a typical natural cationic peptide with excellent antibacterial activity, ε-PL can be adsorbed onto the surface of cell membranes, thereby disrupting the cell membrane structure, inhibiting enzyme activity, interfering with normal gene expression, and ultimately destroying microorganisms. The ACC/DAS hydrogels were used as carriers of ε-PL. With molar ratios of $-NH_2$:-CHO of 1:1, the loading efficiency of the ε-PL in the hydrogels reached up to 99.2% (Figure 6). The Schiff base cross-linked structure exhibits a robust inhibitory effect on the diffusion of ε-PL by means of steric hindrance. However, the ACC_4/DAS hydrogel with a $-NH_2$:-CHO molar ratio of 1:0.6 resulted in a significantly lower loading efficiency. The reduced physical encapsulation rate of the ε-PL in the ACC_4/DAS hydrogel may be attributed to the excessive crosslinks between the polymer chains within the hydrogel.

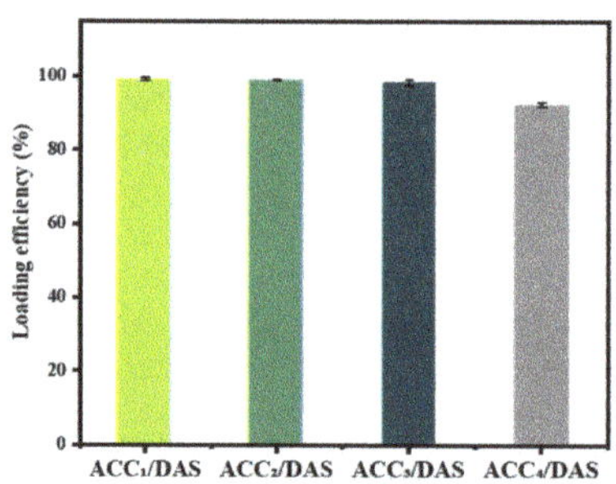

Figure 6. The ε-PL loading efficiency of ACC/DAS hydrogels.

3.5. Antibacterial Activity of the Hydrogels

Hydrogels possess notable appeal as carriers of active compounds owing to their considerable surface area to volume ratio and capacity for structural manipulation [35]. The ACC/DAS hydrogels without ε-PL also exhibited antibacterial properties due to the amino group on the CS chain; however, the antibacterial properties were only temporary: for 24 h against *S. aureus* and 8 h against *E. coli*. When the ACC/DAS/ε-PL hydrogel was added into the medium, water molecules diffused from the medium into the hydrogel network under the osmotic pressure gradient, and the release of ε-PL with antibacterial properties was caused by the swelling of the hydrogel. Figure 7a shows the bacteriostatic effect of the ε-PL and the four ACC/DAS hydrogels loaded with ε-PL against *Staphylococcus aureus*. Among them, the medium with only the addition of ε-PL reached the maximum inhibition rate of $80.67 \pm 0.23\%$ at 24 h; then, the inhibition rate gradually decreased. The inhibition rate was only $31.88 \pm 0.88\%$ at 48 h, which may be due to the depletion of ε-PL. The maximum inhibition rate of the ACC_2/DAS/ε-PL and ACC_3/DAS/ε-PL hydrogels was $71.84 \pm 0.21\%$ and $66.43 \pm 0.92\%$ at 48 h, respectively, and the antibacterial activities still maintained at 60 h, with the values of $49.03 \pm 0.61\%$ and $52.96 \pm 0.20\%$, respectively, suggesting that good swelling behavior is beneficial for the long-time release. The ACC_4/DAS hydrogel had the lowest inhibition rate, indicating that the immoderate crosslinked network was too tight and firm to allow the entry of water molecules, which limited the release of ε-PL. Figure 7b shows the inhibitory effect of the ACC/DAS hydrogels loaded with ε-PL on *E. coli*, and it was found that all of the inhibition rates of the hydrogels on *E. coli* reached the highest at 8 h, with the value of 74.81 ± 1.26%, and the antibacterial activities still maintained at 60 h, with the values of 36.02 ± 1.28%. The bacterial inhibition experiments show that the ACC/DAS/ε-PL hydrogels have a significant effect in inhibiting the growth of both *S. aureus* and *E. coli*. In addition, the hydrogel network formed by the Schiff base reaction is able to prolong the release of ε-PL and extend the time of its action, which has the potential to be used as active food packaging.

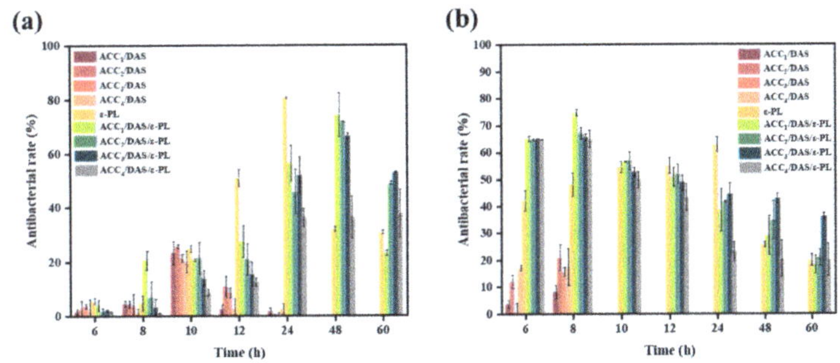

Figure 7. Antibacterial rate of ACC/DAS/ε-PL hydrogels against (**a**) *S. aureus* and (**b**) *E. coli* for 60 h.

4. Conclusions

This study focused on preparing ACC/DAS hydrogels through a Schiff base reaction to attain the optimal loading efficiency and prolonged release of ε-PL. With the increase in the ACC mass concentration, the hydrogel structure became denser, with good thermal stability, and the swelling degree was also enhanced, with the highest value of 67%. The maximum ε-PL loading efficiency of the ACC/DAS hydrogels reached 99.2%. The good swelling behavior of the ACC/DAS hydrogels was beneficial for the long-time release of ε-PL, which prolonged the bacteriostatic time against both *S. aureus* and *E. coli*. However, the excessive Schiff base crosslink of the ACC/DAS hydrogel limited the loading and release of ε-PL. These results demonstrated that the ACC/DAS hydrogels were good candidates for use as a good carrier for active substance delivery and active food packaging for food preservation.

Author Contributions: Conceptualization, J.B. and Y.L.; investigation, Y.Q.; resources, H.H. (Hongman Hou); data curation, Y.L.; writing—original draft preparation, G.Z.; writing—review and editing, H.H. (Hongshun Hao). All authors have read and agreed to the published version of the manuscript.

Funding: This article was supported by the National natural science foundation of China (No. 32102050), the National key research and development foundation (No. 2022YFD2100504), and Liaoning provincial education department basic scientific research foundation (No. LJK MZ20220875).

Data Availability Statement: The data used to support the findings of this study can be made available by the corresponding author upon request.

Acknowledgments: We gratefully appreciate the editors' efficient work in processing our manuscript and the anonymous reviewers' careful work and thoughtful suggestions that have helped improve this paper substantially.

Conflicts of Interest: The authors declare that there is no conflict of interest.

References

1. Liu, T.; Zhao, Y.; Wu, N.; Chen, S.; Xu, M.; Du, H.; Yao, Y.; Tu, Y. Egg white protein-based delivery system for bioactive substances: A review. *Crit. Rev. Food Sci. Nutr.* **2022**, *43*, 1–21. [CrossRef] [PubMed]
2. Zhang, Z.H.; Li, M.F.; Peng, F.; Zhong, S.R.; Huang, Z.; Zong, M.H.; Lou, W.Y. Oxidized high-amylose starch macrogel as a novel delivery vehicle for probiotic and bioactive substances. *Food Hydrocolloid* **2021**, *114*, 106578. [CrossRef]
3. Shima, S.; Sakai, H. Polylysine produced by Streptomyces. *Agric. Biol. Chem.* **1977**, *41*, 1807–1809. [CrossRef]
4. Zahi, M.R.; Hattab, M.; Liang, H.; Yuan, Q. Enhancing the antimicrobial activity of d-limonene nanoemulsion with the inclusion of ε-polylysine. *Food Chem.* **2017**, *221*, 18–23. [CrossRef]
5. Chang, Y.; McLandsborough, L.; McClements, D.J. Interaction of cationic antimicrobial (ε-polylysine) with food-grade biopolymers: Dextran, chitosan, carrageenan, alginate, and pectin. *Food Res. Int.* **2014**, *64*, 396–401. [CrossRef]
6. Song, M.; Lopez-Pena, C.L.; McClements, D.J.; Decker, E.A.; Xiao, H. Safety evaluation and lipid-lowering effects of food-grade biopolymer complexes (ε-polylysine-pectin) in mice fed a high-fat diet. *Food Funct.* **2017**, *8*, 1822–1829. [CrossRef]
7. Saeedi, M.; Vahidi, O.; Moghbeli, M.R.; Ahmadi, S.; Asadnia, M.; Akhavan, O.; Seidi, F.; Rabiee, M.; Saeb, M.R.; Webster, T.J.; et al. Customizing nano-chitosan for sustainable drug delivery. *J. Control. Release* **2022**, *350*, 175–192. [CrossRef]
8. Skopinska-Wisniewska, J.; De la Flor, S.; Kozlowska, J. From Supramolecular Hydrogels to Multifunctional Carriers for Biologically Active Substances. *Nano Micro Mater. Healthc.* **2021**, *22*, 7402. [CrossRef]
9. Abou-Yousef, H.; Sawsan, D.; Mohamed, H. Biocompatible hydrogel based on aldehyde-functionalized cellulose and chitosan for potential control drug release. *Sustain. Chem. Pharm.* **2021**, *21*, 100419. [CrossRef]
10. Guan, X.; Zhang, B.; Li, D.; Ren, J.; Zhu, Y.; Sun, Z.; Chen, Y. Semi-Unzipping of Chitosan-Sodium Alginate Polyelectrolyte Gel for Efficient Capture of Metallic Mineral Ions from Tannery Effluent. *Chem. Eng. J.* **2022**, *452*, 139532. [CrossRef]
11. Pan, Q.; Zhou, C.; Yang, Z.; He, Z.; Wang, C.; Liu, Y.; Song, S.; Gu, H.; Hong, K.; Yu, L.; et al. Preparation and characterization of chitosan derivatives modified with quaternary ammonium salt and quaternary phosphate salt and its effect on tropical fruit preservation. *Food Chem.* **2022**, *387*, 132878. [CrossRef] [PubMed]
12. Upadhyaya, L.; Singh, J.; Agarwal, V.; Tewari, R.P. Biomedical applications of carboxymethyl chitosans. *Carbohyd Polym.* **2013**, *91*, 452–466. [CrossRef] [PubMed]
13. Luo, C.; Li, M.; Yuan, R.; Yang, Y.; Lu, Z.; Ge, L. Biocompatible Self-Healing Coating Based on Schiff Base for Promoting Adhesion of Coral Cells. *Acs Appl. Bio Mater.* **2020**, *3*, 1481–1495. [CrossRef] [PubMed]
14. Dziadek, M.; Dziadek, K.; Salagierski, S. Newly crosslinked chitosan- and chitosan-pectin-based hydrogels with high antioxidant and potential anticancer activity. *Carbohyd Polym.* **2022**, *290*, 119486. [CrossRef] [PubMed]
15. Bi, J.; Tian, C.; Zhang, G.L.; Hao, H.; Hou, H.M. Novel procyanidins-loaded chitosan-graft-polyvinyl alcohol film with sustained antibacterial activity for food packaging. *Food Chem.* **2021**, *365*, 130534. [CrossRef]
16. Xu, Y.; Liu, J.; Guan, S.; Dong, A.; Cao, Y.; Chen, C. A novel Ag/AgO/carboxymethyl chitosan bacteriostatic hydrogel for drug delivery. *Mater. Res. Express* **2020**, *7*, 085431. [CrossRef]
17. Fan, S.; Li, Z.; Fan, C.; Chen, J.; Huang, H. Fast-thermoresponsive carboxylated carbon nanotube/chitosan aerogels with switchable wettability for oil/water separation. *J. Hazard. Mater.* **2022**, *433*, 128808. [CrossRef]
18. Wang, Z.; Su, J.; Ali, A.; Yang, W.; Zhang, R.; Li, Y.; Zhang, L.; Li, J. Chitosan and carboxymethyl chitosan mimic biomineralization and promote microbially induced calcium precipitation. *Carbohyd Polym.* **2022**, *287*, 119335. [CrossRef]
19. Liu, J.; Li, J.; Yu, F.; Zhao, Y.X.; Mo, X.M.; Pan, J.F. In situ forming hydrogel of natural polysaccharides through Schiff base reaction for soft tissue adhesive and hemostasis. *Int. J. Biol. Macromol.* **2020**, *147*, 653–666. [CrossRef]
20. Olanipekun, E.O.; Ayodele, O.; Olatunde, O.C.; Olusegun, S.J. Comparative studies of chitosan and carboxymethyl chitosan doped with nickel and copper: Characterization and antibacterial potential. *Int. J. Biol. Macromol.* **2021**, *183*, 1971–1977. [CrossRef]
21. Mondal, M.d.; Ibrahim, H.; Firoz, A. Cellulosic fibres modified by chitosan and synthesized ecofriendly carboxymethyl chitosan from prawn shell waste. *J. Text. Inst.* **2020**, *111*, 49–59. [CrossRef]
22. Chen, Y.; Miao, W.; Li, X.; Xu, Y.; Gao, H.; Zheng, B. The structure, prties, synthesis method and antimicrobial mechanism of ε-polylysine with the preservative effects for aquatic products. *Trends Food Sci. Technol.* **2023**, *139*, 104131. [CrossRef]
23. Zuo, Y.; Liu, W.; Xiao, J.; Zhao, X.; Zhu, Y.; Wu, Y. Preparation and characterization of dialdehyde starch by one-step acid hydrolysis and oxidation. *Int. J. Biol. Macromol.* **2017**, *103*, 1257–1264. [CrossRef] [PubMed]
24. Li, Y.; Wang, J.H.; Han, Y.; Yue, F.H. The effects of pulsed electric fields treatment on the structure and physicochemical properties of dialdehyde starch. *Food Chem.* **2023**, *408*, 135231. [CrossRef] [PubMed]
25. Li, W.; Wan, Y.; Wang, L.; Zhou, T. Preparation, characterization and releasing property of antibacterial nano-capsules composed of ε-PL-EGCG and sodium alginate-chitosan. *Int. J. Biol. Macromol.* **2022**, *204*, 652–660. [CrossRef]
26. Wang, L.; Wang, M.; Zhou, Y.; Wu, Y.; Ouyang, J. Influence of ultrasound and microwave treatments on the structural and thermal properties of normal maize starch and potato starch: A comparative study. *Food Chem.* **2022**, *377*, 131990. [CrossRef]
27. Ren, J.; Li, M.; Yuan, R.; Pang, A.; Lu, Z.; Ge, L. Adherent self-healing chitosan/dialdehyde starch coating. *Mat. Sci. Eng.* **2020**, *586*, 124203. [CrossRef]
28. Raveendran, R.L.; Anirudhan, T.S. Development of macroscopically ordered liquid crystalline hydrogels from biopolymers with robust antibacterial activity for controlled drug delivery applications. *Polym. Chem.* **2021**, *12*, 3992–4005. [CrossRef]
29. Vlasceanu, G.M.; Crica, L.E.; Pandele, A.M.; Ionita, M. Graphene Oxide Reinforcing Genipin Crosslinked Chitosan-Gelatin Blend Films. *Coatings* **2020**, *10*, 189. [CrossRef]

30. Jaramillo-Quiceno, N.; Rueda-Mira, S.; Felipe Santa Marín, J.; Álvarez-López, C. Development of a novel silk sericin-based hydrogel film by mixture design. *J. Polym. Res.* **2023**, *30*, 120. [CrossRef]
31. Wang, J.; Yang, Z.; Zhou, C.; Qiao, C.; Yuan, F.; Liu, Q.; Luo, X. Preparation and Properties of Composite Hydrogels Based on Microgels Containing Chitosan. *J. Macromol. Sci. B* **2022**, *61*, 557–570. [CrossRef]
32. Taghreed, H.A.; Abir, S.; Ghada, B.; David, R.; Nadia, G. Fabrication of sustainable hydrogels-based chitosan Schiff base and their potential applications. *Arab. J. Chem.* **2022**, *15*, 103511.
33. Jeong, J.P.; Kim, K.; Kim, J.; Kim, Y.; Jung, S. New Polyvinyl Alcohol/Succinoglycan-Based Hydrogels for pH-Responsive Drug Delivery. *Polymers* **2023**, *15*, 3009. [CrossRef]
34. Khan, S.; Anwar, N. Gelatin/carboxymethyl cellulose based stimuli-responsive hydrogels for controlled delivery of 5-fluorouracil, development, in vitro characterization, in vivo safety and bioavailability evaluation. *Carbohyd Polym.* **2021**, *257*, 117617. [CrossRef] [PubMed]
35. Kopka, B.; Kost, B.; Rajkowska, K. A simple strategy for efficient preparation of networks based on poly(2-isopropenyl-2-oxazoline), poly(ethylene oxide), and selected biologically active compounds: Novel hydrogels with antibacterial properties. *Soft Matter* **2021**, *17*, 10683–10695. [CrossRef] [PubMed]

foods

MDPI

Article

Effects of a Novel Starter Culture on Quality Improvement and Putrescine, Cadaverine, and Histamine Inhibition of Fermented Shrimp Paste

Xinyu Li [1,2], Yang Zhang [1,2], Xinxiu Ma [1,2], Gongliang Zhang [1,2] and Hongman Hou [1,2,*]

[1] School of Food Science and Technology, Dalian Polytechnic University, No. 1 Qinggongyuan, Ganjingzi District, Dalian 116034, China; lixinyu990519@163.com (X.L.); zhangyang981223@163.com (Y.Z.); maxinxiu@hotmail.com (X.M.); zgl_mp@163.com (G.Z.)

[2] Liaoning Key Lab for Aquatic Processing Quality and Safety, No. 1 Qinggongyuan, Ganjingzi District, Dalian 116034, China

* Correspondence: houhongman@dlpu.edu.cn; Tel.: +86-411-8632-2020

Abstract: Fermented shrimp paste is a popular food in Asian countries. However, biogenic amines (BAs) are a typically associated hazard commonly found during the fermentation of shrimp paste and pose a food-safety danger. In this work, an autochthonic salt-tolerant *Tetragenococcus muriaticus* TS (*T. muriaticus* TS) strain was used as a starter culture for grasshopper sub shrimp paste fermentation. It was found that with the starter culture, putrescine, cadaverine, and histamine concentrations were significantly lower ($p < 0.05$) with a maximal reduction of 19.20%, 14.01%, and 28.62%, respectively. According to high-throughput sequencing data, *T. muriaticus* TS could change the interactions between species and reduce the abundance of bacterial genera positively associated with BAs, therefore inhibiting the BA accumulation during shrimp paste fermentation. Moreover, the volatile compounds during the fermentation process were also assessed by HS-SPME-GC-MS. With the starter added, the content of pyrazines increased, while the off-odor amines decreased. The odor of the shrimp paste was successfully improved. These results indicate that *T. muriaticus* TS can be used as an appropriate starter culture for improving the safety and quality of grasshopper sub shrimp paste.

Keywords: *Tetragenococcus muriaticus*; microbial community regulation; biogenic amine reduction; volatile compound; grasshopper sub shrimp paste

Citation: Li, X.; Zhang, Y.; Ma, X.; Zhang, G.; Hou, H. Effects of a Novel Starter Culture on Quality Improvement and Putrescine, Cadaverine, and Histamine Inhibition of Fermented Shrimp Paste. *Foods* **2023**, *12*, 2833. https://doi.org/10.3390/foods12152833

Academic Editors: Maria Martuscelli and Evandro Leite de Souza

Received: 27 May 2023
Revised: 6 July 2023
Accepted: 24 July 2023
Published: 26 July 2023

1. Introduction

Shrimp paste is a traditional fermented aquatic food that is popular as a seasoning or side dish in several Southeast Asian and Chinese coastal areas due to its unique flavor and excellent nutritional value [1,2]. Among the various fermented shrimp pastes, grasshopper sub shrimp paste is special due to the particularity of its raw materials. Grasshopper sub shrimps, the most important ingredient, are krill-like shrimps with lengths of 8–10 mm that grow around the confluence of fresh and seawater. Therefore, grasshopper sub shrimp paste is a local specialty of regions around the Bohai Sea in China. Compared with pastes made of other shrimps, grasshopper sub shrimp paste is rich in astaxanthin and calcium contents, and low in fat and cholesterol levels due to the tiny size of the grasshopper sub shrimp [3]. However, the shrimp paste is usually produced at a small scale with a long fermentation period, and the quality of the shrimp paste is easily affected by environmental factors including temperature, the experience of operators, and microbiota [4]. In addition, the abundant amino acids present in shrimp paste provide sufficient substrates to produce some undesirable substances, especially biogenic amines (BAs).

As a kind of low-molecular-weight compound often found in fermented foods, BAs are usually created by microorganisms decarboxylating free amino acids and reducing ketones and aldehydes [5]. The consumption of dietary items containing high concentrations of BAs has a toxic effect on the human body, with symptoms such as headache, heart palpitations, vomiting, diarrhea, and hypertensive crises [6]. Among the common BAs in fermented foods, histamine (HIS) and tyramine (TRY) are considered to be the most dangerous substances, and they can cause toxic symptoms such as "scombroid fish poisoning" and "cheese reaction", respectively [7,8]. The guidance level stated by the FDA for histamine is 50 mg/kg, and the maximum established by the European Food Safety Authority (EFSA) in fresh fish is 200 mg/kg [6]. Moreover, carcinogenic nitrosamines can be produced when cadaverine (CAD) and putrescine (PUT) combine with nitrite, and the toxicity of HIS can also be increased by the two BAs [9,10]. Therefore, the control and reduction of BAs are crucial for shrimp paste production.

Microbes involved in fermentation are vital to the development of the quality of fermented foods. As fermented aquatic food, rather complex microbial communities are engaged during the fermentation of shrimp paste [11]. These microorganisms produce a variety of enzymes and essential nutrients during the fermentation process, imparting the shrimp paste with unique flavor and nutrient characteristics [12]. According to earlier studies, the bacterial communities and interactions between species contribute to the quality and BA content of fermented aquatic foods [13,14]. Bacteria have served as starter cultures in several fermented foods to repress the accumulation of BAs and improve the flavor of the products [15–19]. It has been reported that the *Tetragenococcus* genus consists of moderately halophilic lactic acid bacteria (LAB) and significantly contributes to amino-acid production in several salted fermented foods [20]. Additionally, some *Tetragenococcus* species have been successfully used as starter cultures to reduce the accumulation of BAs, especially HIS [21,22]. Therefore, more knowledge about the mechanism of BA reduction is essential for providing theoretical references for the manufacture of safe, high-quality grasshopper sub shrimp paste products.

In our previous study, *Tetragenococcus muriaticus* TS (*T. muriaticus* TS), which predominates in grasshopper sub shrimp paste and produces fewer BAs, was isolated and subjected to safety assessment [23]. The strain displayed weak resistance to 15 known antibiotics based on the *Enterococcus* breakpoint values and exhibited no hemolytic activity and biofilm formation. Moreover, it performed well in salt tolerance, acid formation, and protease and lipase activities. Moreover, the strain showed positive amine oxidase activity and was able to degrade histamine in in vitro tests, therefore was considered to be a potential starter culture for grasshopper sub shrimp paste fermentation. Although the strain could produce PUT, CAD, and TRY (about 50 mg/kg in total) when supplemented with precursors in MRS broth, it was found to repress BA accumulation in grasshopper sub shrimp paste. In this study, *T. muriaticus* TS was used as a starter culture for the fermentation of grasshopper sub shrimp paste to reduce BA accumulation and improve the quality of shrimp paste in a short fermentation time. The contents of putrescine, cadaverine, and histamine were measured since they carry high risks for food safety. Moreover, the impacts of *T. muriaticus* TS on microbial communities were studied to further explore how the starter could reduce BA content from the perspective of the microbial community. The findings could provide a theoretical basis for the quality and safety control of shrimp paste products.

2. Materials and Methods

2.1. Chemicals and Reagents

Grasshopper sub shrimp was purchased from a local market in Panjin City, Liaoning Province, China. *Tetragenococcus muriaticus* TS was provided by this laboratory. Dansyl chloride was purchased from Dalian Meilun Bio-Technology Co., Ltd. (Dalian, China). Acetonitrile (chromatographic pure) was purchased from Sigma (Shanghai, China). All biogenic amine standards and derivatizing agents (dansyl chloride) were purchased from Sigma (St. Louis, MO, USA).

2.2. Preparation of Starter Culture

First, *Tetragenococcus muriaticus* TS (*T. muriaticus* TS) was cultured in a De Man Rogosa Sharpe (MRS) broth medium at 30 °C for 48 h. Then, the culture was centrifuged at 8000× *g* and 4 °C for 5 min to obtain bacterial cells (Centrifuge: Tomy MX-307, Tokyo, Japan). The cell pellets were then washed twice with sterile saline solution and resuspended in 0.05 mol/L phosphate buffer (pH 7.0). Subsequently, the suspension was adjusted to OD_{600} = 1.2 as the starter culture (5%, *v*/*w*).

2.3. Preparation of Grasshopper Sub Shrimp Paste Samples

The grasshopper sub shrimps caught in April 2022 were mixed with 15% (*w*/*w*) of salt. A total of 40 mL starter cell suspension was placed in a sealed jar containing 800 g of shrimp paste, and the same amount of shrimp paste combined with an equal volume of sterilized water was used as a control sample [14].

The shrimp paste placed in sealed jars was fermented at 30 °C for 35 days and stirred every day. Samples were collected on days 1, 7, 14, 21, 28, and 35. Considering the uniformity and representativeness of the paste, a total of 40 g was collected from the upper, middle, and lower parts of three different fermenters. Collected samples were combined for viable bacteria count analysis immediately, and the remaining samples were stored at −80 °C for further analysis.

2.4. Physicochemical and Microbial Analysis

Samples (10 g) were homogenized in sterilized water (90 mL). Then, the homogenates were subjected to the determination of pH value using a pH meter (FiveEasy Plus^{TM} FE28, Mettler Toledo, Shanghai, China). The total volatile base nitrogen (TVB-N) and amino-acid nitrogen (AAN) contents were determined according to Chinese Standard GB5009.228 (2016) and GB5009.235 (2016), respectively. For the TVB-N, samples (5 g) were mixed with distilled water (75 mL) and allowed to stand for 30 min at room temperature. Thereafter, magnesium oxide (1 g) was added to the distillation tube containing the treated samples and immediately attached to the still of the Kjeldahl analyzer. The same volume of distilled water served as a control group. After that, the distillate was titrated to the endpoint with 0.1 mol/L hydrochloric acid solution. For the AAN, the mixed sample was quickly ground in a mortar, and then 5 g of the treated sample was diluted with distilled water to 100 mL, and the mixture was filtered. A total of 10 mL of filtrate mixed with 60 mL of distilled water was titrated to pH 8.2 with a sodium hydroxide standard solution (0.05 mol/L). Then, 10 mL of formaldehyde solution was added and the mixture was titrated to pH 9.2 with the sodium hydroxide standard solution.

Total aerobic mesophilic bacteria count and Lactic acid bacteria (LAB) plate count were conducted according to the method reported in previous research [16]. Briefly, 5 g samples were mixed with 45 mL sterilized saline in aseptic sampling bags and then homogenized by a stomacher (Stomacher 400 Circulator, Seward, Shanghai, China) for 2 min. Thereafter, decimal dilution was carried out and the suitable diluted solutions (200 μL) were added into the plate count agar (PCA) medium for the total aerobic mesophilic bacterial count, and MRS agar plates for lactic acid bacteria (LAB) count, and they were incubated in an aerobic environment at 37 °C for 48 h and 30 °C for 48 h, respectively. All experiments were repeated in triplicate.

2.5. Biogenic Amine Measurement by HPLC

The previously reported method was used to determine the level of putrescine, cadaverine, and histamine in the grasshopper sub shrimp paste samples [13]. In brief, 5 g of shrimp paste was mixed with 20 mL 10% trichloroacetic acid (TCA), and the mixture was homogenized by vortex mixing and stored at 4 °C to react for 2 h. Subsequently, the mixture was centrifuged at 3000× *g* and 4 °C for 10 min. Then, 200 μL of sodium hydroxide (2 M) and 300 μL of saturated sodium bicarbonate were added into 1 mL of the supernatant, followed by the addition of 2 mL of dansyl chloride solution (10 mg/mL), and the mixture was incubated at 45 °C for 40 min. A saturated sodium bicarbonate solution was obtained by adding a sufficient amount of sodium bicarbonate solid to distilled water until it could no longer be dissolved. After that, 125 μL of ammonia was added and the mixture was incubated at room temperature for 30 min to remove the residual dansyl chloride. The volume of the mixture was then adjusted to 5 mL with acetonitrile. Finally, the mixture was centrifuged at 3000× *g* for 5 min, and the supernatant was filtered twice through a 0.22 μm organic phase ultrafiltration membrane and stored at −80 °C for further HPLC analysis.

BAs were quantified using an Agilent 1260 HPLC unit (Agilent Technologies Inc., Santa Clara, CA, USA) consisting of an Agilent Zorbax SB-C 18 column (4.6 × 150 mm) with a quaternary pump and a diode array detector. A total of 20 μL of the sample was injected into the column at a flow rate of 1.0 mL/min. A binary solution consisting of ultrapure water (solvent A) and acetonitrile (solvent B) was used to elute the column using the following optimized gradient: 0–10 min, 55% B; 10–15 min, 55–65% B; 15–20 min, 65–80% B; 20–25 min, 80% B; 25–30 min, 80–90% B; 30–33 min, 90% B; 33–35 min, 90–55% B. The column temperature was set at 30 °C and the eluent was monitored by absorbance at 254 nm. LOD, RSD, and R^2 are shown in Table 1.

Table 1. Validation parameters of the HPLC method.

Biogenic Amines	LOD/(mg/kg)	RSD/%	Regression Equation	R^2
PUT	0.80	2.36	y = 29.554x − 5.7576	0.9991
CAD	0.80	2.74	y = 30.636x − 32.384	0.9993
HIS	1.20	1.24	y = 26.431x − 26.907	0.9990

The quantification was accomplished by the external standard method, and the BA content was represented as mg/kg. All experiments were repeated in triplicate.

2.6. *Microbial Community Analysis by Illumina High-Throughput Sequencing*

Microbial community genomic DNA of grasshopper sub shrimp paste samples was extracted using the E.Z.N.A. soil DNA kit (Omega Bio-tek, Norcross, GA, USA) according to the instructions. The V3-V4 regions of bacterial 16S rRNA genes were amplified using the primer pair (338F and 806R) according to a previous method [3]. After PCR and purification, a DNA library was constructed and sequenced on the Miseq Illumina platform at Majorbio Bio-Pharm Technology Co., Ltd. (Shanghai, China). The raw sequences were spliced, and quality controlled by Flash software (version 1.2.11) and Fastp (version 0.19.6), and then OTU clustering (similarity 97%) was performed through Uparse (version 11) software. Species taxonomic annotation was conducted by an RDP classifier (version 2.13) based on the SILVA rRNA database (version 138). All samples were run with three repetitions.

2.7. *Identification of Volatile Compounds by HS-SPME-GC-MS*

Solid-phase microextraction (SPME; Supelco, Bellefonte, PA, USA) combined with 7890A-5975C chromatography–mass spectrometry (GC–MS; Agilent, Santa Clara, CA, USA) was used to determine the volatile compounds of samples collected at Day 35. After equilibration of 9 g samples in a 15 mL extraction flask at 70 °C for 15 min, 50/30 μm PDMS/CAR/DVB fiber (Supelco) was inserted into the flask (1 cm from the sample) for extraction of the volatiles at 70 °C for 45 min. Finally, the SPME device was inserted into the injection port of the GC, and desorption was carried out by exposing the fiber in splitless mode to 250 °C for 3 min. GC conditions were as follows: HP-5MS capillary column (30 m × 250 μm × 250 μm; Agilent); column oven temperature: 40 °C for 3 min, 3 °C/min to 120 °C, and finally increased at a rate of 5 °C/min to 240 °C, kept for 5 min; carrier gas: He; flow rate: 1.0 mL/min. MS conditions were listed as follows: electron ionization (EI) mode, voltage 70 eV, ion source temperature 230 °C, mass scanning range *m*/*z* 30–550. The volatile compounds (VCs) were identified by referring to their mass spectra in the NIST 14.0 database and by comparison with their retention index (RI). The relative percentages of each VC were calculated by dividing the peak area of the VC by the sum of the peak areas of all VCs.

2.8. Statistical Analysis

IBM SPSS Statistics 19 software was used for data analysis. The results were subjected to a one-way ANOVA analysis, with $p < 0.05$ considered to be statistically significant. Data were expressed as mean ± standard deviation. A part of the data was analyzed on the online platform of Majorbio Cloud Platform (http://www.majorbio.com (accessed on 15 July 2022)). All experiments were repeated in triplicate.

2.9. Nucleotide Sequence Accession Numbers

The bacterial 16S rRNA gene sequences obtained have been made publicly available in the NCBI Short Read Archive under the accession number PRJNA934728.

3. Results

3.1. Analysis of Bacterial Community in Shrimp Paste Samples

The bacterial composition in shrimp paste samples was determined by high-throughput sequencing and the results are shown in Figure 1. There were 11 genera identified in the control group and 7 genera in the inoculation group, respectively. The dominant genera in both groups were *Marinilactibacillus* and *Tetragenococcus*, with relative abundances of 85.13–38.47% and 8.83–54.16% in the control group and 19.85–22.17% and 78.13–74.82% in the test group, respectively (Figure 1A). After the starter culture inoculation, the relative abundance of *Marinilactibacillus* at 7 and 35 days decreased by 65.28% and 16.30%, respectively. In addition to the dominant genera, some genera with less abundance (<4%), such as *Virgibacillus*, *Atopostipes*, *Salinicoccus*, *Moritella*, *Jeotgalicoccus*, and *Staphylococcus*, were further reduced in relative abundance with the inoculation of starter culture (<2%).

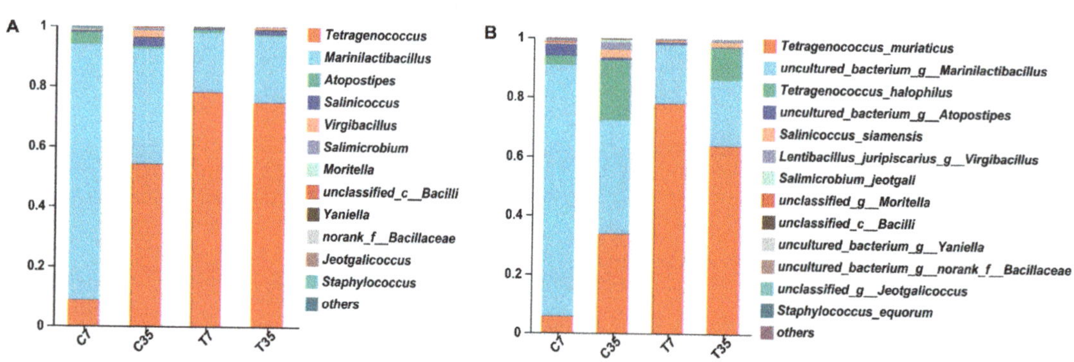

Figure 1. Comparative analysis of bacterial composition in control and test groups on genus (**A**) and species (**B**) level. Relative abundance of bacteria is presented as mean values, measured in triplicate. C: samples without starter culture; T: samples with starter culture. 7, 35: fermentation time (in days).

Eight species were identified in both the control and test group. The predominant species in both groups were uncultured *Marinilactibacillus*, *T. muriaticus*, and *Tetragenococcus halophilus* (*T. halophilus*), with relative abundances of 85.13–38.47%, 5.63–33.78%, and 3.20–20.38% in the control group and 19.85–22.17%, 77.91–63.83%, and 0.22–10.99% in the test group, respectively (Figure 1B). After the starter culture inoculation, the relative abundance of *T. muriaticus* increased by 72.28% and 30.05% by Day 7 and Day 35, respectively. The relative abundance of uncultured *Marinilactibacillus* decreased by 65.28% and 16.30% by Day 7 and Day 35, respectively. The relative abundance of *T. halophilus* decreased by 2.98% and 9.39% by Day 7 and Day 35. In addition, some relatively low-abundance species, such as *Lentibacillus juripiscarius* and *Salinicoccus siamensis*, also decreased in relative abundance after the inoculation of the starter culture. The inoculation resulted in a significant difference in the abundance of the bacterial genera and species.

Principal coordinate analysis (PCoA), determined by the Bray–Curtis and Euclidean distance matrices, was conducted to explore the difference in bacterial diversity between the control and test groups, and the results are shown in Figure 2A. There was a large distance between the control and test group, while a small distance between the samples fermented for 7 and 35 days of the test group was found. The difference between groups was calculated to assess whether it was bigger than the difference within groups using analysis of similarities (ANOSIM), and it was discovered that the former was larger than the latter (Figure 2B). With operational taxonomic units (OTUs) identified at 97% sequence similarity, clustering was carried out to investigate the number of species in each group. A total of 87 OTUs were found at the genus level, as shown by the Venn diagram (Figure 2C). Totals of 83 and 68 OTUs were found in the control and test groups, representing 95.40% and 78.16% of the total OTUs, respectively. Linear discriminant analysis effect size (LEfSe) was conducted to find the bacteria with significant differences in relative abundance between control and test groups, and the result is shown in Figure 2D. The microbial biomarker in the test group was *Tetragenococcus*, while the microbial biomarkers found in the control group were *Marinilactibacillus*, *Lentibacillus*, and *Pseudoalteromonas*. The inoculation of *T. muriaticus* might have caused a significant change in the microbial diversity in the shrimp paste.

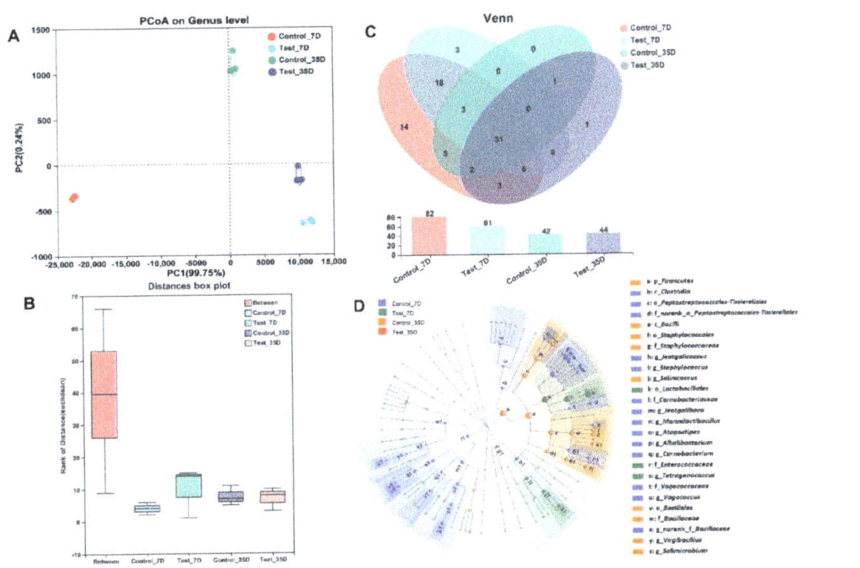

Figure 2. Comparative analysis of bacterial diversity at genus level between control and test groups. (**A**) PCoA of Euclidean. (**B**) Venn chart. (**C**) ANOSIM analysis. (**D**) LEfSe analysis of the microbial community in shrimp paste samples. PCoA: Different samples are represented as points with different colors and shapes. The closer the points are, the more similar the bacterial diversity of the samples. ANOSIM: The "Between" represents the difference between groups, and others represent the differences within each group. LEfSe: Bacteria significantly enriched in the corresponding group are represented by nodes with different colors, and the pale-yellow nodes represent bacteria with no significant difference in relative abundance among difference groups. Control and Test represent shrimp paste inoculated with/without starter, respectively. 7D and 35D: fermentation time (in days).

3.2. *Effect of Starter Culture on the Interspecies Correlation*

The interactions between species were calculated to further investigate the impact of *T. muriaticus* on the microbial community in grasshopper sub shrimp paste. The interaction between different genera in shrimp paste samples changed significantly with the starter culture inoculation, and the number of genera correlated with *Tetragenococcus* increased from 5 to 8 (Figure 3A,B). By the 7th day of fermentation, the new genera positively correlated with *Tetragenococcus* in the test group were *Moritella*, *Staphylococcus*, and unclassified *Planococcaceae*, while the correlation of *Vagooccus* with *Tetragenococcus* disappeared after the starter inoculation. The new genus negatively correlated with *Tetragenococcus* was *Atopostipes*, and the correlation between *Virgibacillus* and *Tetragenococcus* disappeared. After the inoculation of *Tetragenococcus*, the negative correlation between *Tetragenococcus* and *Marinilactibacillus* was stronger. The inoculation of *T. muriaticus* changed the correlation among the dominant bacterial genera in shrimp paste.

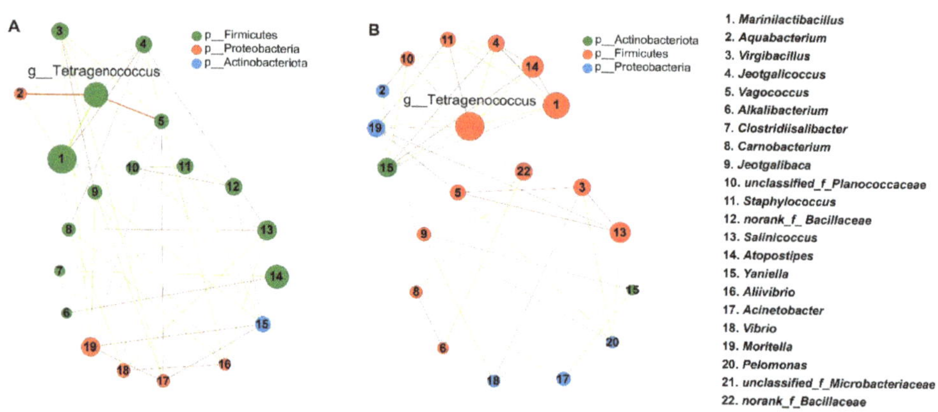

Figure 3. Comparative analysis of interspecies correlation of bacterial genus in control (**A**) and test (**B**) group. Correlations between bacteria are represented by lines between nodes. The red and green lines refer to positive and negative correlations, respectively.

3.3. *Contribution of the Starter Culture to BA Reduction*

Changes in the content of putrescine (PUT), cadaverine (CAD), and histamine (HIS) were measured to explore the BA-degradation ability of *T. muriaticus*, and their contents are shown in Figure 4A–C, respectively. The contents of the three BAs increased continuously in both control and test groups during fermentation, and the inoculation of the starter culture was found to reduce their contents. The highest reduction of PUT and CAD were obtained as 19.20% and 14.01% by Day 7 and 35, respectively. The highest reduction of HIS occurred by Day 7 at 28.62%, and the content of HIS was reduced by 22.81% by Day 35. *T. muriaticus* was able to reduce the BA content when added to grasshopper sub shrimp paste. To further understand how *T. muriaticus* influenced the content of BAs from a microbial perspective, correlations between bacterial genera and the three BAs were conducted, and results are shown in Figure 4D,E. Interestingly, *Tetragenococcus* showed a significantly negative correlation with BAs in the test group while demonstrating a significant positive correlation with BAs in the control group.

3.4. *Effects of Starter Culture on Physicochemical Properties and Microbial Counts of Shrimp Paste*

The results of microbial counts are shown in Figure 5A,B. The total aerobic mesophilic bacterial counts on PCA agar plates grew gradually during the first 7 days and decreased gradually until the end of fermentation in both control and test groups. The highest bacterial counts in the control (7.47 $\log_{10}$ CFU/g) and test (7.75 $\log_{10}$ CFU/g) groups were both obtained on Day 7. The lactic acid bacteria (LAB) count of both control and test groups increased to the maximum value of 6.14 $\log_{10}$ CFU/g and 6.37 $\log_{10}$ CFU/g on Days 28 and 14, respectively, and then gradually decreased to 5.98 $\log_{10}$ CFU/g on Day 35. From Day 0 to Day 21, both the total aerobic mesophilic bacteria and LAB counts of the test group were consistently higher than those of the control group, with a difference of no more than 0.5 $\log_{10}$ CFU/g.

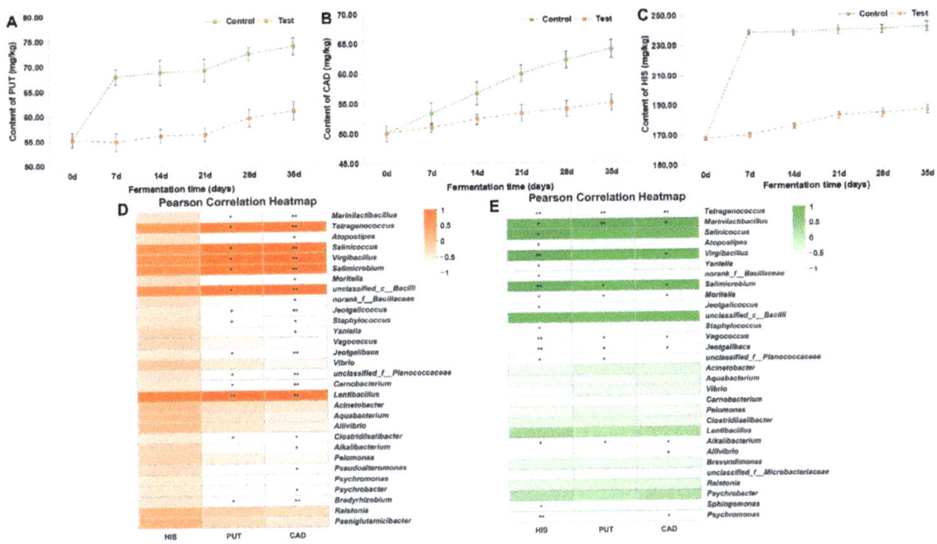

Figure 4. Content of putrescine (**A**), cadaverine (**B**), histamine (**C**) in shrimp paste samples and Pearson correlation analysis between bacteria genus and BAs in control (**D**) and test (**E**) groups. * $0.01 < p \leq 0.05$, ** $0.001 < p \leq 0.01$.

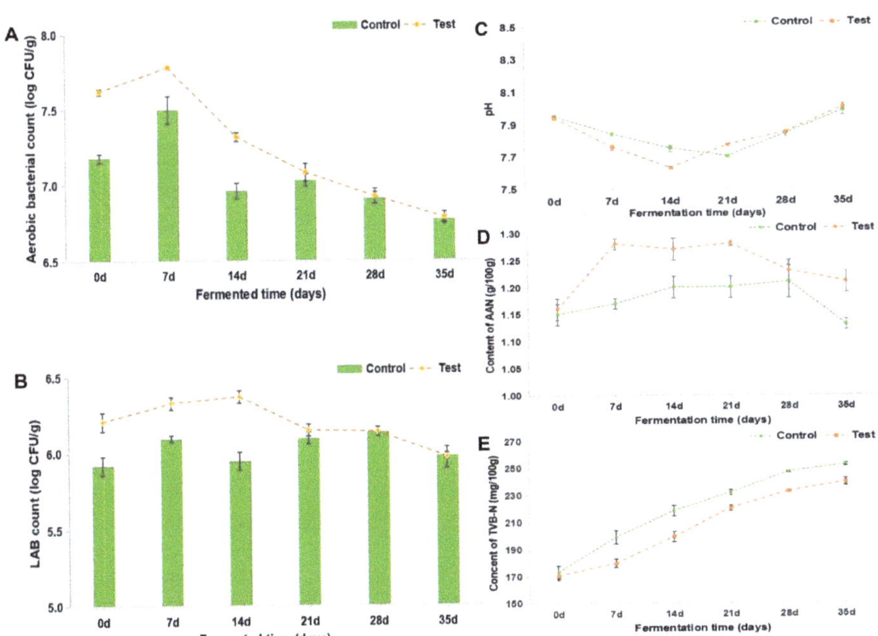

Figure 5. Microbiological and physicochemical characteristics of shrimp paste samples. (**A**) total aerobic bacterial count. (**B**) lactic acid bacteria count. (**C**) pH values. (**D**) Amino-acid nitrogen (AAN) content. (**E**) Total volatile base nitrogen (TVB-N) content.

To investigate the impact of *T. muriaticus* on the physicochemical quality of shrimp paste, the pH values, total volatile base nitrogen (TVB-N), and amino-acid nitrogen (AAN) concentrations of shrimp paste throughout the fermentation were measured, and the results are presented in Figure 5C–E. The initial pH values of the control and test groups were both 7.9 (Figure 5C). The pH levels of shrimp paste first dropped to 7.6–7.7 and then quickly rose to 8.0. There was consistently more AAN present in the test group than in the control group throughout the whole fermentation. The content of AAN was measured since it was an indicator of the degree of fermentation of shrimp paste. The initial content of AAN in both groups was about 1.15 g/100 g (Figure 5D). The AAN content first increased to a relatively stable level and then decreased to 1.13 and 1.21 g/100 g in the control and test group, respectively. When compared to the control group, the AAN level in the test group was consistently higher. To assess the degree of amino-acid degradation, the content of TVB-N was examined throughout the fermentation process. The TVB-N concentrations of shrimp paste samples were about 170 mg/100 g before the fermentation began (Figure 5E). The TVB-N accumulated gradually in both batches throughout the fermentation processes and upon 35 days of fermentation, the values in the control and test group reached 252 and 240 mg/100 g, respectively. The TVB-N concentrations in both batches were lower than the limit (450 mg/100g) of the Chinese commercial standard for shrimp sauce (SB/T 10525-2009).

3.5. Impacts of Starter Culture on Volatile Compounds of Shrimp Paste

The variations in volatile components of shrimp paste samples collected on Day 35 were determined using the HS-SPME-GC-MS. Overall, a total of 25 volatile compounds (VCs), including 6 pyrazines, 1 hydrocarbon, 2 esters, 1 ether, 1 acid, 1 ketone, 2 alcohols, and 9 other compounds, were found in shrimp paste by GC–MS (Table 2). The most prominent VC type found in fermented shrimp paste was pyrazine, which comprised around 91.9% and 96% of the total VCs in the control and test groups, respectively. The esters and ethers were only identified in the control group, while acids, ketones, and alcohols were only found in the test group. It is worth noting that among the 9 other substances, amines were identified with an abundance of about 7% and 0.6% in the control and test group, respectively.

Table 2. Volatile compounds identified by HS-SPME-GC-MS in shrimp paste samples.

No.	RT (min)	Compound Name	Relative Percentage (%)	
			Control	Test
		Pyrazines		
A1	15.09	3-Ethyl-2,5-dimethyl-pyrazine	55.40	56.31
A2	11.41	Trimethyl-pyrazine	17.08	18.73
A3	7.11	2,5-Dimethyl-pyrazine	16.50	17.45
A4	10.45	2,6-Dimethyl-pyrazine	0.27	0.20
A5	19.48	3,5-Diethyl-2-methyl-pyrazine	2.64	ND
A6	19.31	2,3,5-Trimethyl-6-ethylpyrazine	ND	3.27
		Hydrocarbons		
B1	1.89	Oxirane, 2-butyl-3-methyl-, cis-	0.22	ND
B2	37.35	Hexa-t-butylthiatrisiletane	ND	0.03
B3	38.29	Silane, [[4-[1,2-bis[(trimethylsilyl)oxy] ethyl]-1,2-phenylene] bis(oxy)] bis [trimethyl-]	ND	0.04
		Esters		
C1	5.08	Arsenous acid, tris(trimethylsilyl) ester	0.49	ND
C2	32.58	Benzenesulfonic acid, 4-methyl-, bicyclo [3.2.1] oct-6-yl ester	0.05	ND
		Ethers		
D1	38.29	18-Methyl-nonadecane-1,2-dio, trimethylsilyl ether	0.06	ND
		Acids		
E1	1.59	12-Methylaminolauric acid	ND	1.36

Table 2. *Cont.*

No.	RT (min)	Compound Name	Relative Percentage (%)	
			Control	Test
		Ketones		
F1	25.76	Thebacon	ND	0.07
		Alcohols		
G1	1.40	Cyclobutanol	ND	1.62
G2	51.51	trans-Farnesol	ND	0.18
		Other compounds		
H1	1.42	2-Propanamine	3.82	0.59
H2	1.61	Octodrine	3.20	ND
H3	3.78	1,2,4-trimethyl-Piperazine	0.17	ND
H4	49.20	4H-1-Benzopyran-2-carbonyl chloride, 4-oxo-	0.04	ND
H5	25.76	2-Chloro-4-(4-methoxyphenyl)-6-(4-nitrophenyl) pyrimidine	0.08	ND
H6	3.75	1,3,3-Trimethyl-diaziridine	ND	0.07
H7	29.84	1-Ethoxy-2-methyl-benzene	ND	0.06
H8	47.68	5-Fluoroorotyl-N, N-dimethylhydrazide	ND	0.01
H9	52.08	1H-Indole, 2-(1,1-dimethyl-2-propenyl)-6-(3-methyl-2-butenyl)-	ND	0.01

ND: Not detected. RT: Retention time in the capillary column. Control and Test represent shrimp paste fermented without/with starter culture for 35 days, respectively.

4. Discussion

Bacterial communities involved in the fermentation process are crucial for the quality of fermented shrimp paste. Previous studies have suggested that the dominant bacteria in fermented shrimp paste are *Pseudognonas*, *Staphylococcus*, and lactic acid bacteria such as *Tetragenococcus* and *Marinilactibacillus* [24,25]. In this study, *Tetragenococcus* and *Marinilactibacillus* with quite high abundance played a crucial role during shrimp paste fermentation, while *Staphylococcus* was low in abundance and contributed little to the fermentation process. It has been shown that *Tetragenococcus* is an important and predominant genus in grasshopper sub shrimp paste [13]. As a common lactic acid bacterium (LAB) in fermented foods, *Tetragenococcus* can grow well in environments with high salt concentrations, so they are crucial for the synthesis of several nutrients and flavor substances not only in shrimp paste but also in many fermented foods with high salt level [26,27]. *Marinilactibacillus* is a representative group of basophilic marine LAB and has been found in fermented herring and fermented rays [28,29]. A previous study has demonstrated that halophilic and alkaliphilic strains of *Marinilactibacillus* played a part in the fermentation of surface-ripened soft cheese and significantly influenced the quality of the product [30]. *Marinilactibacillus* might also play an important part in the quality formation of grasshopper sub shrimp paste, which was as rich in protein as cheese. The high-throughput sequencing data showed that the microbial composition in grasshopper sub shrimp paste has changed significantly with the inoculation of the starter culture. For the result of PCoA, the large distance between the control and test groups showed that the starter culture dramatically changed the microbial diversity in the shrimp paste, while the small distance between the samples fermented for 7 and 35 days of the test group indicated that the microbial community became more stable with the starter culture inoculation. Moreover, the results of ANOSIM and LEfSe also showed that the microbial diversity in the grasshopper sub shrimp paste changed significantly with the inoculation of the starter culture. For the species relevance network map, the more a node links to other nodes, the more important the node is [14]. The inoculation of *T. muriaticus* TS enhanced the impacts of *Tetragenococcus* during the fermentation process since there were more bacterial genera linked to the *Tetragenococcus* node in the test group than in the control group. The bacterial community changes might result in changes in the quality of the grasshopper sub shrimp paste during the fermentation process.

Excessive histamine (HIS) intake may result in food intoxication, which can cause symptoms including edemas, migraines, and low blood pressure. Although putrescine

(PUT) and cadaverine (CAD) have far less toxicity, their interactions with amine oxidases may slow down the metabolism and increase the toxicity of histamine [31]. Controlling their content in food is therefore crucial for improving food safety. In our previous study, *T. muriaticus* TS showed histamine degradation of 12.60% in vitro. In this study, the histamine content decreased by 22.81% with the starter culture inoculation. A previous study has found that when the *Tetragenococcus halophilus* strain MJ4 was inoculated into fermented shrimp paste, the formation of cadaverine during the fermentation was repressed [20]. Our study demonstrated that *T. muriaticus* could enhance the food safety of shrimp paste by lowering the content of BA content produced during fermentation. LAB starters have been reported to inhibit the growth of bacteria with high BA-producing activities such as strains of enterococci, lactobacilli, streptococci, and lactococci [7]. Therefore, *T. muriaticus* might suppress the development of other LAB, therefore reducing the BA accumulation in the grasshopper sub shrimp paste.

The primary mechanisms for the synthesis of BAs in fermented foods are microbial amino-acid decarboxylation or amination and transamination of aldehydes and ketones. Therefore, correlations between bacterial genera and the three BAs were conducted. According to the results shown in the Pearson correlation heatmap, *T. muriaticus* could affect the correlation between the dominant bacterial genera and BAs (Figure 4D,E). *Tetragenococcus* was a dominating bacterium with a very high abundance in shrimp paste samples (Figure 1A), and the findings revealed that it positively correlated with BAs in the control group while negatively correlated with BAs in the test group. According to a previously published paper, *Tetragenococcus* was a dominant bacterium and positively correlated with BAs in fermented shrimp paste [23]. On the contrary, in our study, *Tetragenococcus* was negatively correlated with BAs in the test group. BA-producing capacity has been reported to be a strain-specific characteristic [5], and the *T. muriaticus* strain used in our work has been proven to suppress the accumulation of BAs in both vitro and grasshopper sub shrimp paste [23]. In the test group, *T. muriaticus* TS with BA-reducing ability grew rapidly and became the dominant strain, therefore the correlation between *Tetragenococcus* and BAs became inversely correlated. Moreover, although *Marinilactibacillus* was positively correlated with the three BAs in the test group, its abundance fell dramatically after the inoculation of the starter culture (Figure 1A).

The impacts of starter culture on the physicochemical quality of shrimp paste were also studied. Both the total aerobic mesophilic bacterial and LAB counts increased with the starter culture inoculation during the early fermentation phase, which might suggest that the starter culture could adapt efficiently to the fermentation environment. In addition, the increase in total aerobic mesophilic bacterial and LAB population was conducive to the formation of shrimp paste. The slight decrease in pH values during the early period of the fermentation process might be caused by the accumulation of lactic acid produced by LAB in the shrimp paste, consistent with the result of the LAB count (Figure 5B). The increase in pH values during the late fermentation stage might be due to the BAs or other basic substances produced by the excessive hydrolysis of proteins in shrimp paste [24]. In general, the inoculation of *T. muriaticus* TS had no significant effect on the pH values of the fermented shrimp paste. TVB-N is often employed as a physicochemical indicator to assess the freshness of aquatic items [32]. In general, a high TVB-N level means great destruction of amino acids. The data shows that *T. muriaticus* TS could significantly repress the accumulation of TVB-N and increase the content of AAN during the fermentation of grasshopper sub shrimp paste ($p < 0.05$). This could be brought on by the powerful capacity of *Tetragenococcus* for the enzymatic degradation of proteins during the fermentation of shrimp paste [24], and more proteins were broken down into nitrogen compounds existing in the form of amino acids rather than volatile base nitrogen compounds.

As an important characteristic, the odor changes in the shrimp paste were investigated by GC–MS. A strain of *T. muriaticus* has been used as a starter culture to influence the kind and abundance of key volatile components to enhance the volatile flavor of low-salt fish sauce [19]. In this work, the indigenous *T. muriaticus* TS strain was used for the grasshopper sub shrimp paste fermentation. According to the data of GC–MS, pyrazine, hydrocarbon, esters, ether, acid, ketone, and alcohol were identified as the volatile compounds in the grasshopper sub shrimp paste. These compounds have also been identified with different compositions and relative content in other fermented shrimp pastes. Shrimp species and fermentation processes may be responsible for the significant difference in the volatile compounds [25,33,34]. Pyrazines were the predominant volatile compound in shrimp paste samples and the starter culture increased the content of pyrazines in shrimp paste (Table 2). In this study, 3-ethyl-2,5-dimethyl-Pyrazine, trimethyl-Pyrazine, and 2,5-dimethyl-Pyrazine were the dominant pyrazine compounds with high content, consistent with previous studies [25,35]. Pyrazines are a crucial component of food flavor, and they have a significant impact on the sensory qualities of food. Due to their low taste threshold, they are crucial in the development of shrimp paste flavor and often provide a meaty, nutty, and roasted fragrance like roast potatoes [34]. Esters, ethers, acids, ketones, and alcohols were also identified in shrimp paste samples in this study (Table 2). However, they contributed less to the flavor of shrimp paste because they were relatively low in abundance ($\leq$1.7%). It has been reported that some off-odor amine compounds were commonly produced by spoilage bacteria [36]. In this work, 2-Propanamine was identified in both control and test groups and Octodrine was found only in the control group, and the content of these two amines in grasshopper sub shrimp paste decreased by about 6.4% with the inoculation of the starter culture. These results demonstrated that *T. muriaticus* TS might improve the volatile flavor of grasshopper sub shrimp paste by enhancing the aroma brought on by pyrazines and reducing the unpleasant odor caused by amines.

5. Conclusions

Autochthonic salt-tolerant *Tetragenococcus muriaticus* TS (*T. muriaticus* TS) was successfully used as a starter culture to decrease BAs content. Contents of putrescine (PUT), cadaverine (CAD), and histamine (HIS) decreased significantly with the inoculation of *T. muriaticus* TS, reaching the maximal reduction at 19.20%, 14.01%, and 28.62%, respectively. The overall odor of the grasshopper sub shrimp paste was improved by the increase of pyrazines and the decrease of amines. Results of high-throughput sequencing demonstrated that the inoculation of *T. muriaticus* TS could interfere with the bacterial structure and interspecies correlation. Correlation analysis found that the inoculation of *T. muriaticus* TS could suppress the growth of bacteria that were positively correlated with BAs, thus reducing the BA accumulation in grasshopper sub shrimp paste. In summary, inoculation of starters such as *T. muriaticus* TS was an effective method to produce grasshopper sub shrimp paste with higher quality and lower food-safety risks caused by high levels of BAs.

Author Contributions: Conceptualization, X.L.; Methodology X.L., Y.Z. and G.Z.; Investigation, X.L., Y.Z. and X.M.; Data curation, H.H.; Writing—original draft preparation, X.L.; Writing—review and editing, H.H.; Supervision, H.H.; Project administration, H.H.; Funding acquisition, H.H. All authors have read and agreed to the published version of the manuscript.

Funding: This research was funded by the National Key Technology Research and Development Program, grant number 2017YFC1600403; The APC was funded by H.H.

Data Availability Statement: The data presented in this study are available on request from the corresponding author. The data are not publicly available due to the need for the first author to the apply for a master's degree.

Conflicts of Interest: The authors declare no conflict of interest. The funders had no role in the design of the study; in the collection, analyses, or interpretation of data; in the writing of the manuscript; or in the decision to publish the results.

References

1. Li, W.; Mi, S.; Liu, X.; Sang, Y.; Sun, J. Variations in the physicochemical properties and bacterial community composition during fermentation of low-salt shrimp paste. *Food Res. Int.* **2022**, *154*, 111034. [CrossRef] [PubMed]
2. Yu, J.; Lu, K.; Dong, X.; Xie, W. *Virgibacillus* sp. SK37 and *Staphylococcus nepalensis* JS11 as potential starters to improve taste of shrimp paste. *LWT Food Sci. Technol.* **2022**, *154*, 112657. [CrossRef]
3. Sang, X.; Li, K.; Zhu, Y.; Ma, X.; Hao, H.; Bi, J.; Zhang, G.; Hou, H. The Impact of Microbial Diversity on Biogenic Amines Formation in Grasshopper Sub Shrimp Paste during the Fermentation. *Front. Microbiol.* **2020**, *11*, 782. [CrossRef] [PubMed]
4. Yu, J.; Lu, K.; Zi, J.; Yang, X.; Zheng, Z.; Xie, W. Halophilic bacteria as starter cultures: A new strategy to accelerate fermentation and enhance flavor of shrimp paste. *Food Chem.* **2022**, *393*, 133393. [CrossRef]
5. Park, Y.K.; Lee, J.H.; Mah, J.H. Occurrence and reduction of biogenic amines in traditional Asian fermented soybean foods: A review. *Food Chem.* **2019**, *278*, 1–9. [CrossRef]
6. Anal, A.K.; Perpetuini, G.; Petchkongkaew, A.; Tan, R.; Avallone, S.; Tofalo, R.; Nguyen, H.V.; Chu-Ky, S.; Ho, P.H.; Phan, T.T.; et al. Food safety risks in traditional fermented food from South-East Asia. *Food Control* **2020**, *109*, 106922. [CrossRef]
7. Barbieri, F.; Montanari, C.; Gardini, F.; Tabanelli, G. Biogenic Amine Production by Lactic Acid Bacteria: A Review. *Foods* **2019**, *8*, 17. [CrossRef]
8. Jaguey-Hernandez, Y.; Aguilar-Arteaga, K.; Ojeda-Ramirez, D.; Anorve-Morga, J.; Gonzalez-Olivares, L.G.; Castaneda-Ovando, A. Biogenic amines levels in food processing: Efforts for their control in foodstuffs. *Food Res. Int.* **2021**, *144*, 110341. [CrossRef]
9. Doeun, D.; Davaatseren, M.; Chung, M.S. Biogenic amines in foods. *Food Sci. Biotechnol.* **2017**, *26*, 1463–1474. [CrossRef]
10. Jin, Y.H.; Lee, J.H.; Park, Y.K.; Lee, J.H.; Mah, J.H. The Occurrence of Biogenic Amines and Determination of Biogenic Amine-Producing Lactic Acid Bacteria in Kkakdugi and Chonggak Kimchi. *Foods* **2019**, *8*, 73. [CrossRef]
11. Shen, Y.; Wu, Y.; Wang, Y.; Li, L.; Li, C.; Zhao, Y.; Yang, S. Contribution of autochthonous microbiota succession to flavor formation during Chinese fermented mandarin fish (*Siniperca chuatsi*). *Food Chem.* **2021**, *348*, 129107. [CrossRef]
12. Wang, Y.; Li, C.; Li, L.; Yang, X.; Chen, S.; Wu, Y.; Zhao, Y.; Wang, J.; Wei, Y.; Yang, D. Application of UHPLC-Q/TOF-MS-based metabolomics in the evaluation of metabolites and taste quality of Chinese fish sauce (Yu-lu) during fermentation. *Food Chem.* **2019**, *296*, 132–141. [CrossRef]
13. Sang, X.; Ma, X.; Hao, H.; Bi, J.; Zhang, G.; Hou, H. Evaluation of biogenic amines and microbial composition in the Chinese traditional fermented food grasshopper sub shrimp paste. *LWT Food Sci. Technol.* **2020**, *134*, 109979. [CrossRef]
14. Ma, X.; Zhang, Y.; Li, X.; Bi, J.; Zhang, G.; Hao, H.; Hou, H. Impacts of salt-tolerant *Staphylococcus nepalensis* 5-5 on bacterial composition and biogenic amines accumulation in fish sauce fermentation. *Int. J. Food Microbiol.* **2022**, *361*, 109464. [CrossRef]
15. Saelao, S.; Maneerat, S.; Thongruck, K.; Watthanasakphuban, N.; Wiriyagulopas, S.; Chobert, J.M.; Haertle, T. Reduction of tyramine accumulation in Thai fermented shrimp (*kung-som*) by nisin Z-producing *Lactococcus lactis* KTH0-1S as starter culture. *Food Control* **2018**, *90*, 249–258. [CrossRef]
16. Hua, Q.; Gao, P.; Xu, Y.; Xia, W.; Sun, Y.; Jiang, Q. Effect of commercial starter cultures on the quality characteristics of fermented fish-chili paste. *LWT Food Sci. Technol.* **2020**, *122*, 109016. [CrossRef]
17. Zhang, Y.; Qin, Y.; Wang, Y.; Huang, Y.; Li, P.; Li, P. Lactobacillus plantarum LPL-1, a bacteriocin producing strain, changed the bacterial community composition and improved the safety of low-salt fermented sausages. *LWT Food Sci. Technol.* **2020**, *128*, 109385. [CrossRef]
18. Hua, Q.; Sun, Y.; Xu, Y.; Gao, P.; Xia, W. Bacterial community succession and biogenic amine changes during fermentation of fish-chili paste inoculated with different commercial starter cultures. *Int. J. Food Sci. Technol.* **2021**, *56*, 6752–6764. [CrossRef]
19. Li, C.; Li, W.; Li, L.; Chen, S.; Wu, Y.; Qi, B. Microbial community changes induced by a newly isolated salt-tolerant *Tetragenococcus muriaticus* improve the volatile flavor formation in low-salt fish sauce. *Food Res. Int.* **2022**, *156*, 111153. [CrossRef]
20. Kim, K.H.; Lee, S.H.; Chun, B.H.; Jeong, S.E.; Jeon, C.O. *Tetragenococcus halophilus* MJ4 as a starter culture for repressing biogenic amine (cadaverine) formation during *saeu-jeot* (salted shrimp) fermentation. *Food Microbiol.* **2019**, *82*, 465–473. [CrossRef]
21. Lee, J.H.; Heo, S.; Jeong, K.; Lee, B.; Jeong, D.W. Genomic insights into the non-histamine production and proteolytic and lipolytic activities of *Tetragenococcus halophilus* KUD23. *FEMS Microbiol. Lett.* **2017**, *365*, fnx252. [CrossRef] [PubMed]
22. Jeong, D.W.; Heo, S.; Lee, J.H. Safety assessment of *Tetragenococcus halophilus* isolates from doenjang, a Korean high-salt-fermented soybean paste. *Food Microbiol.* **2017**, *62*, 92–98. [CrossRef] [PubMed]
23. Sang, X.; Ma, X.; Zhang, Y.; Hao, H.; Bi, J.; Zhang, G.; Hou, H. Assessment of the Distribution and Safety of *Tetragenococcus muriaticus* for Potential Application in the Preparation of Chinese Grasshopper Sub Shrimp Paste. *Front. Microbiol.* **2021**, *12*, 628838. [CrossRef] [PubMed]
24. Li, W.; Lu, H.; He, Z.; Sang, Y.; Sun, J. Quality characteristics and bacterial community of a Chinese salt-fermented shrimp paste. *LWT Food Sci. Technol.* **2021**, *136*, 110358. [CrossRef]
25. Yao, Y.; Zhou, X.; Hadiatullah, H.; Zhang, J.; Zhao, G. Determination of microbial diversities and aroma characteristics of Beitang shrimp paste. *Food Chem.* **2021**, *344*, 128695. [CrossRef]
26. Ma, X.; Bi, J.; Li, X.; Zhang, G.; Hao, H.; Hou, H. Contribution of Microorganisms to Biogenic Amine Accumulation during Fish Sauce Fermentation and Screening of Novel Starters. *Foods* **2021**, *10*, 2572. [CrossRef]
27. Ryu, J.-A.; Kim, E.; Yang, S.-M.; Lee, S.; Yoon, S.-R.; Jang, K.-S.; Kim, H.-Y. High-throughput sequencing of the microbial community associated with the physicochemical properties of meju (dried fermented soybean) and doenjang (traditional Korean fermented soybean paste). *LWT Food Sci. Technol.* **2021**, *146*, 111473. [CrossRef]

28. Belleggia, L.; Aquilanti, L.; Ferrocino, I.; Milanovic, V.; Garofalo, C.; Clementi, F.; Cocolin, L.; Mozzon, M.; Foligni, R.; Haouet, M.N.; et al. Discovering microbiota and volatile compounds of *surströmming*, the traditional Swedish sour herring. *Food Microbiol.* **2020**, *91*, 103503. [CrossRef]
29. Zhao, C.C.; Eun, J.B. Shotgun metagenomics approach reveals the bacterial community and metabolic pathways in commercial *hongeo* product, a traditional Korean fermented skate product. *Food Res. Int.* **2020**, *131*, 109030. [CrossRef]
30. Suzuki, T.; Matsutani, M.; Matsuyama, M.; Unno, R.; Matsushita, H.; Sugiyama, M.; Yamasato, K.; Koizumi, Y.; Ishikawa, M. Growth and metabolic properties of halophilic and alkaliphilic lactic acid bacterial strains of *Marinilactibacillus psychrotolerans* isolated from surface-ripened soft cheese. *Int. Dairy J.* **2021**, *112*, 104840. [CrossRef]
31. Ruiz-Capillas, C.; Herrero, A.M. Impact of Biogenic Amines on Food Quality and Safety. *Foods* **2019**, *8*, 62. [CrossRef]
32. Ma, Q.; Lu, X.; Wang, W.; Hubbe, M.A.; Liu, Y.; Mu, J.; Wang, J.; Sun, J.; Rojas, O.J. Recent developments in colorimetric and optical indicators stimulated by volatile base nitrogen to monitor seafood freshness. *Food Packag. Shelf Life* **2021**, *28*, 100634. [CrossRef]
33. Yu, J.; Lu, K.; Sun, J.; Xie, W.; Song, L. The Flavor and Antioxidant Activity Change Pattern of Shrimp Head Paste during Fermentation. *J. Ocean Univ. China* **2022**, *21*, 195–203. [CrossRef]
34. Li, Y.; Yuan, L.; Liu, H.; Liu, H.; Zhou, Y.; Li, M.; Gao, R. Analysis of the changes of volatile flavor compounds in a traditional Chinese shrimp paste during fermentation based on electronic nose, SPME-GC-MS and HS-GC-IMS. *Food Sci. Hum. Wellness* **2023**, *12*, 173–182. [CrossRef]
35. Yu, J.; Lu, K.; Zi, J.; Yang, X.; Xie, W. Characterization of aroma profiles and aroma-active compounds in high-salt and low-salt shrimp paste by molecular sensory science. *Food Biosci.* **2022**, *45*, 101470. [CrossRef]
36. Odeyemi, O.A.; Burke, C.M.; Bolch, C.C.J.; Stanley, R. Seafood spoilage microbiota and associated volatile organic compounds at different storage temperatures and packaging conditions. *Int. J. Food Microbiol.* **2018**, *280*, 87–99. [CrossRef]

foods

MDPI

Article

Combined Effects of Cold and Hot Air Drying on Physicochemical Properties of Semi-Dried *Takifugu obscurus* Fillets

Ye Zhu [1,2], Xiaoting Chen [1,3], Kun Qiao [1], Bei Chen [1], Min Xu [1,3], Shuilin Cai [1], Wenzheng Shi [2,*] and Zhiyu Liu [1,*]

1 Key Laboratory of Cultivation and High-Value Utilization of Marine Organisms in Fujian Province, National Research and Development Center for Marine Fish Processing (Xiamen), Fisheries Research Institute of Fujian, Xiamen 361013, China; yezhuzy@163.com (Y.Z.); xtchen@jmu.edu.cn (X.C.); qiaokun@xmu.edu.cn (K.Q.); chenbeifjfri@foxmail.com (B.C.); xumin@jmu.edu.cn (M.X.); caishuilin@hqu.edu.cn (S.C.)

2 College of Food Sciences & Technology, Shanghai Ocean University, Shanghai 201306, China

3 College of Ocean Food and Biological Engineering, Jimei University, Xiamen 361021, China

* Correspondence: wzshi@shou.edu.cn (W.S.); 13906008638@163.com (Z.L.); Tel.: +86-156-9216-5859 (W.S.); +86-139-0600-8638 (Z.L.)

Abstract: The physicochemical properties of semi-dried *Takifugu obscurus* fillets in cold air drying (CAD), hot air drying (HAD), and cold and hot air combined drying (CHACD) were analyzed based on pH, water state, lipid oxidation, protein degradation, and microstructure, using a texture analyzer, low-field nuclear magnetic resonance, thiobarbituric acid, frozen sections, sodium dodecyl sulfate polyacrylamide gel electrophoresis, and differential scanning calorimetry. Water binding to the samples was enhanced by all three drying methods, and the immobilized water content of CHACD was between that of HAD and CAD. The pH of the semi-dried fillets was improved by CHACD. When compared to HAD and CAD, CHACD improved the springiness and chewiness of the fillets, especially cold air drying for 90 min (CAD-90), with values of 0.97 and 59.79 g, respectively. The muscle fibers were arranged compactly and clearly in CAD-90, having higher muscle toughness. CHACD reduced the drying time and degree of lipid oxidation compared to HAD and CAD. CAD better preserved protein composition, whereas HAD and CHACD promoted actin production; CHACD had a higher protein denaturation temperature (74.08–74.57 °C). CHACD results in better physicochemical properties than HAD or CAD, including shortened drying time, reduced lipid oxidation, enhanced protein stability, and denser tissue structure. These results provide a theoretical basis for selecting the appropriate drying method for *T. obscurus* in industrial applications.

Keywords: *Takifugu obscurus*; drying; physicochemical properties; microstructure; protein degradation

Citation: Zhu, Y.; Chen, X.; Qiao, K.; Chen, B.; Xu, M.; Cai, S.; Shi, W.; Liu, Z. Combined Effects of Cold and Hot Air Drying on Physicochemical Properties of Semi-Dried *Takifugu obscurus* Fillets. *Foods* **2023**, *12*, 1649. https://doi.org/10.3390/foods12081649

Academic Editor: Jingran Bi

Received: 10 March 2023
Revised: 10 April 2023
Accepted: 12 April 2023
Published: 14 April 2023

1. Introduction

Pufferfish (*Tetraodontidae*), belonging to *Tetraodontiformes*, is famous worldwide for its tetrodotoxin. Pufferfish abound in China with more than 40 species widely distributed in the East China Sea, Bohai Sea, Yellow Sea, and Yangtze River. *Takifugu obscurus* and *Takifugu rubripes* are the only two species of pufferfish that have been legally processed in China since 2016 [1]. Pufferfish meat is low in fat and high in protein, mineral elements, a variety of essential amino acids, and taurine, making it a fish with high nutritional value that is popular with consumers in Japan, Korea, and China [2]. According to the 2022 China Fishery Statistical Yearbook, the total amount of farmed pufferfish reached 29,950 tons in 2021, and the overall production will continue to increase [3]. However, the current processing capacity of pufferfish is higher than the supply, and pufferfish meat is not easily stored, as it only has a shelf life of about 4 d when refrigerated at 4 °C [4]. Therefore, extending the shelf life of pufferfish is important for improving production. In addition, the current processing method is too simple, as fish is mainly stored salt-dried and at low temperatures; this problem can be addressed by developing market-adapted pufferfish products.

Drying is an essential step in the aquatic products processing because it can extend shelf life, improve quality, and reduce storage and transportation costs by reducing the internal moisture content, thus inhibiting the growth of microorganisms and enzyme activity [5]. Traditional sun drying exhibits poor color, texture, and odor quality [6]. Hot air drying (HAD) in the processing of aquatic products—although higher temperatures can improve drying speed—involves the risk of destroying nutrients and physicochemical properties during the drying process [7]. Cold air drying (CAD) technology is appropriate for drying aquatic products with high protein content, as it can minimize the thermal denaturation of proteins, lipid oxidation, color change, and loss of flavor substances; however, the drying speed is slow [8]. Cold and hot air combined drying (CHACD) offers the advantages of both CAD and HAD, and has thus become a more popular method.

Different drying methods have different impacts on the water status, microstructure, lipid oxidation, and protein degradation of food products, leading to changes in their physicochemical properties [9]. Through drying, water evaporates within the aquatic product and the rate of lipid contact with oxygen increases [10]. During processing, moderate oxidation produces a satisfactory flavor, but over-oxidation leads to undesirable flavors [11], such as sourness and a disagreeable smell, and can even affect the health of consumers. The heating temperatures during drying can also lead to different degrees of denaturation of myofibrillar proteins, directly affecting their structure. Myofibrillar proteins are important functional proteins in meat products [12], and their denaturation results in coagulation and shrinkage, leading to significant changes in the texture of aquatic products, including springiness and hardness [13]. With increased drying time, more water evaporates from the surface of the material and muscle fibers contract, leading to the hardening of the tissue while hindering the outward diffusion of water. At the same time, the intermediates generated by lipid oxidation during drying cross-link with proteins through the Maillard and free radical reactions, causing protein denaturation and therefore a significant change in color after drying [14]. The selection of the appropriate drying method is necessary for maintaining the physicochemical properties of food products.

Hence, this study aimed to comprehensively analyze the physicochemical properties of semi-dried *T. obscurus* fillets by CAD, HAD, or CHACD. The effects of pH, water state, lipid oxidation, protein degradation, and microstructure were analyzed using texture analyzer, low-field nuclear magnetic resonance (LF-NMR), thiobarbituric acid (TBA), frozen sections, sodium dodecyl sulfate polyacrylamide gel electrophoresis (SDS-PAGE), and differential scanning calorimetry (DSC). The results obtained by these techniques provide a theoretical basis for selecting the most appropriate drying method for pufferfish fillets.

2. Materials and Methods

2.1. Sample Preparation

Two-year-old farmed *T. obscurus*, with a weight of 430 ± 30 g and a length of 26 ± 1 cm, was purchased from Fujian Tunzixian Aquatic Products Co., Ltd. (Zhangzhou, China). Living pufferfish were humanely slaughtered, gutted, skinned, and headed according to the guidelines issued by the Ministry of Agriculture of the P. R. of China (SC/T 3033–2016). The dorsal meat was washed with tap water and cut into pieces (20 × 15 × 5 mm, weight 2.0 ± 0.1 g) for drying. The pieces had an initial moisture content of 80 ± 1%.

2.2. Drying Conditions

The parameters of the different drying methods were optimized based on previous research [15], and the following methods were used to dry the fillets. HAD was performed using a hot air oven (BGZ-240, Boxun, Shanghai, China) for 37.5 min at 70 °C, with a wind speed of 1.5 m/s. CAD was operated using a cold air oven (HFD-2, Ouchen, Nanjing, China) for 146 min at 20 °C, with a relative humidity of 41% and a wind speed of 3 m/s. CHACD was performed using the above specified CAD and HAD parameters, with CAD performed for 30, 60, 90, and 120 min followed by HAD performed for 29, 28, 19, and 17 min. The final moisture content of all dried fillets was below 40 ± 1%.

2.3. Determination of pH

The pH was determined according to the method described by Fan et al. [16], with slight modifications. Approximately 2 g of fillets were dispersed in 18 mL of distilled water and homogenized using a T25 ULTRA-TURRAX® (IKA Werke GmbH & Co., KG, Staufen, Germany). Using an Eppendorf 5810R high-speed freezing centrifuge (Eppendorf AG, Hamburg, Germany), the sample was centrifuged at 10,621× *g* for 10 min at 4 °C. The pH of the filtrate was measured using a digital pH meter (Mettler Toledo FE28, Shanghai, China) and measured three times in parallel.

2.4. Determination of Texture

The samples were investigated using an A/MORS P/5S probe (5 mm diameter, TA-XTplus, Stable Micro Systems, Godalming, UK), and a slight modification was made by referring to Guo et al. [17]. Two consecutive cycles at 50% sample deformation were applied at 1.5 mm/s test speed. Other parameters including pre-test speed, post-test speed and trigger force, were set at 2 mm/s, 2 mm/s, and 5 g, respectively. TPA analysis results were expressed in terms of hardness, chewiness, springiness, and resilience. Six parallel measurements were performed for each sample group.

2.5. LF-NMR Analysis

The protocol for LF-NMR analysis was slightly modified from that reported by Wang et al. [18]. A MesoMR LF-NMR analyzer (Shanghai Niumag Analytical Instrument, Shanghai, China) with a magnetic field strength of 0.5 T, corresponding to a spectrometer frequency (SF) of 21 MHz, was employed. The samples were transferred to the center of cylindrical glass tubes (25-mm radiofrequency coil) and equilibrated to 25 °C before detection. The transverse relaxation time (T_2) was measured using a Carr–Purcell–Meiboom–Gill pulse sequence (CPMG). Data were acquired from 8000 echoes (NECH) over 16 scans (NS). The repetition time between scans was 3500 ms, and the 90 (P1) and 180 (P2) pulse times were 5 and 10 μs, respectively. The echo time (TE) was 0.2 ms, the number of sampling points (TD) was 320,026, the magnet frequency (SW) was 200 and regulate analog gain 1 (RG1) was 20 db. The MultiExp Inv analysis software Version 4.0 (Shanghai Niumag Analytical Instrument Co., Shanghai, China) based on the Simultaneous Iterative Reconstruction technique (SIRT) was used to obtain the single-component relaxation time (T_{2W}) and transverse relaxation time (T_2) distributions through mono-exponential or multi-exponential fitting. The samples were analyzed six times each.

2.6. Determination of Thiobarbituric Acid Value

The thiobarbituric acid (TBA) value was measured according to the method described by Wang et al. [19] with slight modifications. Three grams of minced fillet samples were homogenized in 30 mL of ice cooling 7.5% trichloroacetic acid solution containing 0.1% ethylene diamine tetraacetic acid (EDTA) at 15,000× *g* for 1 min. After filtering the homogenate, 5 mL of 0.02 M thiobarbituric acid was added to 5 mL of filtrate and mixed. The mixture was maintained in a boiling water bath for 40 min. After the solution had cooled, the absorbance of the upper layer was determined at 532 nm using a VICTOR Nivo™ multimode plate reader (PerkinElmer Inc., Waltham, MA, USA). A standard curve was obtained using 1, 1, 3, 3-tetraethoxypropane and the results were expressed as mg of malonaldehyde (MDA)/kg. Each sample group was measured in triplicate.

2.7. Determination of Microstructure

The frozen sections were measured according to Sigurgisladottir et al. [20], with slight modifications. All samples were collected from the same location on each fillet, and the geometric centers of the fillets were cut into 5 × 5 × 5 mm pieces. The samples were embedded in plastic tubes containing O.C.T. compound (SAKURA Tissue Tek®, Torrance, CA, USA) and frozen in liquid nitrogen. Freezing occurred in approximately 40 s below −80 °C, and samples were then stored at −80 °C until sectioning. The specimens were sectioned (10 μm), frozen at −25 °C in a freezing microtome (Leica CM1950, Shanghai, China) for vertical cuts, and then placed on glass slides. After hematoxylin–eosin (HE) staining, the microstructures of the samples were examined using a fluorescent inverted microscope (DMi8, Leica Microsystems, Shanghai, China) using a 20× objective lens.

2.8. Protein Component Analysis

The method described by Yu et al. [21] was slightly modified. Briefly, semi-dried fillets of minced samples (5 g) were homogenized with 25 mL of Tris-HCI (20 mmol/L, pH 7.5) using a T25 ULTRA-TURRAX® (IKA Werke GmbH & Co. KG, Staufen, Germany). The homogenate was centrifuged at 10,621× *g* for 10 min at 4 °C using an Eppendorf 5810R high-speed freezing centrifuge (Eppendorf AG, Hamburg, Germany). The precipitate was then re-homogenized and centrifuged again under the same conditions. The combined supernatants were treated with 10 mL of 50% trichloroacetic acid solution (TCA), left for 2 h, and then centrifuged at 10,621× *g* for 15 min. After centrifugation, the supernatant fractions were non-protein nitrogen (NPN) and precipitated as water-soluble protein (WSP). The precipitate obtained after the first centrifugation was added to 40 mL of 0.6 mol/L NaCl Tris-HCl buffer (20 mmol/L, pH 7.5), homogenized for 1 min, stirred for 1 h, and centrifuged at 10,621× *g* for 15 min at 4 °C. The supernatant was repeated twice for salt-soluble protein (SSP), and the residue was used as insoluble protein (ISP). The nitrogen contents of the different components were determined using the Kjeldahl method. Each group of samples was measured thrice in parallel. All the above operations were performed at 4 °C, and the solutions pre-cooled before use.

2.9. Sodium Dodecyl Sulfate Polyacrylamide Gel Electrophoresis Analysis

Extraction was performed following the method used by Setyabrata et al. [22], with a few modifications. Protein extraction was performed by homogenizing 0.1 g of fillet samples in 20 mL of whole protein extraction buffer (ratio of RIPA buffer (high) to PMSF was 100:1). The protein concentration of each sample was adjusted to 1 mg/mL using extraction buffer. Prior to sample loading, each sample was mixed well with its respective sample buffer. The samples were heated at 100 °C for 5 min in a loading buffer, cooled to 25 °C, and then 10 μL of sample solution was loaded into each well. The gels were prepared using 12% separating gel and 5% stacking gel. The gels were run on a Bio-Rad® Criterion cell system (Bio-Rad® Laboratories, Hercules, CA, USA) equipped with a DYY-6C electrophoresis apparatus (Beijing Liuyi Biotechnology Co., Ltd., Beijing, China) using 200 V constant voltage for 50–60 min. After running the gel, it was stained with 0.25% Coomassie brilliant blue R-250 (Bio-Rad, Richmond, CA, USA) and destained in a solution containing 5% ethanol and 10% acetic acid to visualize the gel. Gel visualization and image export were performed using an EPSON perfection V700 photo (Seiko Epson Co., Nagano, Japan). The protein bands were identified based on the determined molecular weight in parallel with the relative mobility of the protein bands to the dual-color pre-stained protein standard (10–250 kDa, EpiZyme, Shanghai, China).

2.10. Determination of Denaturation Temperature

Thermal transition properties were determined using differential scanning calorimetry (DSC; DSC 3, Mettler-Toledo International Trading Co., Ltd., Zurich, Switzerland). The method was performed according to Shang et al. [23]. After accurate weighing, 16–17 mg of each semi-dried sample was placed in a DSC aluminum pan, the added mass was recorded (accurate to 0.0001 g), and the pan was hermetically sealed. The initial temperature was set to 20 °C and the constant temperature time was 10 min. Then scanned from 20 to 100 °C, with a following temperature rise rate was of 5 °C/min at a nitrogen flow rate of

100 mL/min, and a reaction gas flow rate of 50 mL/min. A hermetically sealed empty pan was used as a reference. Three replicates were performed for each sample.

2.11. Statistical Analysis

One-way analysis of variance (ANOVA) followed by LSD and Duncan's test were used to check for significant differences ($p < 0.05$) using SPSS Statistics 26 (SPSS Inc., Chicago, IL, USA). All experiments were performed three times, except for TBA and LF-NMR analyses, which were conducted in six replicates. Data are expressed as the mean ± standard deviation (SD) (n = 3). The data were plotted using OriginPro2018 (OriginLab Corp., Northampton, MA, USA). DSC data were analyzed using STARe Evaluation Software Version 16.20 (Mettler-Toledo International Trading Co., Ltd., Zurich, Switzerland) to create temperature plots.

3. Results and Discussion

3.1. Fillet pH

The pH is an important index for evaluating the quality of meat. Figure 1 shows the changes in the pH of pufferfish fillets under different drying methods. The pH of all fillets from all three treatments was weakly acidic, with that of fillets treated with HAD and CAD measuring 6.07 and 6.12, respectively, whereas that of CHACD-treated fillets was increased to 6.15–6.27. A strongly acidic environment leads to protein denaturation and decreases the water-holding capacity of muscle. Acidity is an important factor in flabby meat quality, and an increase in pH within a certain range results in improved meat quality [24]. However, increases in pH may be due to microbial production of nitrogenous substances and the production of substances, such as peptides and amines, by protein degradation during the drying process [25].

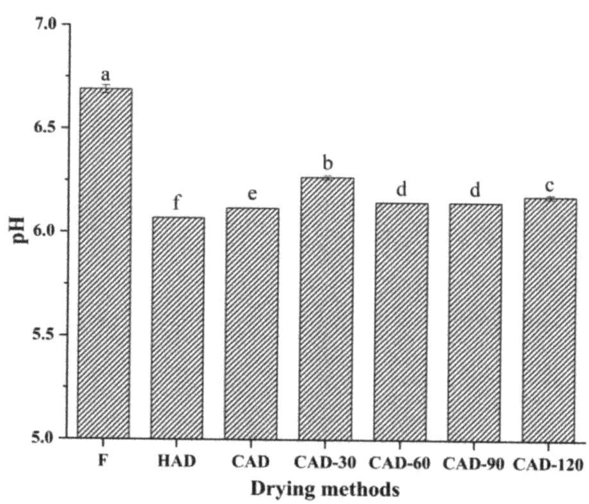

Figure 1. Changes in pH of semi-dried pufferfish fillets under different drying methods. Values with different superscripts are significantly different ($p < 0.05$). F, fresh undried fillets; HAD, hot air drying; CAD, cold air drying; CAD-30, cold air drying for 30 min; CAD-60, cold air drying for 60 min; CAD-90, cold air drying for 90 min; CAD-120, cold air drying for 120 min.

3.2. Fillet Texture

Texture is an important sensory attribute for consumer acceptance of aquatic products and influences the choice of subsequent processing methods [26]. Variations in pufferfish fillet textural properties, such as hardness, springiness, chewiness, and resilience under different drying methods, are shown in Table 1. CAD and CHACD helped maintain the hardness of the fillets, whereas HAD had the highest hardness (218.14 g) and chewiness

(70.42 g). Hardening of tissue texture could be attributed to the rapid loss of surface water at higher temperatures and the appearance of hardened layers on the surface, along with increased protein denaturation and aggregation, contraction of myofibril and connective tissues, and reduction of extracellular space, intracellular lumens, and channels [27]. Similar results were reported by Vega-Gálvez et al. [28], where the hardness of giant squid meat increased dramatically after drying under different high-temperature conditions. Chewiness refers to the energy required to chew a solid to a swallowable state—the larger the value of elasticity, the better the taste of the fish. Appropriate increases in springiness and chewiness are beneficial for improving taste. Compared with HAD and CAD, CHACD improved springiness and chewiness, in which CAD-90 resulted in higher springiness and chewiness, with values of 0.97 and 59.79 g, respectively. The other three CHACD groups all had some correlation with CAD. This result may be due to the fact that CAD was performed first during CHACD, in that the connective tissue was less damaged, and the muscle fibers gradually contracted and tightly connected while water evaporated and dried to a certain extent before undergoing HAD, thus reducing the adverse effect of hardening of the sample surface during high-temperature drying. However, the resilience of all three drying methods was lower than that of the fresh samples, and there was no significant difference ($p > 0.05$). Changes in fillet texture after drying can be attributed to protein denaturation due to solute concentration, heat, and enzymatic denaturation [29].

Table 1. Texture characteristics of pufferfish fillets under different drying methods.

Drying Methods	Hardness (g)	Springiness	Chewiness (g)	Resilience
F	133.60 ± 23.55^{b}	0.91 ± 0.02^{ab}	53.57 ± 8.05^{b}	0.28 ± 0.03^{a}
HAD	218.14 ± 28.28^{a}	0.60 ± 0.04^{c}	70.42 ± 11.95^{a}	0.16 ± 0.01^{c}
CAD	93.96 ± 10.35^{b}	0.88 ± 0.07^{b}	56.62 ± 6.74^{ab}	0.20 ± 0.03^{b}
CAD-30	101.04 ± 22.89^{b}	0.87 ± 0.06^{b}	55.68 ± 10.53^{ab}	0.21 ± 0.02^{b}
CAD-60	105.54 ± 23.72^{b}	0.88 ± 0.05^{b}	59.57 ± 14.48^{ab}	0.19 ± 0.01^{b}
CAD-90	102.27 ± 11.70^{b}	0.97 ± 0.02^{a}	59.79 ± 6.26^{ab}	0.19 ± 0.02^{b}
CAD-120	90.48 ± 19.07^{b}	0.88 ± 0.10^{b}	49.88 ± 14.53^{b}	0.18 ± 0.01^{bc}

Data are expressed as the mean ± SD ($n = 6$). Values within a column with different superscripts are significantly different ($p < 0.05$). F, fresh undried fillets; HAD, hot air drying; CAD, cold air drying; CAD-30, cold air drying for 30 min; CAD-60, cold air drying for 60 min; CAD-90, cold air drying for 90 min; CAD-120, cold air drying for 120 min.

3.3. LF-NMR of Fillets

The state and migration of water were further characterized using LF-NMR transverse relaxation time (T_2). The relaxation time T_2 is related to the degree of binding of the sample and hydrogen protons, and a higher binding of hydrogen protons indicates a lower mobility of water in the sample, which is expressed as a decrease in relaxation time T_2 [30]. Bound water (T_{20} and T_{21}) has a short relaxation time of 1–10 ms and represents water bound to macromolecules; immobilized water (T_{22}) in the range of 10–100 ms indicates the water inside the myofibril and reticular tissues, which accounts for most of the total water content of the samples; and free water (T_{23}) with the longest relaxation time mainly occurs between 100 and 1000 ms [31].

Figure 2A shows the relaxation time T_2 curves of fresh pufferfish fillets obtained using the three drying methods, which resulted in a left shift of the peak positions of T_{21} and T_{22}, a decrease in the peak area, and a shortening of the relaxation time T_2. Across all three drying methods, the T_{22} of the fillets decreased when compared to T_{21} and T_{23}, especially from 49.77 ms to 24.77 ms after HAD, indicating that the water binding to the sample increased and the mobility of immobilized water gradually decreased. This is consistent with the findings of Zhang et al. [32], in which the myofibril structure gradually shrank during drying, displaying a tighter network structure and enhanced water absorption capacity, resulting in a shorter T_2 relaxation time. The relative percentages of water in the three states can be seen in Figure 2B, and the content of immobilized water in the undried fresh samples was higher, accounting for 95.09% of the total water content; the fillets mainly contained bound and immobilized water, and the content of immobilized water was reduced in all four groups of CHACD, and between HAD and CAD. The increase in the T_{21} peak area was mainly due to the conversion of immobilized water into free water and its subsequent outward diffusion during the drying process. The T_{23} peak area increased in all drying methods, with the highest after CAD-30 containing 1.14%, which may be due to the destruction of the protein structure and the release of immobilized water from the outer parts of myofibrils [33].

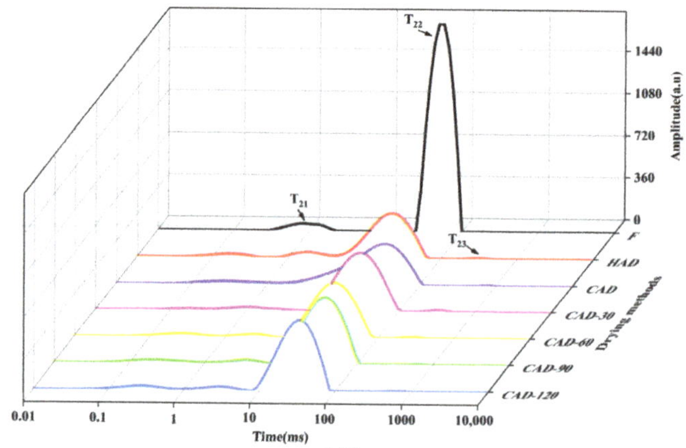

(A)

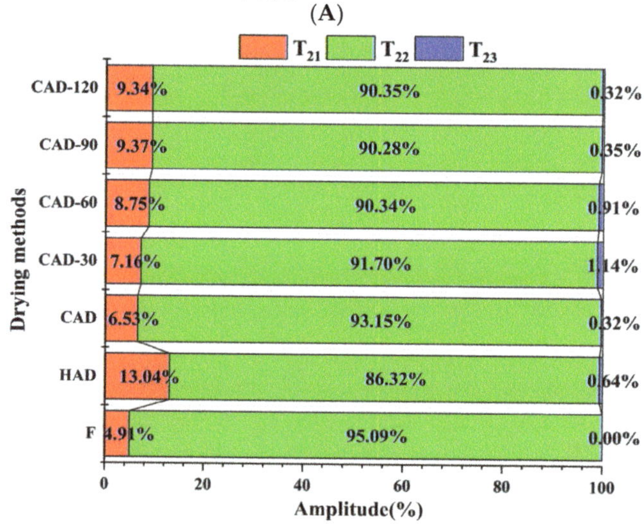

(B)

Figure 2. Curve of T_2 transverse relaxation time (**A**) and proportion of various peak areas (**B**) in semi-dried pufferfish fillets. F, fresh undried fillets; HAD, hot air drying; CAD, cold air drying; CAD-30, cold air drying for 30 min; CAD-60, cold air drying for 60 min; CAD-90, cold air drying for 90 min; CAD-120, cold air drying for 120 min.

3.4. Oxidation of Fillet Lipids

The TBA value corresponds with the content of malondialdehyde (MDA) produced during lipid oxidation, which is a natural and final product of lipid peroxidation [34]; therefore, the degree of lipid oxidation in aquatic products can be expressed as TBA [14]. Figure 3 shows the variation in TBA values of semi-dried pufferfish fillets treated with different drying methods. The three drying methods had some effect on the lipid oxidation of pufferfish fillets. CAD had the largest TBA value of 0.024 mg MDA/kg among the three drying methods, indicating that the effect of CAD on the lipid oxidation of fillets was more obvious, which may be due to the relatively longer exposure of fish fillets to air [35] and increased oxidation of unsaturated fatty acids during dehydration [36]. Compared with CAD, CHACD reduced the degree of lipid oxidation by reducing the drying time. The degree of lipid oxidation in CHACD was similar to that in HAD ($p < 0.05$).

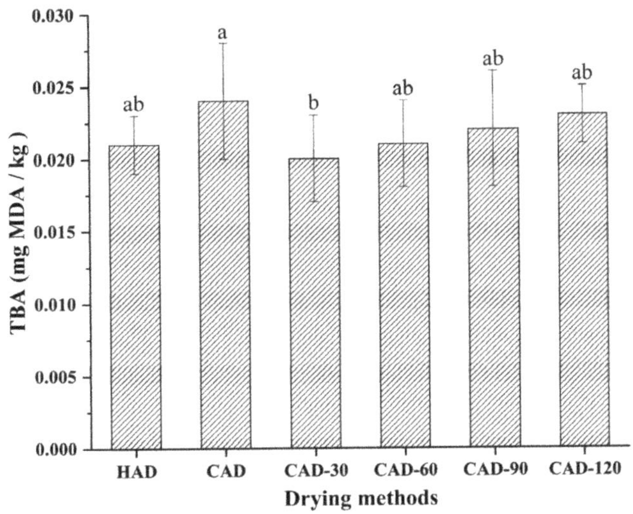

Figure 3. Changes in thiobarbituric acid (TBA) values in semi-dried pufferfish fillets. Values with different superscripts are significantly different ($p < 0.05$). HAD, hot air drying; CAD, cold air drying; CAD-30, cold air drying for 30 min; CAD-60, cold air drying for 60 min; CAD-90, cold air drying for 90 min; CAD-120, cold air drying for 120 min.

3.5. Fillet Microstructure

The changes in fillet microstructure (vertical section) under different drying methods are shown in Figure 4. The microstructure is closely related to the quality of the fillets after drying and can indicate the degree of protein degradation [37]. As shown in Figure 4, the muscle fibers of the undried fresh fish samples were intact and well defined. Using different drying methods, the muscle fibers showed different degrees of shrinkage. After HAD, the muscle fibers were firmer but showed fiber breakage and significant degradation. This may be due to the higher temperature, which accelerates protein denaturation, and the smaller size of myofibrils and collagen, resulting in a reduction in muscle fiber diameter and length [27]. Fillet microstructure after CAD had smaller and well-defined muscle fiber gaps, which may be due to the lower temperature and longer drying duration, which allows sufficient time for muscle fiber contraction and reduced protease activity, thus mitigating protein degradation [38]. In contrast, CHACD-treated fillets more and more resembled the microstructure of CAD-treated fillets with increasing CAD treatment time, and the muscle fibers changed from partially broken to intact, well-arranged, and well-defined. Among them, the muscle fiber structure of CAD-90-treated fillets was the most arranged, compact, well-defined, and complete in structure among the four CHACD groups, showing high muscle toughness. This indicated that CAD-90 resulted in better texture properties, which was consistent with the results of the texture analysis.

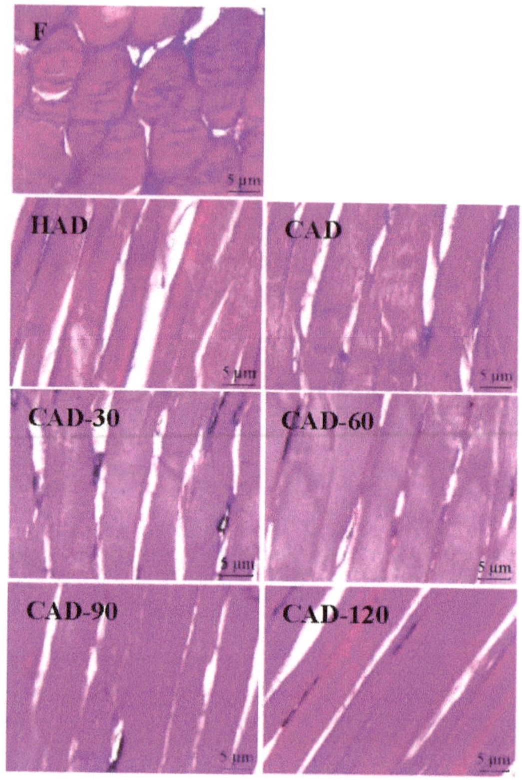

Figure 4. Microstructure (vertical section) of semi-dried pufferfish fillets under different drying methods. F, fresh undried fillets; HAD, hot air drying; CAD, cold air drying; CAD-30, cold air drying for 30 min; CAD-60, cold air drying for 60 min; CAD-90, cold air drying for 90 min; CAD-120, cold air drying for 120 min.

3.6. Fillet Protein Composition

According to the difference in fish protein solubility, pufferfish proteins were divided into WSP, SSP, ISP, and NPN. The changes in the protein composition of pufferfish fillets subjected to different drying methods are summarized in Table 2. The WSP content in fresh fillets was the highest, accounting for 58.09% of the total protein, followed by ISP, accounting for 26.76%, whereas the content of SSP and NPN was lower. The NPN content increased but varied less among the three drying methods. NPN is composed of free amino acids, small peptides, and nucleic acids [39]. With the three drying methods, the moisture content in the fillets was reduced, which helped to inhibit microbial and enzymatic activities. However, drying degraded proteins with a large molecular mass to produce free amino acids and peptides, thereby increasing the NPN content. When compared to the fresh samples, the WSP content of the two drying methods decreased and was not significantly different ($p > 0.05$), except for CAD, which could better maintain WSP. When comparing the four CHACD groups, the WSP content tended to increase slightly with an increase in CAD time. This may be due to a decrease in WSP solubility as a result of protein concentration. For CAD, this may be due to a decrease in WSP solubility as a result of protein concentration due to altered water loss at low temperatures. For HAD, this may be due to WSP denaturation caused by high temperature, which leads to a decrease in WSP solubility. However, CHACD combines the characteristics of both drying methods, leading to a decrease in solubility [40]. When compared to the undried fresh samples, the SSP of CAD significantly increased from 6.93 to 8.15 mgN/g. However, the SSP content of HAD and CHACD significantly decreased ($p < 0.05$), with the four CHACD groups increasing with increasing CAD time. When comparing the three drying methods, ISP increased significantly ($p < 0.05$), and the content in CHACD decreased with increasing CAD time, with the largest increases in CAD-30 and CAD-60, with 19.32 and 18.69 mgN/g, respectively. The significant increase in ISP in HAD and CHACD may be related to the aggregation of WSP and SSP by thermal denaturation and the fact that protein fractions in fish fillets are more unstable during HAD, leading to a significant increase in ISP content [21].

Table 2. Protein composition of pufferfish fillets under different drying methods (mgN/g).

Drying Methods	NPN	WSP	SSP	ISP
F	0.20 ± 0.01^{c}	27.33 ± 0.71^{a}	6.93 ± 0.40^{b}	12.59 ± 0.86^{c}
HAD	0.26 ± 0.01^{a}	19.37 ± 0.43^{b}	3.56 ± 0.32^{e}	15.57 ± 1.50^{b}
CAD	0.23 ± 0.01^{b}	26.80 ± 1.73^{a}	8.15 ± 0.14^{a}	15.53 ± 0.69^{b}
CAD-30	0.27 ± 0.01^{a}	18.63 ± 0.32^{b}	3.55 ± 0.04^{e}	19.32 ± 0.94^{a}
CAD-60	0.27 ± 0.01^{a}	19.60 ± 1.77^{b}	3.78 ± 0.11^{de}	18.69 ± 0.98^{a}
CAD-90	0.27 ± 0.01^{a}	20.30 ± 1.46^{b}	4.33 ± 0.33^{d}	17.60 ± 0.53^{ab}
CAD-120	0.26 ± 0.01^{a}	20.51 ± 0.39^{b}	5.39 ± 0.35^{c}	15.93 ± 0.39^{b}

Data are expressed as the mean ± SD ($n = 6$). Values within a column with different superscripts are significantly different ($p < 0.05$). NPN, non-protein nitrogen; WSP, water-soluble protein; SSP, salt-soluble protein; ISP, insoluble protein; F, fresh undried fillets; HAD, hot air drying; CAD, cold air drying; CAD-30, cold air drying for 30 min; CAD-60, cold air drying for 60 min; CAD-90, cold air drying for 90 min; CAD-120, cold air drying for 120 min.

Proteins play a key role in determining the physical properties of meat products. Han et al. [41] also found that heat induction caused a negative correlation between the degree of pork myofibrillar protein aggregation and immobilized water content and showed that protein aggregation tends to form network structures. This is consistent with the LF-NMR and microstructure analysis, where the relative percentage of the immobilized water content was lower in HAD than in CAD and CHACD, and the myofibrils were more tightly arranged in the HAD and CHACD groups.

3.7. Degradation of Fillet Proteins

SDS-PAGE can be used to determine molecular weight and to analyze the number of subunits in protein molecules [42]. We analyzed three common proteins: myosin heavy chain (MHC, 200 kDa), myosin (100 kDa), and actin (45 kDa) [43]. Figure 5 shows the changes in SDS-PAGE of fillets treated using different drying methods. The MHC and myosin bands were clearly visible after CAD when compared to fresh samples, and the actin band remained unaltered. In HAD and CHACD, the MHC band disappeared and myosin remained unaltered, whereas the actin band became thicker and the color deepened to be more clearly visible. CHACD showed an increase, then a decrease, and then an increase in the actin band with the increase in CAD time, indicating that heating could reduce the thermal stability of myosin and actin and accelerate protein degradation. Zheng et al. [44] also found that actin bands disappeared more slowly than myosin bands during heating of chicken meat myofibrillar proteins.

3.8. Thermal Stability of Fillet Proteins

Thermal stability is an important measure of protein stability in aquatic products. In general, the treated fish had two heat absorption peaks near 50 and 75 °C for myosin (T_{P1}) and actin (T_{P2}) [45], respectively, where myosin is the most unstable of the structural proteins and completes the entire denaturation process between 40–60 °C [46]; actin is the most thermally stable protein, starting to denature at 71 °C, endoscopically transitioning between 78 °C, and denaturing completely when heated to approximately 83 °C [23]. Figure 6 shows the two peak values of endothermic transitions in samples treated with different drying methods. The T_{P1} value after HAD (52.50 °C) was higher than that after CAD, whereas all the CHACD myosins underwent different degrees of thermal denaturation. While the T_{P2} after HAD (73.94 °C) was higher than that after CAD (73.70 °C), it was lower than that after all CHACD treatments (74.08–74.57 °C), except for CAD-60 (73.85 °C), indicating that CHACD could also effectively reduce the degradation of actin, which was consistent with the results of SDS-PAGE. The enthalpy of denaturation (ΔH) of the samples treated with the three drying methods varied from 0.11–0.42 (peak I) and 0.21–0.29 (peak II), indicating that the increase in heating temperature was associated with a decrease in Δ [47].

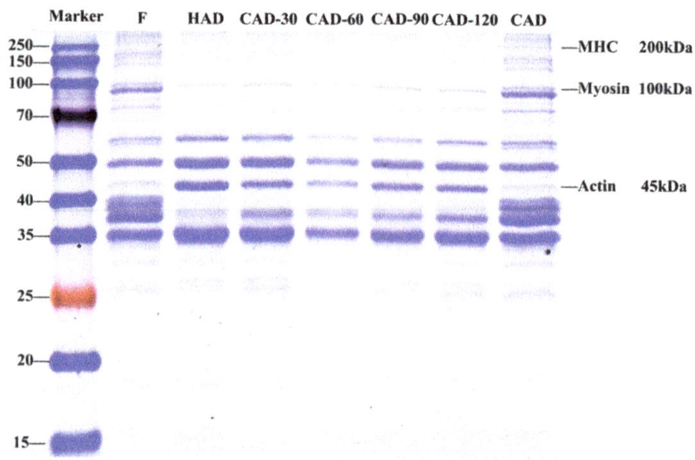

Figure 5. SDS-PAGE of myofibrillar proteins from semi-dried pufferfish fillets dried using different methods. F, fresh, undried fillets; HAD, hot air drying; CAD, cold air drying; CAD-30, cold air drying for 30 min; CAD-60, cold air drying for 60 min; CAD-90, cold air drying for 90 min; CAD-120, cold air drying for 120 min.

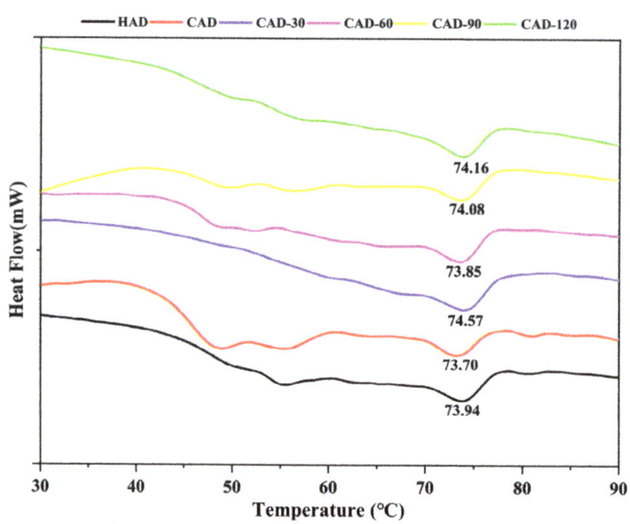

Figure 6. Differential scanning calorimetry of semi-dried pufferfish fillets under different drying methods. F, fresh, undried fillets; HAD, hot air drying; CAD, cold air drying; CAD-30, cold air drying for 30 min; CAD-60, cold air drying for 60 min; CAD-90, cold air drying for 90 min; CAD-120, cold air drying for 120 min.

4. Conclusions

We compared the effects of CAD, HAD, and CHACD on the physicochemical properties of *T. obscurus*, including the pH, water state, lipid oxidation, protein degradation, and microstructure. All three drying methods enhanced the binding of water to the sample, existing mainly in a bound or immobilized form that was converted into free water to diffuse outward. The content of CHACD-immobilized water was between that of HAD and CAD. The pH after HAD and CAD was low, whereas CHACD increased the pH to some extent, but the fish treated with all three drying methods were weakly acidic. HAD-treated fillets had the highest hardness and chewiness, but fiber breakage and degradation were observed. When compared to HAD and CAD, CHACD improved the springiness and chewiness of the fillets, among which CAD-90 resulted in higher springiness and chewiness and showed higher muscle toughness among the four CHACD groups. All three drying methods resulted in some degree of lipid oxidation. The TBA value of CAD was the highest among the three drying methods, whereas CHACD reduced the drying time and decreased the degree of lipid oxidation through combined drying. In terms of protein degradation, CAD better maintained the protein composition, while HAD and CHACD promoted the formation of actin, and CHACD had a higher protein denaturation temperature. We found that CHACD combines the advantages of CAD and HAD with respect to effects on the physicochemical characteristics of semi-dried pufferfish fillets, including shortened drying time, reduced degree of lipid oxidation, enhanced protein stability, and denser tissue structure. This study provides a theoretical basis to improve research on and development of the most suitable drying method for semi-dried *T. obscurus* in industrial applications.

Author Contributions: Conceptualization, Y.Z.; methodology, Y.Z. and X.C.; validation, X.C.; formal analysis, Y.Z., X.C. and K.Q.; investigation, Y.Z., B.C. and M.X.; resources, Y.Z., K.Q. and S.C.; data curation, Y.Z. and X.C.; writing—original draft preparation, Y.Z.; writing—review and editing, W.S. and Z.L.; visualization, Y.Z.; supervision, W.S. and Z.L.; project administration, W.S. and Z.L.; funding acquisition, W.S. and Z.L. All authors have read and agreed to the published version of the manuscript.

Funding: This research was funded by the National Key R&D Program of China, grant number 2018YFD0901102 and the National Key R&D Program of China, grant number 2019YFD0902003.

Institutional Review Board Statement: Not applicable.

Informed Consent Statement: Not applicable.

Data Availability Statement: Data is contained within the article.

Conflicts of Interest: The authors declare no conflict of interest.

References

1. Zhang, D.; Yang, N.; Fisk, I.D.; Li, J.; Liu, Y.; Wang, W. Impact of cooking on the sensory perception and volatile compounds of Takifugu rubripes. *Food Chem.* **2022**, *371*, 131165. [CrossRef] [PubMed]
2. Bi, H.; Cai, D.; Zhang, R.; Zhu, Y.; Zhang, D.; Qiao, L.; Liu, Y. Mass spectrometry-based metabolomics approach to reveal differential compounds in pufferfish soups: Flavor, nutrition, and safety. *Food Chem.* **2019**, *301*, 125261. [CrossRef] [PubMed]
3. China Fishery Bureau. *China Fishery Statistical Yearbook*; Chinese Agriculture Express: Beijing, China, 2022; pp. 22–25. (In Chinese)
4. Yang, F.; Jing, D.; Yu, D.; Xia, W.; Jiang, Q.; Xu, Y.; Yu, P. Differential roles of ice crystal, endogenous proteolytic activities and oxidation in softening of obscure pufferfish (*Takifugu obscurus*) fillets during frozen storage. *Food Chem.* **2019**, *278*, 452–459. [CrossRef] [PubMed]
5. Liu, J.; Zhao, Y.; Shi, Q.; Wu, X.; Fang, Z. Water distribution, physicochemical and microstructural properties of scallop adductors as affected by different drying methods. *J. Food Compos. Anal.* **2023**, *115*, 104966. [CrossRef]
6. Rasul, M.G.; Majumdar, B.C.; Afrin, F.; Bapary, M.A.J.; Shah, A.K.M.A. Biochemical, Microbiological, and Sensory Properties of Dried Silver Carp (*Hypophthalmichthys molitrix*) Influenced by Various Drying Methods. *Fishes* **2018**, *3*, 25. [CrossRef]
7. Zhao, X.; Cheng, X.; Zang, M.; Wang, L.; Li, X.; Yue, Y.; Liu, B. Insights into the characteristics and molecular transformation of lipids in *Litopenaeus vannamei* during drying from combined lipidomics. *J. Food Compos. Anal.* **2022**, *114*, 104809. [CrossRef]
8. Aykın-Dinçer, E.; Erbaş, M. Cold dryer as novel process for producing a minimally processed and dried meat. *Innov. Food Sci. Emerg. Technol.* **2019**, *57*, 102113. [CrossRef]

9. Rasul, M.; Kabir, I.E.; Yuan, C.; Shah, A.K.M.A. Effects of drying methods on physicochemical, microbiological and sensory properties of torpedo scad (*Megalaspis cordyla*). *J. Microbiol. Biotechnol. Food Sci.* **2021**, *10*, e2796. [CrossRef]
10. Zhao, B.; Zhou, H.; Zhang, S.; Pan, X.; Li, S.; Zhu, N.; Wu, Q.; Wang, S.; Qiao, X.; Chen, W. Changes of protein oxidation, lipid oxidation and lipolysis in Chinese dry sausage with different sodium chloride curing salt content. *Food Sci. Hum. Wellness* **2020**, *9*, 328–337. [CrossRef]
11. Rasul, M.G.; Yuan, C.; Shah, A.K.M.A. Chemical and Microbiological Hazards of Dried Fishes in Bangladesh: A Food Safety Concern. *Food Nutr. Sci.* **2020**, *11*, 523–539. [CrossRef]
12. Kalem, I.K.; Bhat, Z.F.; Kumar, S.; Desai, A. Terminalia arjuna: A novel natural preservative for improved lipid oxidative stability and storage quality of muscle foods. *Food Sci. Hum. Wellness* **2017**, *6*, 167–175. [CrossRef]
13. Shi, S.; Zhao, M.; Li, Y.; Kong, B.; Liu, Q.; Sun, F.; Yu, W.; Xia, X. Effect of hot air gradient drying on quality and appearance of beef jerky. *LWT* **2021**, *150*, 111974. [CrossRef]
14. Fu, X.; Lin, Q.; Xu, S.; Wang, Z. Effect of drying methods and antioxidants on the flavor and lipid oxidation of silver carp slices. *LWT* **2015**, *61*, 251–257. [CrossRef]
15. Zhu, Y.; Chen, X.; Pan, N.; Liu, S.; Su, Y.; Xiao, M.; Shi, W.; Liu, Z. The Effects of Five Different Drying Methods on the Quality of Semi-dried *Takifugu obscurus* Fillets. *LWT* **2022**, *161*, 113340. [CrossRef]
16. Fan, H.; Luo, Y.; Yin, X.; Bao, Y.; Feng, L. Biogenic amine and quality changes in lightly salt- and sugar-salted black carp (*Mylopharyngodon piceus*) fillets stored at 4 °C. *Food Chem.* **2014**, *159*, 20–28. [CrossRef] [PubMed]
17. Guo, X.; Wang, Y.; Lu, S.; Wang, J.; Fu, H.; Gu, B.; Lyu, B.; Wang, Q. Changes in proteolysis, protein oxidation, flavor, color and texture of dry-cured mutton ham during storage. *LWT* **2021**, *149*, 111860. [CrossRef]
18. Wang, X.; Xie, X.; Zhang, T.; Zheng, Y.; Guo, Q. Effect of edible coating on the whole large yellow croaker (*Pseudosciaena crocea*) after a 3-day storage at −18 °C: With emphasis on the correlation between water status and classical quality indices. *LWT* **2022**, *163*, 113514. [CrossRef]
19. Ying, W.; Ya-Ting, J.; Jin-Xuan, C.; Yin-Ji, C.; Yang-Ying, S.; Xiao-Qun, Z.; Dao-Dong, P.; Chang-Rong, O.; Ning, G. Study on lipolysis-oxidation and volatile flavour compounds of dry-cured goose with different curing salt content during production. *Food Chem.* **2016**, *190*, 33–40. [CrossRef]
20. Sigurgisladottir, S.; Sigurdardottir, M.S.; Torrissen, O.; Vallet, J.L.; Hafsteinsson, H. Effects of different salting and smoking processes on the microstructure, the texture and yield of Atlantic salmon (*Salmo salar*) fillets. *Food Res. Int.* **2000**, *33*, 847–855. [CrossRef]
21. Yu, D.; Feng, T.; Jiang, Q.; Yang, F.; Gao, P.; Xu, Y.; Xia, W. The change characteristics in moisture distribution, physical properties and protein denaturation of slightly salted silver carp (*Hypophthalmichthys molitrix*) fillets during cold/hot air drying processing. *LWT* **2021**, *137*, 110466. [CrossRef]
22. Setyabrata, D.; Kim, Y.H.B. Impacts of aging/freezing sequence on microstructure, protein degradation and physico-chemical properties of beef muscles. *Meat Sci.* **2019**, *151*, 64–74. [CrossRef] [PubMed]
23. Shang, S.; Wu, B.; Fu, B.; Jiang, P.; Liu, Y.; Qi, L.; Du, M.; Dong, X. Enzyme treatment-induced tenderization of puffer fish meat and its relation to physicochemical changes of myofibril protein. *LWT* **2022**, *155*, 112891. [CrossRef]
24. Zhang, Z.Y.; Limbu, S.M.; Zhao, S.H.; Chen, L.Q.; Luo, Y.; Zhang, M.L.; Qiao, F.; Du, Z.Y. Dietary l-carnitine supplementation recovers the increased pH and hardness in fillets caused by high-fat diet in Nile tilapia (*Oreochromis niloticus*). *Food Chem.* **2022**, *382*, 132367. [CrossRef] [PubMed]
25. Delbarre-Ladrat, C.; Chéret, R.; Taylor, R.; Verrez-Bagnis, V. Trends in Postmortem Aging in Fish: Understanding of Proteolysis and Disorganization of the Myofibrillar Structure. *Crit. Rev. Food Sci. Nutr.* **2006**, *46*, 409–421. [CrossRef]
26. Yu, D.; Jiang, Q.; Xu, Y.; Xia, W. The shelf life extension of refrigerated grass carp (*Ctenopharyngodon idellus*) fillets by chitosan coating combined with glycerol monolaurate. *Int. J. Biol. Macromol.* **2017**, *101*, 448–454. [CrossRef]
27. Deng, Y.; Luo, Y.; Wang, Y.; Yue, J.; Liu, Z.; Zhong, Y.; Zhao, Y.; Yang, H. Drying-induced protein and microstructure damages of squid fillets affected moisture distribution and rehydration ability during rehydration. *J. Food Eng.* **2014**, *123*, 23–31. [CrossRef]
28. Vega-Gálvez, A.; Miranda, M.; Clavería, R.; Quispe, I.; Vergara, J.; Uribe, E.; Paez, H.; Di Scala, K. Effect of air temperature on drying kinetics and quality characteristics of osmo-treated jumbo squid (*Dosidicus gigas*). *LWT* **2011**, *44*, 16–23. [CrossRef]
29. Ortiz, J.; Lemus-Mondaca, R.; Vega-Gálvez, A.; Ah-Hen, K.; Puente-Diaz, L.; Zura-Bravo, L.; Aubourg, S. Influence of air-drying temperature on drying kinetics, colour, firmness and biochemical characteristics of Atlantic salmon (*Salmo salar L.*) fillets. *Food Chem.* **2013**, *139*, 162–169. [CrossRef]
30. Liang, Y.; Xie, Y.; Li, D.; Luo, Y.; Hong, H. Dynamics of water mobility, salt diffusion and hardness changes in bighead carp fillets during low-salting. *LWT* **2021**, *135*, 110033. [CrossRef]
31. Han, Z.; Zhang, J.; Zheng, J.; Li, X.; Shao, J.H. The study of protein conformation and hydration characteristics of meat batters at various phase transition temperatures combined with Low-field nuclear magnetic resonance and Fourier transform infrared spectroscopy. *Food Chem.* **2019**, *280*, 263–269. [CrossRef]
32. Zhang, L.; Xue, Y.; Xu, J.; Li, Z.; Xue, C. Effects of high-temperature treatment (≥100 °C) on Alaska Pollock (*Theragra chalcogramma*) surimi gels. *J. Food Eng.* **2013**, *115*, 115–120. [CrossRef]
33. Cheng, S.; Zhang, T.; Yao, L.; Wang, X.; Song, Y.; Wang, H.; Wang, H.; Tan, M. Use of low-field-NMR and MRI to characterize water mobility and distribution in pacific oyster (*Crassostrea gigas*) during drying process. *Dry Technol.* **2018**, *36*, 630–636. [CrossRef]

34. Barbosa, V.; Camacho, C.; Oliveira, H.; Anacleto, P.; Maulvault, A.L.; Delgado, I.; Ventura, M.; Dias, J.; Ribeiro, L.; Pousão-Ferreira, P.; et al. Physicochemical properties of iodine and selenium biofortified Sparus aurata and Cyprinus carpio during frozen storage. *Food Chem.* **2022**, *397*, 133780. [CrossRef] [PubMed]
35. Díaz, P.; Linares, M.B.; Egea, M.; Auqui, S.M.; Garrido, M.D. TBARs distillation method: Revision to minimize the interference from yellow pigments in meat products. *Meat Sci.* **2014**, *98*, 569–573. [CrossRef]
36. Zhang, Q.; Chen, X.; Ding, Y.; Ke, Z.; Zhou, X.; Zhang, J. Diversity and succession of the microbial community and its correlation with lipid oxidation in dry-cured black carp (*Mylopharyngodon piceus*) during storage. *Food Microbiol.* **2021**, *98*, 103686. [CrossRef]
37. Wang, Q.; Liu, B.; Cao, J.; Li, C.; Duan, Z. The impacts of vacuum microwave drying on osmosis dehydration of tilapia fillets. *J. Food Process Eng.* **2019**, *42*, e12956. [CrossRef]
38. Su, Y.; Zhang, M.; Bhandari, B.; Zhang, W. Enhancement of water removing and the quality of fried purple-fleshed sweet potato in the vacuum frying by combined power ultrasound and microwave technology. *Ultrason. Sonochem.* **2018**, *44*, 368–379. [CrossRef]
39. Zhao, C.C.; Benjakul, S.; Eun, J.B. Changes in protein compositions and textural properties of the muscle of skate fermented at 10 °C. *Int. J. Food Prop.* **2019**, *22*, 173–185. [CrossRef]
40. Yu, P.; Zhang, Z.; Tang, X.; Yu, D.; Jiang, Q.; Gao, P.; Yang, F. Effects of acidification and sterilisation on the quality of channel catfish (*Ietalurus punctatus*) fillets. *Int. J. Food Sci. Technol.* **2022**, *57*, 7296–7306. [CrossRef]
41. Han, M.; Wang, P.; Xu, X.; Zhou, G. Low-field NMR study of heat-induced gelation of pork myofibrillar proteins and its relationship with microstructural characteristics. *Int. Food Res. J.* **2014**, *62*, 1175–1182. [CrossRef]
42. Wei, H.; Rasul, M.G.; Sun, Z.; Yang, W.; Huang, T.; Yamashita, T.; Takaki, K.; Yuan, C. Study on nucleotide, myofibrillar protein, biochemical properties and microstructure of freeze-dried Scallop striated muscle during storage and rehydration. *Food Res. Int.* **2022**, *158*, 111461. [CrossRef] [PubMed]
43. Jiang, Q.; Han, J.; Gao, P.; Yu, L.; Xu, Y.; Xia, W. Effect of heating temperature and duration on the texture and protein composition of Bighead Carp (*Aristichthys nobilis*) muscle. *Int. J. Food Prop.* **2018**, *21*, 2110–2120. [CrossRef]
44. Zheng, H.; Han, M.; Bai, Y.; Xu, X.; Zhou, G. Combination of high pressure and heat on the gelation of chicken myofibrillar proteins. *Innov. Food Sci. Emerg. Technol.* **2019**, *52*, 122–130. [CrossRef]
45. Wang, L.; Shi, L.; Jiao, C.; Wu, W.; Li, X.; Wang, J.; Qiao, Y.; Liao, L.; Ding, A.; Xiong, G.; et al. Effects of different sugars on the thermal properties and microstructures of Mandarin fish (*Siniperca chuats*). *LWT* **2019**, *99*, 84–88. [CrossRef]
46. Bendall, J.R.; Restall, D.J. The cooking of single myofibres, small myofibre bundles and muscle strips from beef M. psoas and M. sternomandibularis muscles at varying heating rates and temperatures. *Meat Sci.* **1983**, *8*, 93–117. [CrossRef]
47. Cao, M.; Cao, A.; Wang, J.; Cai, L.; Regenstein, J.; Ruan, Y.; Li, X. Effect of magnetic nanoparticles plus microwave or far-infrared thawing on protein conformation changes and moisture migration of red seabream (*Pagrus major*) fillets. *Food Chem.* **2018**, *266*, 498–507. [CrossRef]

foods
MDPI

Article

Effects of Modified Atmosphere Packaging with Varied CO_2 and O_2 Concentrations on the Texture, Protein, and Odor Characteristics of Salmon during Cold Storage

Yun-Fang Qian [1,2,3,4], Cheng-Cheng Liu [1], Jing-Jing Zhang [1], Per Ertbjerg [4,*] and Sheng-Ping Yang [1,2,3,*]

1 College of Food Science & Technology, Shanghai Ocean University, Shanghai 201306, China
2 Shanghai Engineering Research Center of Aquatic Product Processing & Preservation, Shanghai Ocean University, Shanghai 201306, China
3 National Experimental Teaching Demonstration Center for Food Science and Engineering, Shanghai Ocean University, Shanghai 201306, China
4 Department of Food and Nutrition, University of Helsinki, 00014 Helsinki, Finland
* Correspondence: per.ertbjerg@helsinki.fi (P.E.); spyang@shou.edu.cn (S.-P.Y.); Tel.: +358-50-318909 (P.E.); +86-21-6190-0400 (S.-P.Y.)

Abstract: The effect of gas ratio on the growth of bacteria has been well demonstrated, but some adverse effects of modified atmosphere packaging (MAP) on seafoods have also been found. To provide a better understanding of the effects of CO_2 and O_2 concentrations (CO_2 from 40% to 100% and O_2 from 0% to 30%) in MAP on the texture and protein contents and odor characteristics of salmon during cold storage, the physiochemical, microbial, and odor indicators were compared with those without treatment (CK). Generally, MAP treatments hindered the increase of microbial counts, total volatile basic nitrogen, and TCA-soluble peptides, and decreased the water-holding capacity, hardness, springiness, and sarcoplasmic and myofibrillar protein contents. The results also indicated that $60\%CO_2/10\%O_2/30\%N_2$ was optimal and decreased the total mesophilic bacterial counts by 2.8 log cfu/g in comparison with CK on day 12. In agreement, the concentration of CO_2 of 60% showed the lowest myofibrillar protein degradation, and less subsequent loss of hardness. The electronic nose characteristics analysis indicated that $60\%CO_2/20\%O_2/20\%N_2$ and $60\%CO_2/10\%O_2/30\%N_2$ had the best effect to maintain the original odor profiles of salmon. The correlation analysis demonstrated that microbial growth had a strong relationship with myofibrillar and sarcoplasmic protein content. It can be concluded that $60\%CO_2/10\%O_2/30\%N_2$ displayed the best effect to achieve the goal of preventing protein degradation and odor changes in salmon fillets.

Keywords: salmon; modified atmosphere packaging; protein content; odor characteristics; correlation analysis

Citation: Qian, Y.-F.; Liu, C.-C.; Zhang, J.-J.; Ertbjerg, P.; Yang, S.-P. Effects of Modified Atmosphere Packaging with Varied CO_2 and O_2 Concentrations on the Texture, Protein, and Odor Characteristics of Salmon during Cold Storage. *Foods* **2022**, *11*, 3560. https://doi.org/10.3390/foods11223560

Academic Editor: Jingran Bi

Received: 30 September 2022
Accepted: 7 November 2022
Published: 9 November 2022

1. Introduction

Salmon (*Salmo salar*) is an anadromous fish that is mainly produced in the Arctic and Atlantic Oceans and is favored by consumers worldwide. In recent years, salmon has become increasingly popular for its high nutritional value and good taste [1]. However, salmon is extremely vulnerable to spoilage during the period of storage and transportation. The spoilage is mainly dominated by microbial activities, consequently resulting in organoleptic rejection and economic loss [2].

To maintain the quality of salmon and prolong the shelf life, techniques are required with safety, eco-friendly, and convenient properties. Modified atmosphere packaging (MAP) is one of the most used physical techniques to prolong the shelf life of aquatic products by inhibiting microbial growth and delaying protein degradation [3,4]. The gas mixture in MAP is usually composed of carbon dioxide (CO_2), nitrogen (N_2), and oxygen (O_2), in which CO_2 is considered to have bacteriostatic and fungistatic properties, N_2 to balance the gas composition and to prevent packaging collapse, and O_2 to inhibit the growth of anaerobic microorganisms and reduce the formation of trimethylamine in fresh fish [5,6]. Most studies are focused on bacterial growth and microbial community [7–9], and it is considered that the bacteriostatic effect of MAP increases with an increasing concentration of CO_2 [10]. Therefore, MAP with high concentrations of CO_2 is generally applied for the chilled storage of animal products [11]. A modified atmosphere containing 96% CO_2 was reported to limit microbial growth in Atlantic salmon [7]. In another study, freezing treatment with an atmosphere of 60% CO_2 and 40% N_2 could not only inhibit the growth of *Photobacterium phosphoreum*, but could also inhibit the production of biogenic amines [2]. Recent studies also revealed that high concentrations of CO_2 displayed some adverse effects of MAP on the sensory and physical characteristics of seafoods [12]. However, only few studies focused on the mechanisms of how the atmosphere composition affects the texture, protein, and odor characteristics of salmon fillets. Due to the importance of sensory quality affecting consumers' decisions, it is necessary to systematically study the effect of different concentrations of CO_2 and O_2 on the muscle and odor quality of salmon.

Therefore, a better understanding of how the atmosphere affects the quality deterioration of salmon and how to optimize the gas mixture of MAP is needed. The aim of this study was to investigate the influences of different concentrations of CO_2 and O_2 on the texture, muscle protein, and odor characteristics of salmon at 4 °C, and the correlation between texture, protein content, and microbial growth was also analyzed.

2. Materials and Methods

2.1. Sample Preparation

The original salmon (*Salmo salar*) samples (each one about 6 kg) were purchased from a local supermarket (place of origin: Norway), where the salmon samples were beheaded, skinned, and deboned by skillful staff in the workshops of the supermarket. The fish was cut into pieces each weighing about 100 g, and 49 pieces were selected randomly for further study. Then they were transferred to the laboratory in an incubator with ice slurry within one hour. After arrival, the samples were washed in cold water, and then the drips were drained off before putting them into polyamide/polyethylene pouches (PA/PE, Xilong Packaging Co., Ltd., Shijiazhuang, China; size: 32×22 cm; thickness: 0.32 mm; oxygen permeability (O_2P): 7 $cm^3/(m^2 \cdot 24\ h \cdot 0.1\ MPa)$) randomly. To investigate the effect of CO_2 and O_2 on the texture, protein, and odor characteristics of salmon, the pouches were separated into eight different batches, settled as following (Figure 1): (1) CK: air, ambient; (2) 100/0/0: 100%CO_2; (3) 80/0/20: 80%CO_2/20%N_2; (4) 60/0/40: 60%CO_2/40%N_2; (5) 40/0/60: 40%CO_2/60%N_2; (6) 60/10/30: 60%CO_2/10%O_2/30%N_2; (7) 60/20/20: 60%CO_2/20%O_2/20%N_2; (8) 60/30/10: 60%CO_2/30%O_2/10%N_2.

The MAP pouches were emptied by vacuum, and then the gas mixtures were flushed into them, and, finally, they were heat-sealed by gas-flushing equipment (Model BQ-360W, Shanghai Qingba Food Packaging Machinery Co., Ltd., Shanghai, China) with a gas volume-to-product (G/P) ratio of 3:1. Finally, all samples were stored at 4 °C for 12 days.

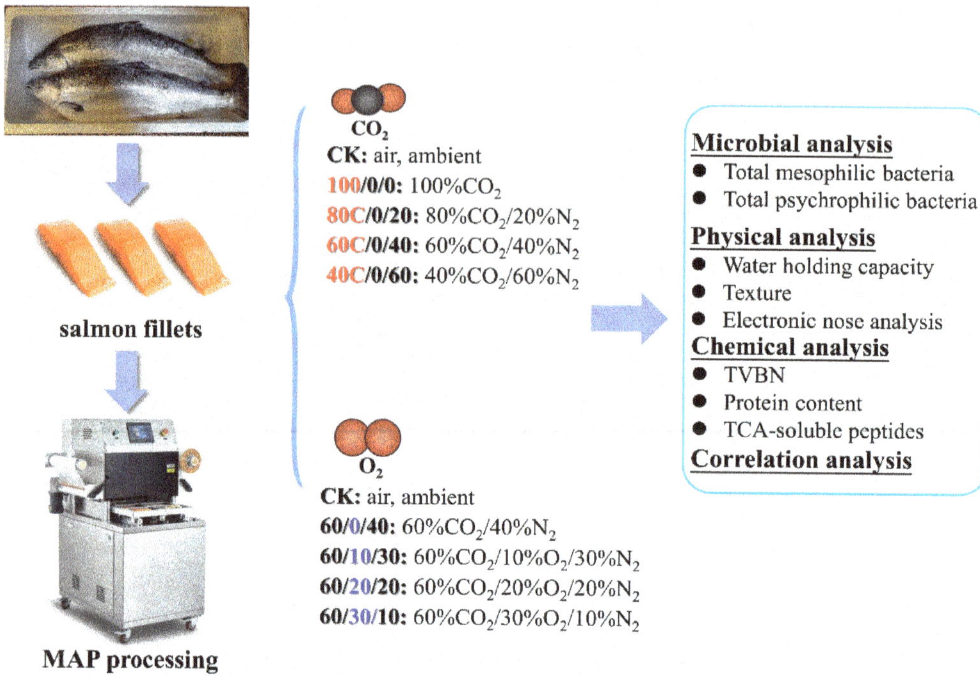

Figure 1. A schematic overview of the experimental study.

2.2. Microbiological Analysis

The bacterial counts were measured according to the method described by Yu et al. [13] with some modifications. A sample (25 g) was mixed with 225 mL sterilized salines in a sterile plastic bag and homogenized for 1 min to 2 min. After dilution in 10-folded series, 1 mL of an appropriate dilution was added into Plate Count Agar (PCA, No. HB0101, Qingdao Hope Biol-Technology Co., Ltd., Qingdao, China) with 5% NaCl (Sangon Biotech Co., Ltd. Shanghai, China). The total mesophilic bacterial count (TMB) was incubated on PCA at 30 °C for two days. Next, the total psychrophilic bacteria (TPB) were cultivated at 4 °C for ten days.

2.3. Determination of Total Volatile Basic Nitrogen (TVB-N)

The total volatile basic nitrogen (TVB-N) value was measured according to the previous studies by an Automatic Kjeldahl Apparatus (KjeltecTM8400; FOSS Quality Assurance Co., Ltd., Copenhagen, Denmark) [14]. The value of the TVB-N was expressed as mg N/100 g salmon. Each sample was tested in triplicate.

2.4. Determination of Water-Holding Capacity (WHC)

The measurement method in this study was developed by Bao et al. [15] and Liu et al. [16]. In brief, the sample was cut into small pieces (weight about 2 g, recorded as M_1), and then the sample was weighed again after centrifugation at 1590 g for 10 min, and the value was recorded as M_2. Triplicate samples were calculated by the following formula:

$$WHC = \frac{M_2}{M_1} \times 100\% \quad (1)$$

2.5. Texture Profile Analysis (TPA)

The texture test was conducted according to Wang et al. [17] by a TA.XT Plus analyzer (Stable Micro Systems, Ltd., Surrey, UK) equipped with a P/5 probe. The fresh sample was cut into small squares (20 mm × 20 mm × 10 mm). The parameters of hardness, chewiness, springiness, and gumminess were acquired at a speed of 1 mm/s and a sample deformation of 50%. Each sample was tested six times.

2.6. Determination of TCA-Soluble Peptide Content

The TCA-soluble peptide content determination was performed according to the method of Zhuang et al. [18] with some modifications. The minced sample (2 g) was homogenized with 18 mL 5% (*w/v*) trichloroacetic acid (TCA) solution, and incubated at 4 °C for 1 h. Then the sample was centrifuged at 10,000× *g* for 5 min. The supernatant was collected. The content of TCA-soluble peptide was determined by Lowry's method, and was expressed as μmol tyrosine/g muscle. Each sample was tested three times.

2.7. Determination of Protein Content

The myofibrillar protein was extracted according to the method of Lv et al. [19]. About 2 g minced fish sample was homogenized with 20 mL Tris-maleate buffer A (20 mmol/L pH 7.0, 0.05 mol/L KCl) at 4 °C. Then, the solution was centrifuged for 15 min at a speed of 10,000× *g*. The supernatant containing the sarcoplasmic protein was collected. The precipitate was homogenized with 20 mL Tris-maleate buffer B (20 mmol/L, pH 7.0, containing 0.6 mol/L KCl). The solution was incubated for 1 h at 4 °C. After centrifugation at 10,000× *g* for 15 min, the supernatant containing the salt-soluble myofibrillar protein was obtained. The Biuret method was used to determine the content of protein using bovine serum albumin as standard, and the result was expressed as mg/g. All samples were repeated three times.

2.8. Electronic Nose Analysis

The electronic nose (Alpha MOS, Toulouse, France) consisted of an auto-sampling device, a detector unit containing 18 metal oxide sensors (MOSs), and pattern recognition software for data recording, and interpretation was used to provide flavor attributes of the salmon during storage. The electronic nose sensor signals were measured by the method of Zhang et al. [20]. Each sample was mixed with 0.18 g/mL NaCl solution at a mass ratio of 1:1. Before the test, the salmon sample was put into a 20 mL airtight vial and incubated at 45 °C for 50 min. The test parameters were settled with the measurement time for 120 s and the standby procedure for 500 s. All samples were repeated six times, and only the last five pieces of data of each sample were used for further analysis to eliminate the impact from the previous group with different treatments. The signals were further analyzed by the system (Alpha MOS, Toulouse, France) to distinguish discrepancies between the odor profiles of the groups with different gas ratios on different days, and the principal component analysis diagrams (PCA) were outputted by the system.

2.9. Statistic Analysis

The analyses described above were carried out on days 0, 2, 4, 6, 8, 10, and 12. The one-way ANOVA procedure with the package type ((1) CK: air, ambient; (2) 100/0/0: 100%CO_2; (3) 80/0/20: 80%CO_2/20%N_2; (4) 60/0/40: 60%CO_2/40%N_2; (5) 40/0/60: 40%CO_2/60%N_2; (6) 60/10/30: 60%CO_2/10%O_2/30%N_2; (7) 60/20/20: 60%CO_2/20%O_2/20%N_2; (8) 60/30/10: 60%CO_2/30%O_2/10%N_2.), followed by Duncan's test, was carried out for multiple comparisons by the SPSS 20.0 software (SPSS Version 20.0, Inc., Chicago, IL, USA), and significance was judged at 0.05. The diagrams of the changes in the indicators of samples under different atmospheres against storage time were designed by the Origin2018 software (OriginLab, Northampton, MA, USA). The results are expressed as means ± standard deviation. The relationship between indicators was assessed by Pearson correlation analysis, and the diagram was also designed by the Origin2018 software (OriginLab, Northampton, MA, USA).

3. Results and Discussion

3.1. Microbial Growth

The effect of different concentrations of CO_2 and O_2 on the growth of mesophilic bacteria in salmon is shown in Figure 2A. The initial TMB count was 3.79 log cfu/g, indicating that the sample was fresh. The TMB counts increased continuously during storage in all packaging conditions. All MAP groups had significantly lower TMB counts than the sample under air after 4 days of storage. The TMB counts of the control were close to 7 log cfu/g on day 6 and reached 9.20 log cfu/g at the end of storage, whereas most samples with 60% CO_2 did not exceed the threshold during the whole period, except for 60%CO_2/30%O_2/10%N_2. It is found in Figure 2A that MAP with CO_2 higher than 40% effectively inhibited the growth of the mesophilic bacteria, which was also stated by Fernandez et al. [10]. Similarly, it is reported that MAP with CO_2 higher than 40% could prolong the shelf life of Pacific white shrimp effectively [21]. However, higher CO_2 did not always lead to lower bacterial counts, because 100%CO_2 and 80%CO_2/20%N_2 had higher TMB counts than 60%CO_2/40%N_2. It was hypothesized that high concentrations of CO_2 might help to increase the solubility of the muscle protein, consequently providing nutrients for bacterial growth [22]. As shown in Figure 2A, different concentrations of O_2 also affected the increase of TMB counts. It is known that the existence of O_2 can slow the growth of anaerobic bacteria [23]. However, our study showed that the TMB counts of 60%CO_2/30%O_2/10% N_2 were higher than 60%CO_2/40%N_2, 60%CO_2/10%O_2/30%N_2, and 60%CO_2/20%O_2/20%N_2, indicating that 30%O_2 could promote bacterial growth. Therefore, a MAP with O_2 higher than 20% was not recommended according to the results.

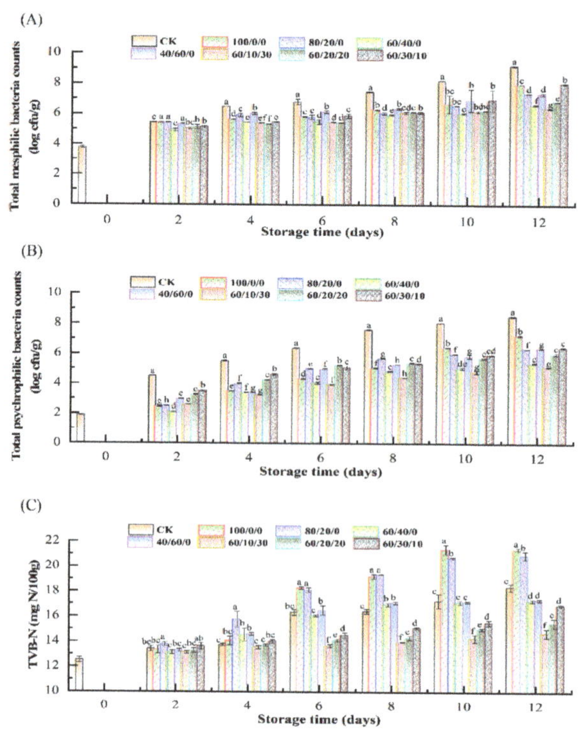

Figure 2. Changes of total mesophilic bacteria (**A**), the total psychrophilic bacteria (**B**), and TVB-N (**C**) in salmon fillets under MAP containing different CO_2/O_2/N_2 ratios during storage at 4 °C. Different lowercase letters (a–h) represent significant differences ($p < 0.05$) between different groups at the same storage period.

The changes of total psychrophilic bacteria count of salmon packaged under different concentrations of CO_2 and O_2 are shown in Figure 2B. The TPB counts of the control rose from 1.87 log cfu/g to 8.52 log cfu/g during storage. Compared with the results of TMB, it is demonstrated that the proportion of psychrophilic bacteria to mesophilic bacteria was low initially, but increased during storage, in agreement with the study of Yang et al. [24]. The pattern change of TPB was similar to TMB, and $60\%CO_2/10\%O_2/30\%N_2$ had the lowest TPB counts. Therefore, $60\%CO_2/10\%O_2/30\%N_2$ was recommended according to the results of bacterial growth.

3.2. Changes of TVB-N Content

Animal products rich in protein, amino acid, and other nutrients can easily be decomposed into ammonia and other basic volatile substances, which could be detected as total volatile basic nitrogen (TVB-N) content [25]. Figure 2C display the changes of TVB-N content in salmon under different modified atmosphere packaging. The TVB-N of all groups increased gradually during storage. The samples of $100\%CO_2$ and $80\%CO_2/20\%N_2$ had higher TVB-N contents than other groups after 6 days of storage. The results indicated that CO_2, when higher than 80%, might promote the accumulation of TVB-N. Compared with the changes of bacterial counts, it is found that the changes of TVB-N were consistent with the bacterial counts. It is also observed that $60\%CO_2/10\%O_2/30\%N_2$, $60\%CO_2/20\%O_2/20\%N_2$, and $60\%CO_2/30\%O_2/10\%N_2$ had lower TVB-N contents than CK and $60\%CO_2/40\%N_2$, indicating that O_2 was also beneficial for inhibiting the generation of TVB-N. However, the inhibitory effect of O_2 decreased with its increasing concentration. This phenomenon might be contributed to the microbial growth [26].

3.3. Changes of Water-Holding Capacity

The water-holding capacity (WHC) of all samples displayed a decreasing tendency as the storage time increased (Figure 3A). The WHC of CK decreased from 83.01% to 74.31%. It was found that CK had lower WHC than MAP groups except for $100\%CO_2$ and $80\%CO_2/20\%N_2$. An adverse effect of CO_2 on WHC was observed in this study, which might be associated with the solubilization of CO_2 in the muscle, leading to muscle shrinkage and a decreased ability to retain water [26]. Another reason might be that CO_2 promoted proteolytic degradation. The samples of $60\%CO_2/10\%O_2/30\%N_2$ and $60\%CO_2/20\%O_2/20\%N_2$ had the highest level of WHC, indicating low concentrations of O_2 would be beneficial for maintaining the WHC of salmon fillets. However, atmospheres with O_2 concentrations higher than 20% might accelerate protein oxidation and result in decreased WHC [27].

3.4. Changes of Texture Properties

The texture profiles of salmon during MAP storage are shown in Figure 3B–E. As shown in Figure 3B, the hardness of salmon reduced gradually during storage time, which may be attributed to the activity of endogenous and microbial proteolytic enzymes [28]. The hardness of CK samples was at the lowest level at the later storage times. As shown in Figure 3B, it was found that the hardness of sample 100/0/0 was lower than other MAP groups, especially on day 6, indicating that 100% CO_2 had an adverse effect on the hardness of salmon. However, CO_2 at 80% and 60% inhibited the decrease of hardness. The modified atmosphere with different concentrations of O_2 also affected the hardness (Figure 3B). The samples of $60\%CO_2/10\%O_2/30\%N_2$ and $60\%CO_2/20\%O_2/20\%N_2$ had higher hardness than other groups, and the hardness of $60\%CO_2/30\%O_2/10\%N_2$ was relatively lower, which was coherent with the results of microbial growth.

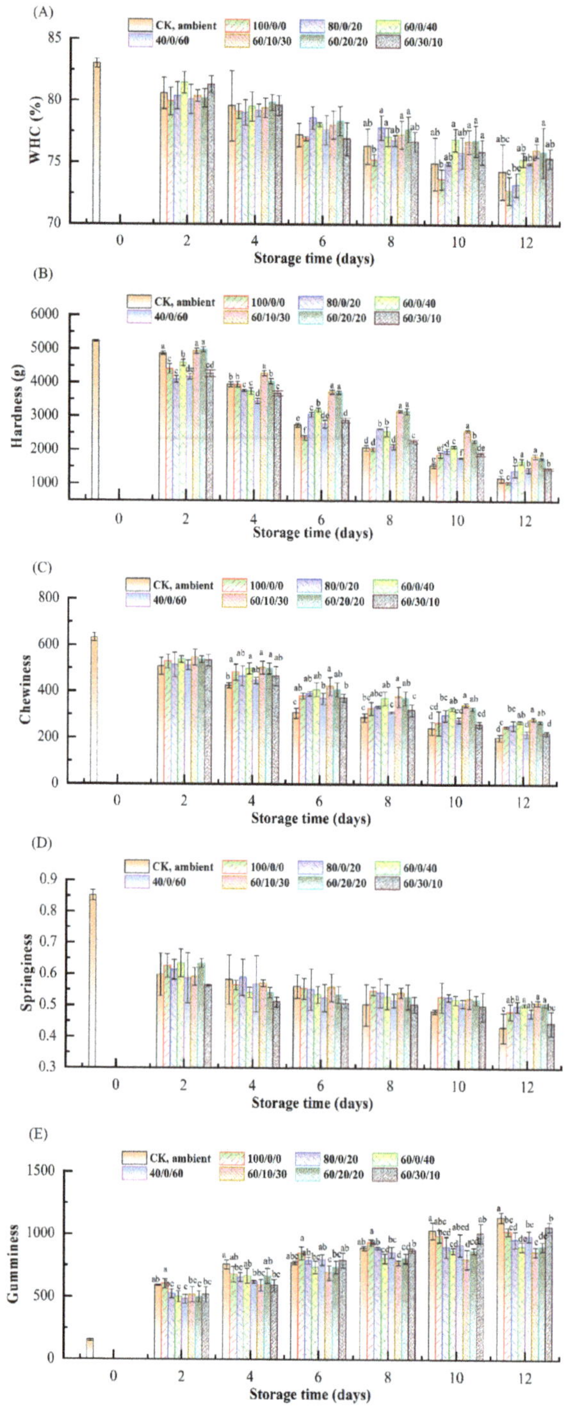

Figure 3. Changes of WHC (**A**), hardness (**B**), chewiness (**C**), springiness (**D**), and gumminess (**E**) of salmon fillets under MAP containing different $CO_2/O_2/N_2$ ratios at 4 °C. Different lowercase letters (a–d) represent significant differences ($p < 0.05$) between different groups at the same storage period.

The results of the chewiness and springiness of salmon during storage are displayed in Figure 3C,D, respectively. It was observed that both the chewiness and springiness decreased during storage. The samples of 60%CO_2/10%O_2/30%N_2 and 60%CO_2/20%O_2/20%N_2 had the highest values of chewiness, meaning a lower level of O_2 would be suitable for maintaining the chewiness of salmon. The initial springiness of fresh salmon was 0.85, and the control decreased to 0.56 on day 6, and even reached 0.43 on day 12. A sharp decrease of springiness could be observed on day 2, and the samples of 60%CO_2/10%O_2/30%N_2 and 60%CO_2/10%O_2/30%N_2 held the highest springiness on day 12. The results indicated that MAP with 60% CO_2 and O_2 below 20% maintained the chewiness and springiness of salmon fillets most effectively, and higher CO_2 and O_2 provided an adverse effect.

The gumminess of salmon fillets increased continuously with the extension of storage time (Figure 3E), which likely was mainly due to the microbial growth [29]. Significant differences were observed between CK and MAP groups at the end of storage, probably because MAP effectively inhibited microbial growth and reproduction. Both CO_2 and O_2 had some impact on the gumminess of salmon fillets. Compared with the groups with different gas mixtures, the results indicated that 60% CO_2 was more appropriate to slow the increase of gumminess, whereas 10% O_2 had a better effect than 0%, 20%, and 30%. Therefore, the concentrations of CO_2 should not be higher than 60%, and a low level of O_2 (~10%) was more beneficial for maintaining the gumminess of salmon fillets.

3.5. Changes of TCA-Soluble Peptide Content

The content of TCA-soluble peptides is an important indicator of protein degradation in fish flesh. As shown in Figure 4A, the TCA-soluble peptide content of the fresh sample was 1.44 μmol tyrosine/g. With increased storage time, the content of TCA-soluble peptides of CK increased to 10.66 μmol tyrosine/g on day 12. The content of TCA-soluble peptides in 100C was higher than other MAP groups at the end of storage. This phenomenon is probably related to the disruption of muscle cells by CO_2, leading to an increase of proteolytic activity. The content of TCA-soluble peptides of MAP groups with different concentrations of O_2 increased much more slowly. Among them, the sample of 60%CO_2/10%O_2/30%N_2 had the lowest TCA-soluble peptide content, and, thus, only reached 6.12 μmol tyrosine/g on day 12. It is reported that the protein degradation is not only induced by endogenous enzymes [30,31], but is also caused by spoilage bacteria [32]. Therefore, the lower content of TCA-soluble peptides found in the sample of 60%CO_2/10%O_2/30%N_2 could be mainly attributed to its antimicrobial effect, consistent with the results of microbial growth.

3.6. Changes of Protein Contents

Salmon flesh is mainly composed of sarcoplasmic and myofibrillar protein, which account for almost 85% of the total protein contents [33,34]. Therefore, the changes of sarcoplasmic and myofibrillar protein contents were measured and the results are displayed in Figure 4B,C.

The results show that the sarcoplasmic protein (SP) contents in all packaging systems decreased during storage. It is well known that sarcoplasmic proteins, to a large extent, are composed of small and intermediate size proteins, such as some water-soluble enzymes [35]. Our results showed that samples stored without oxygen under 100%, 80%, and 40% CO_2 had lower contents of sarcoplasmic protein than those under 60% CO_2, indicating that 60% CO_2 balanced the anti-bacterial activity and anti-hydrolysis activity. Moreover, the contents of sarcoplasmic protein also decreased with the increasing O_2 concentrations, and 60%CO_2/10%O_2/30%N_2 and 60%CO_2/40%N_2 had the highest contents of sarcoplasmic protein compared to other groups. This phenomenon should be attributed to the lower microbial metabolic activity in these samples.

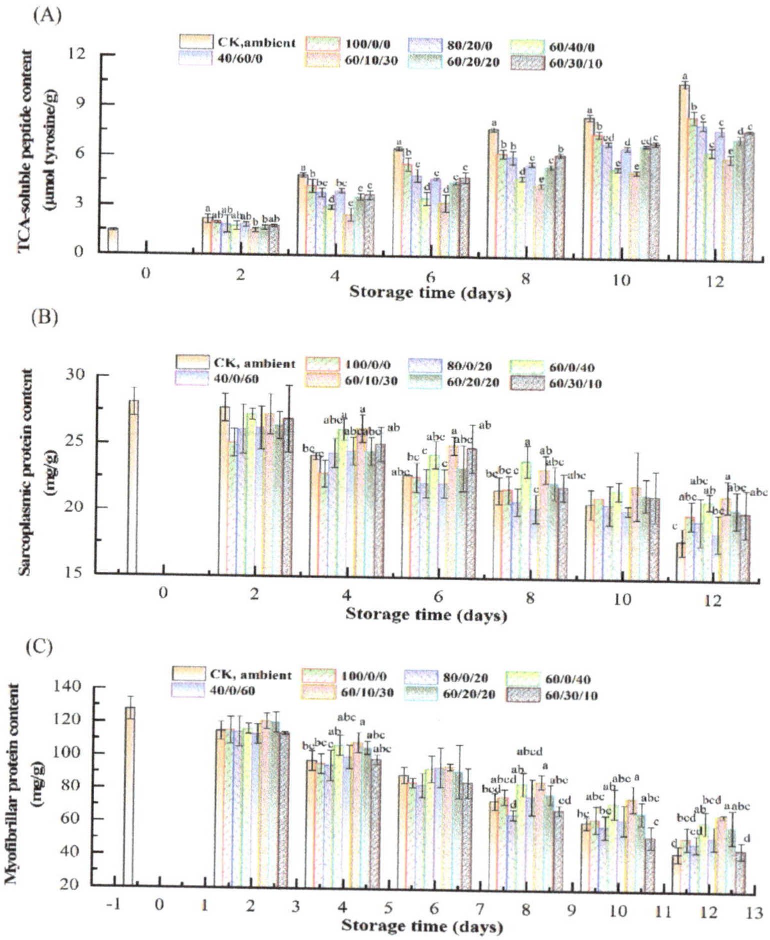

Figure 4. Changes of TCA-soluble peptides (**A**), sarcoplasmic protein content (**B**), and myofibrillar protein content (**C**) of salmon under MAP containing different $CO_2/O_2/N_2$ ratios at 4 °C. Different lowercase letters (a–c) represent significant differences ($p < 0.05$) between different groups at the same storage period.

As shown in Figure 4C, the content of myofibrillar protein (MP) soluble at high ionic strength in all groups decreased gradually, especially in CK. Myofibrillar protein is known as an important component of the muscle fiber, which is composed of salt-soluble proteins such as myosin, actin, and troponins [36]. The initial content of myofibrillar protein in fresh salmon was 128 mg/g, but it decreased sharply to 89 mg/g after 6 days of storage, and even to 43 mg/g at the end of storage in CK. The decrease of myofibrillar protein content should be attributed to the aggregation, degradation, and oxidation processes [37]. The content of myofibrillar protein in samples stored under 60% CO_2 was maintained at a higher level. This was probably mainly because this MAP system could effectively maintain the protein content by inhibiting microbial growth [38].

3.7. Changes of Odor Characteristics

In this study, the electronic nose was used to analyze the odor characteristics of salmon fillets during the MAP storage period. The principal component analysis (PCA) was used to distinguish electronic nose sensor response signals of the samples on day 6 and day 12 (Figure 5). The results are presented in the form of a two-dimensional scatter diagram on the basis of the original information system and feature vectors in linear classification after dimensionality reduction. The different index (DI) was used to reflect the overall differentiation effect of odor characteristics, and its range was generally from 0 to 100, with 80 as the distinguishing threshold [39,40]. The figures also indicate that the first two principal component axes (PC1 and PC2) and the spatial distribution areas reflected the differences between samples. The DI of Figure 5A,B were 82 and 84, respectively. PC1 and PC2 for Figure 5A were 99.46% and 0.35%, whereas those for Figure 5B were 99.43% and 0.40%, respectively. Therefore, the PCA results covered almost all of the varied information, and the results could be distinguished effectively. As shown in Figure 5A, the plots of salmon fillets stored under 100%CO_2 after 6 days of storage were distant from the initial samples, the control, and other MAP groups, whereas the sample stored under 60%CO_2/20%O_2/20%N_2 was clustered close to CK-0d (fresh sample), suggesting similar odor profiles. Therefore, 100%CO_2 atmosphere packaging might change the intrinsic odor characteristics of salmon. After 12 days of storage, the plots of all samples were more distant from the fresh sample (CK-0d) (Figure 5B), and the plots of 60/20/20-12d and 60/10/30-12d were much closer to CK-0d than other samples. Therefore, 60%CO_2/20%O_2/20%N_2 and 60%CO_2/10%O_2/30%N_2 displayed a better effect to maintain the original odor profiles of salmon. The changes of odor in samples stored under higher concentrations of CO_2 and lower concentrations of O_2 might be attributed to the growth of lactic bacteria, which are anaerobic, gram-positive, and CO_2-resistant species, and can provide sour off-odors [41].

(A)

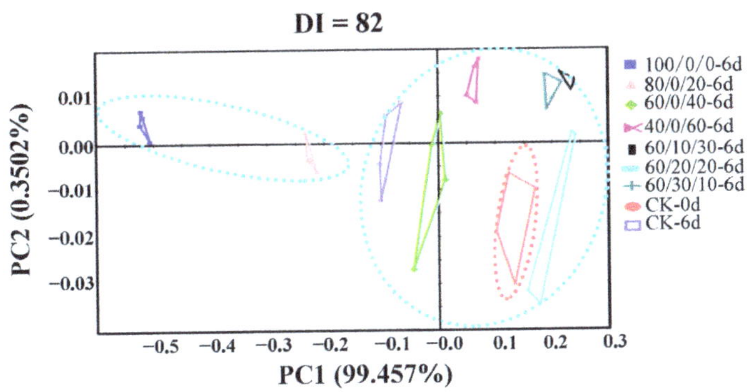

(B)

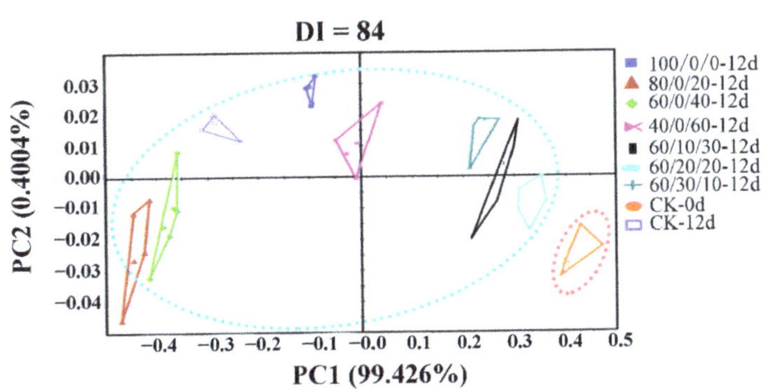

Figure 5. PCA plots of the volatile odor characteristics by electronic analysis of salmon under MAP containing different CO_2/O_2/N_2 ratios at 4 °C on day 6 (**A**) and day 12 (**B**) in comparison with CK-0d.

3.8. Correlation Analysis

The results of the correlation analysis between protein changes, microbial growth, and texture properties are shown in Figure 6. The results show that TMB, TPB, the contents of TCA-soluble peptides, myofibrillar protein, and sarcoplasmic protein all had significant correlations with hardness, springiness, chewiness, gumminess, and WHC. The absolute value of Pearson-related coefficients between TMB/TPB and myofibrillar/sarcoplasmic protein/TCA-soluble peptides was even higher than 0.8, indicating that the microbial growth had a strong relationship with the myofibrillar and sarcoplasmic protein content. Moreover, WHC seemed to have a stronger relationship with chewiness, hardness, and myofibrillar protein content than gumminess, springiness, sarcoplasmic protein, and TCA-soluble peptide content. Therefore, the intact structure of the myofibrillar protein was essential for maintaining WHC, chewiness, and hardness.

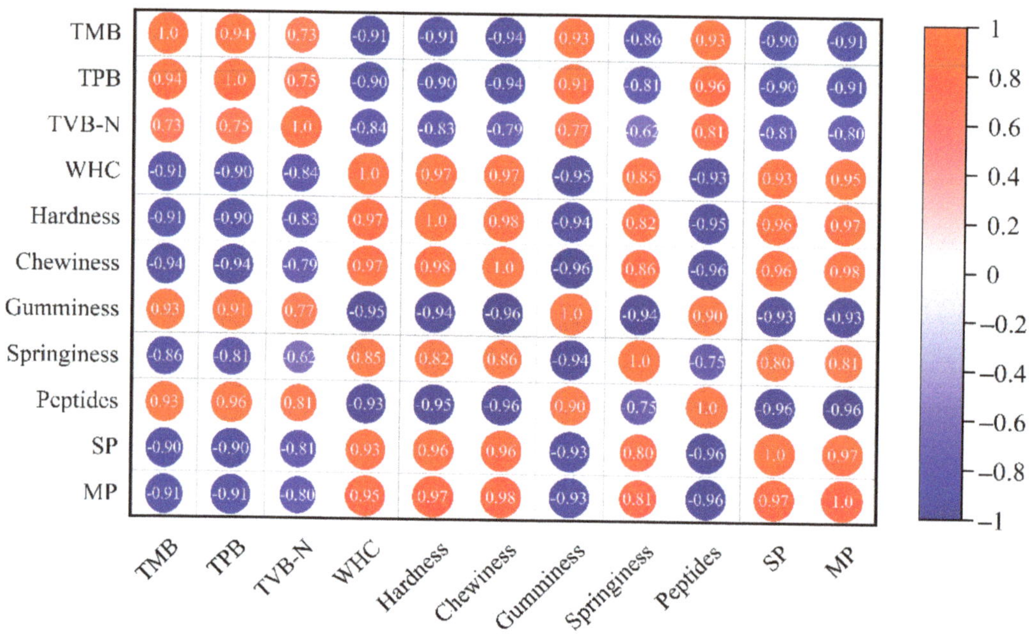

Figure 6. Heatmap of Pearson correlation between the TMB, TPB, TVB-N, WHC, hardness, chewiness, gumminess, springiness, TCA-soluble peptides, sarcoplasmic and myofibrillar protein content in salmon under modified atmosphere packaging. The circle size represents the absolute value of the correlation coefficient, where red represents positive correlation, and blue represents negative correlation.

The effect of CO_2 concentrations on the hardness, WHC, myofibrillar protein content, and TPB of salmon fillets are shown in Figure 7. The values plotted were on 6 d and 12 d with CO_2 concentrations from 40% to 100% and 0% O_2. It could be observed that hardness increased with an increasing CO_2 concentration when the CO_2 concentration was up to 60%. However, when the CO_2 concentration was higher than 60%, increasing CO_2 would result in decreased hardness. A similar tendency was found in WHC and myofibrillar protein content. It is hypothesized that high CO_2 contributed to the changes in the microstructure of the muscle [42]. The increase of CO_2 concentration led to a lower count of TPB when CO_2 was below 60%, which could be attributed to the antibacterial effect of CO_2. However, CO_2 higher than 60% had an adverse effect on TPB. Therefore, 60% CO_2 was recommended to maintain the quality of salmon fillets.

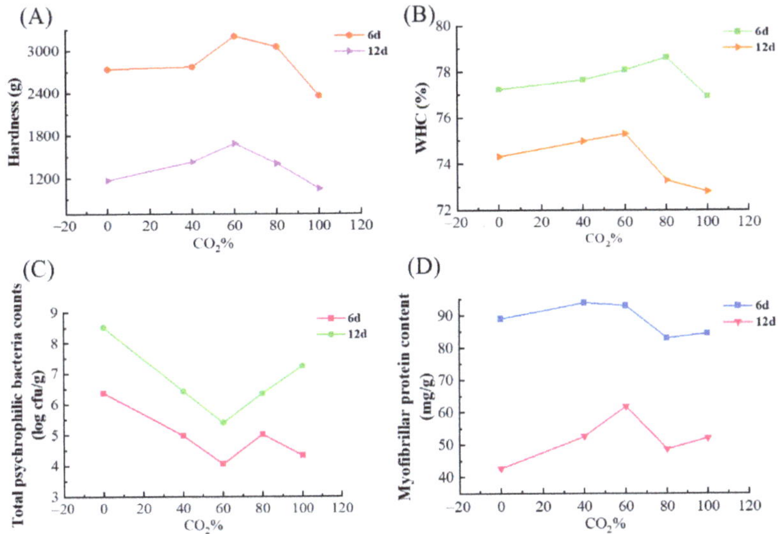

Figure 7. Relationship between CO_2 concentrations and hardness (**A**), WHC (**B**), TPB (**C**), and myofibrillar protein content (**D**) of salmon fillets stored under modified atmosphere packaging containing O_2 of 0% at 4 °C on 6 d and 12 d.

4. Conclusions

In this study, the results showed that increased CO_2 from 40% to 100% and decreased O_2 from 30% to 0% could effectively inhibit the growth of mesophilic and psychrophilic bacteria of salmon fillets from day 4 to day 10. However, 100% CO_2 and 80% CO_2 also led to a reduction of hardness, WHC, and myofibrillar protein content in comparison with 60% CO_2. Moreover, CO_2 higher than 80% might also promote the accumulation of TVB-N in salmon. Additionally, O_2 higher than 20% also possibly affected protein degradation and subsequent muscle texture deterioration. The E-nose analysis indicated that 60%CO_2/20%O_2/20%N_2 and 60%CO_2/10%O_2/30%N_2 displayed the best effect to maintain the original odor profiles of salmon. The microbial growth had a strong negative correlation with myofibrillar and sarcoplasmic protein content. A stronger positive relationship between WHC and chewiness, hardness, and myofibrillar protein content was also found in the results. In conclusion, the content of myofibrillar protein is essential for the texture and water-holding capacity. Accordingly, a high CO_2 content of 60% and low O_2 of 10% in MAP is recommended to maintain the quality of salmon.

Author Contributions: Y.-F.Q.: conceptualization, writing—review and editing, funding acquisition; C.-C.L.: experiment, data curation, writing—original draft; J.-J.Z.: experiment; P.E.: methodology, writing—review and editing; S.-P.Y.: methodology, investigation, project administration. All authors have read and agreed to the published version of the manuscript.

Funding: This work was funded by China Scholarship Council (No: 202008310018), National Natural Science Foundation of China (No: 31501551), and Special Fund for the Development of Science and Technology of Shanghai Ocean University (No: A2-2006-20-200203).

Institutional Review Board Statement: Not applicable.

Informed Consent Statement: Not applicable.

Data Availability Statement: Data are contained within the article.

Acknowledgments: Open access funding provided by University of Helsinki.

Conflicts of Interest: The authors declare that they have no conflict of interest to this work.

References

1. Badiola, M.; Gartzia, I.; Basurko, O.C.; Mendiola, D. Land-based growth of Atlantic salmon (*Salmo salar*) and consumers' acceptance. *Aquacult. Res.* **2017**, *48*, 4666–4683. [CrossRef]
2. Emborg, J.; Laursen, B.; Rathjen, T.; Dalgaard, P. Microbial spoilage and formation of biogenic amines in fresh and thawed modified atmosphere-packed salmon (*Salmo salar*) at 2 °C. *J. Appl. Microbiol.* **2002**, *92*, 790–799. [CrossRef] [PubMed]
3. Zhang, X.; Huang, W.; Xie, J. Effect of different packaging methods on protein oxidation and degradation of grouper (*Epinephelus coioides*) during refrigerated storage. *Foods* **2019**, *8*, 325. [CrossRef]
4. Eduardo, E.; Luis, G.; Jaime, A. Effects of vacuum and modified atmosphere packaging on the quality and shelf-life of gray triggerfish (*Balistes capriscus*) fillets. *Foods* **2021**, *10*, 250. [CrossRef]
5. Daniels, J.A.; Krishnamurthi, R.; Rizvi, S.S.H. A review of effects of carbon dioxide on microbial growth and food quality. *J. Food Prot.* **1985**, *48*, 532–537. [CrossRef] [PubMed]
6. Couvert, O.; Divanac'h, M.-L.; Lochardet, A.; Thuault, D.; Huchet, V. Modelling the effect of oxygen concentration on bacterial growth rates. *Food Microbiol.* **2019**, *77*, 21–25. [CrossRef]
7. Milne, D.; Powell, S.M. Limited microbial growth in Atlantic salmon packed in a modified atmosphere. *Food Control* **2014**, *42*, 29–33. [CrossRef]
8. Powell, S.M.; Tamplin, M.L. Microbial communities on Australian modified atmosphere packaged Atlantic salmon. *Food Microbiol.* **2012**, *30*, 226–232. [CrossRef]
9. Powell, S.; Ratkowsky, D.; Tamplin, M. Predictive model for the growth of spoilage bacteria on modified atmosphere packaged Atlantic salmon produced in Australia. *Food Microbiol.* **2015**, *47*, 111–115. [CrossRef]
10. Fernandez, K.; Aspe, E.; Roeckel, M. Shelf-life extension on fillets of Atlantic Salmon (*Salmo salar*) using natural additives, superchilling and modified atmosphere packaging. *Food Control.* **2009**, *20*, 1036–1042. [CrossRef]
11. Singh, P.; Abas Wani, A.; Saengerlaub, S.; Langowski, H.C. Understanding critical factors for the quality and shelf-life of MAP fresh meat: A review. *Crit. Rev. Food Sci. Nutr.* **2011**, *51*, 146–177. [CrossRef] [PubMed]
12. Bono, G.; Badalucco, C.V.; Cusumano, S.; Palmegiano, G.B. Toward shrimp without chemical additives: A combined freezing-MAP approach. *LWT* **2012**, *46*, 274–279. [CrossRef]
13. Yu, Y.J.; Yang, S.P.; Lin, T.; Qian, Y.F.; Xie, J.; Hu, C. Effect of cold chain logistic interruptions on lipid oxidation and volatile organic compounds of salmon (*Salmo salar*) and their correlations with water dynamics. *Front. Nutr.* **2020**, *7*, 155. [CrossRef] [PubMed]
14. Qian, Y.F.; Cheng, Y.; Ye, J.X.; Zhao, Y.; Xie, J.; Yang, S.P. Targeting shrimp spoiler *Shewanella putrefaciens*: Application of ε-polylysine and oregano essential oil in Pacific white shrimp preservation. *Food Control* **2021**, *123*, 107702. [CrossRef]
15. Bao, Y.; Boeren, S.; Ertbjerg, P. Myofibrillar protein oxidation affects filament charges, aggregation and water-holding. *Meat Sci.* **2018**, *135*, 102–108. [CrossRef] [PubMed]
16. Liu, Y.; Zhang, L.; Gao, S.; Bao, Y.; Tan, Y.; Luo, Y.; Li, X.; Hong, H. Effect of protein oxidation in meat and exudates on the water holding capacity in bighead carp (*Hypophthalmichthys nobilis*) subjected to frozen storage. *Food Chem.* **2022**, *370*, 131079. [CrossRef]
17. Wang, S.; Xiang, W.; Fan, H.; Xie, J.; Qian, Y.F. Study on the mobility of water and its correlation with the spoilage process of salmon (*Salmo solar*) stored at 0 and 4 °C by low-field nuclear magnetic resonance (LF NMR ^{1}H). *J. Food Sci. Technol.* **2018**, *55*, 173–182. [CrossRef]
18. Zhuang, S.; Li, Y.; Hong, H.; Liu, Y.; Shu, R.; Luo, Y. Effects of ethyl lauroyl arginate hydrochloride on microbiota, quality and biochemical changes of container-cultured largemouth bass (*Micropterus salmonides*) fillets during storage at 4 °C. *Food Chem.* **2020**, *324*, 126886. [CrossRef]
19. Lv, M.; Mei, K.; Zhang, H.; Xu, D.; Yang, W. Effects of electron beam irradiation on the biochemical properties and structure of myofibrillar protein from *Tegillarca granosa* meat. *Food Chem.* **2018**, *254*, 64–69. [CrossRef]
20. Zhang, J.; Wang, X.; Shi, W. Odor characteristics of white croaker and small yellow croaker fish during refrigerated storage. *J. Food Biochem.* **2019**, *43*, e12852. [CrossRef]
21. Kimbuathong, N.; Pattarin, L.; Harnkarnsujarit, N. Inhibition of melanosis and microbial growth in Pacific white shrimp (*Litopenaeus vannamei*) using high CO_2 modified atmosphere packaging. *Food Chem.* **2020**, *312*, 126114. [CrossRef]
22. Qian, Y.F.; Xie, J.; Yang, S.P.; Huang, S.; Wu, W.H.; Li, L. Inhibitory effect of a quercetin-based soaking formulation and modified atmospheric packaging (MAP) on muscle degradation of Pacific white shrimp (*Litopenaeus vannamei*). *LWT* **2015**, *63*, 1339–1346. [CrossRef]
23. McSharry, S.; Koolman, L.; Whyte, P.; Bolton, D. An investigation of the survival and/or growth of Clostridioides (*Clostridium*) difficile in beef stored under aerobic, anaerobic and commercial vacuum packaging conditions at 2 °C and 20 °C. *Food Control* **2021**, *119*, 107475. [CrossRef]
24. Yang, S.P.; Xie, J.; Qian, Y.F. Determination of spoilage microbiota of pacific white shrimp during ambient and cold storage using next-generation sequencing and culture-dependent method. *J. Food Sci.* **2017**, *82*, 1178–1183. [CrossRef]
25. Lv, Y.; Xie, J. Effects of freeze-thaw cycles on water migration, microstructure and protein oxidation in cuttlefish. *Foods* **2021**, *10*, 2576. [CrossRef] [PubMed]
26. Xie, J.; Tang, Y.R.; Yang, S.P.; Qian, Y.F. Effects of whey protein films on the quality of thawed bigeye tuna (*Thunnus obesus*) chunks under modified atmosphere packaging and vacuum packaging conditions. *Food Sci. Biotechnol.* **2017**, *26*, 937–945. [CrossRef]

27. Wang, C.; Wang, H.; Li, X.; Zhang, C. Effects of oxygen concentration in modified atmosphere packaging on water holding capacity of pork steaks. *Meat Sci.* **2019**, *148*, 189–197. [CrossRef]
28. Perez-Won, M.L.; Roberto, M.; Herrera-Lavados, C.; Reyes, J.E.; Roco, T.; Palma-Acevedo, A.; Tabilo-Munizaga, G.; Aubourg, S.P. Combined treatments of high hydrostatic pressure and CO_2 in coho salmon (*Oncorhynchus kisutch*): Effects on enzyme inactivation, physicochemical properties, and microbial shelf life. *Foods* **2020**, *9*, 273. [CrossRef]
29. Qi, M.; Yan, H.; Zhang, Y.; Yuan, Y. Impact of high voltage prick electrostatic field (HVPEF) processing on the quality of ready-to-eat fresh salmon (*Salmo salar*) fillets during storage. *Food Control* **2022**, *137*, 108918. [CrossRef]
30. Li, X.; Hu, L.; Zhu, X.; Guo, X.; Deng, X.; Zhang, J. The effect of caspase-3 in mitochondrial apoptosis activation on degradation of structure proteins of *Esox lucius* during postmortem storage. *Food Chem.* **2022**, *367*, 130767. [CrossRef]
31. Ahmed, Z.; Donkor, O.S.; Wayne, A.; Vasiljevic, T. Calpains- and cathepsins-induced myofibrillar changes in post-mortem fish: Impact on structural softening and release of bioactive peptides. *Trends Food Sci. Technol.* **2015**, *45*, 130–146. [CrossRef]
32. Yi, Z.; Xie, J. Assessment of spoilage potential and amino acids deamination & decarboxylation activities of Shewanella putrefaciens in bigeye tuna (*Thunnus obesus*). *LWT* **2022**, *156*, 113016. [CrossRef]
33. Wang, Y.Y.; Tayyab Rashid, M.; Yan, J.K.; Ma, H. Effect of multi-frequency ultrasound thawing on the structure and rheological properties of myofibrillar proteins from small yellow croaker. *Ultrason. Sonochem.* **2021**, *70*, 105352. [CrossRef] [PubMed]
34. Ding, J.; Zhao, X.; Li, X.; Huang, Q. Effects of different recovered sarcoplasmic proteins on the gel performance, water distribution and network structure of silver carp surimi. *Food Hydrocoll.* **2022**, *131*, 107835. [CrossRef]
35. Yang, N.; Liang, X.; Cao, J.; Zhang, Q.; Tan, Y.; Xu, B.; Yang, Y.; Wang, Y.; Yang, Q.; Liu, H.; et al. Denaturation manner of sarcoplasmic proteins in pale, soft and exudative meat determines their positive impacts on myofibrillar water-holding capacity. *Meat Sci.* **2022**, *185*, 108723. [CrossRef]
36. Ertbjerg, P.; Puolanne, E. Muscle structure, sarcomere length and influences on meat quality: A review. *Meat Sci.* **2017**, *132*, 139–152. [CrossRef]
37. Tironi, V.A.; Tomas, M.C.; Anon, M.C. Lipid and protein deterioration during the chilled storage of minced sea salmon (*Pseudopercis semifasciata*). *J. Sci. Food Agric.* **2007**, *87*, 2239–2246. [CrossRef]
38. Zhuang, S.; Tan, Y.; Hong, H.; Li, D.; Zhang, L.; Luo, Y. Exploration of the roles of spoilage bacteria in degrading grass carp proteins during chilled storage: A combined metagenomic and metabolomic approach. *Food Res. Int.* **2022**, *152*, 110926. [CrossRef]
39. Tian, S.; Xiao, X.; Deng, S. Sinusoidal envelope voltammetry as a new readout technique for electronic tongues. *Mikrochim. Acta* **2012**, *178*, 315–321. [CrossRef]
40. Duan, Z.; Dong, S.; Dong, Y.; Gao, Q. Geographical origin identification of two salmonid species via flavor compound analysis using headspace-gas chromatography-ion mobility spectrometry combined with electronic nose and tongue. *Food Res. Int.* **2021**, *145*, 110385. [CrossRef]
41. Gänzle, M.G. Lactic metabolism revisited: Metabolism of lactic acid bacteria in food fermentations and food spoilage. *Curr. Opin. Food Sci.* **2015**, *2*, 106–117. [CrossRef]
42. Sorheim, O.; Ofstad, R.; Lea, P. Effects of carbon dioxide on yield, texture and microstructure of cooked ground beef. *Meat Sci.* **2004**, *67*, 231–236. [CrossRef] [PubMed]

foods

MDPI

Article

Changes in Protein Degradation and Non-Volatile Flavor Substances of Swimming Crab (*Portunus trituberculatus*) during Steaming

Qin Chen [1], Yurui Zhang [1], Lunan Jing [1], Naiyong Xiao [1], Xugan Wu [2,*] and Wenzheng Shi [1,3,*]

[1] College of Food Science and Technology, Shanghai Ocean University, Shanghai 201306, China
[2] College of Fisheries and Life Science, Shanghai Ocean University, Shanghai 201306, China
[3] National R&D Branch Center for Freshwater Aquatic Products Processing Technology (Shanghai), Shanghai 201306, China
* Correspondence: xgwu@shou.edu.cn (X.W.); wzshi@shou.edu.cn (W.S.); Tel.: +86-15692165021 (X.W.); +86-15692165859 (W.S.)

Abstract: To investigate the effect of steaming time (0, 5, 10, 15, 20, and 25 min) on the protein degradation and non-volatile flavor substances of swimming crab (*Portunus trituberculatus*), the moisture content, total nitrogen (TN), non-protein nitrogen (NPN), free amino acids (FAAs), flavor nucleotides, electronic tongue analysis, and sensory evaluation were determined. The results showed that the contents of NPN and total FAAs were the highest after crabs steamed for 10 min. Meanwhile, the AMP (adenosine monophosphate) content reached the maximum value (332.83 mg/100 g) and the taste active value (TAV) reached 6.67, which indicated that AMP contributes the most to the taste of steamed crab at 10 min. The electronic tongue distinguished the taste difference well, and the sensory score was the highest at 15 min. Combined with equivalent umami concentration (EUC) and TAV value, swimming crab (weight = 200 ± 20 g) steamed for 10–15 min tasted best.

Keywords: *Portunus trituberculatus*; protein degradation; non-volatile flavor; free amino acids; flavor nucleotides; steam

Citation: Chen, Q.; Zhang, Y.; Jing, L.; Xiao, N.; Wu, X.; Shi, W. Changes in Protein Degradation and Non-Volatile Flavor Substances of Swimming Crab (*Portunus trituberculatus*) during Steaming. *Foods* **2022**, *11*, 3502. https://doi.org/10.3390/foods11213502

Academic Editor: Jingran Bi and Yeşim Özoğul

Received: 13 September 2022
Accepted: 19 October 2022
Published: 3 November 2022

1. Introduction

Swimming crab (*Portunus trituberculatus*), one of the most critical marine economic crustacean species in the world, is popular for its succulent meat, ample nutrition, and distinctive flavor [1,2], and the marine capture production in 2020 has amounted to 367 thousand tons [3]. Due to the characteristic of easy corruption, swimming crabs are mainly sold fresh or frozen, which leads to low utilization of production and processing. However, hot-processed items have a higher commercial value in global commercial marketplaces, especially when supply is affected by the season [4]. Hence, it is significant to study the processing quality of swimming crab further to supply high-quality products to consumers and establish a higher-quality market.

Thermal processing is defined as heating meat to a temperature high enough to denature proteins [5], which is an effective measurement to enhance food quality. Thermal processing methods, including microwaving, steaming, boiling, and roasting, have been used to process frozen surimi, sturgeon, shrimp, and mangrove crab, which had a significant effect on flavor formation and consumer performances [6–9]. Generally, the most common hot processing procedures for crabs are steaming and boiling. Shi et al. [10] investigated the effects of steaming versus boiling on the flavor of swimming crab and discovered that steaming with hot water produced crabs with more water-soluble flavor substances. Simultaneously, steaming time also has a significant impact on the flavor formation and consumer choice of aquatic products. If the steaming time is insufficient, the meat can't be separated from the exoskeleton adequately. If the steaming time is too long, excessive water loss deteriorates texture; the Maillard reaction reduces the nutritional value. Furthermore, because the quantities of ammonia in crustacean muscles are higher than in vertebrate muscles, superheated steam creates unpleasant ammonia compounds [11]. In several studies, the recommended steaming time for sturgeon and black carp was determined [7,12]. Moreover, the characteristic flavor compounds during steaming have been identified, which provide a theoretical basis for increasing the flavor and promoting the processing of aquatic products. However, the flavor of swimming crab was evaluated mainly in terms of different genders, edible parts, habitat environments, and dietary nutrition [13–16], and there is little data on the effect of steaming time on the flavor formation of swimming crab. Thus, investigating the flavor change of swimming crab depending on steaming time is of great significance.

Flavor includes taste and smell, and taste is one of the most critical characteristics that influence consumer preference. There are two main types of taste compounds in aquatic products: nitrogen-containing compounds (e.g., free amino acids (FAAs), flavor nucleotides, betaine, trimethylamine oxide, etc.) and others without nitrogen [17]. FAAs are the end products of protein degradation, divided into umami, sweet, and bitter amino acids. Flavor nucleotides, including guanosine monophosphate (GMP), inosine monophosphate (IMP), and adenosine monophosphate (AMP); AMP provides umami and sweetness taste profiles and inhibits bitterness; IMP is merely sweet when the concentration is between 0.5–1.0 mg/mL and shows intense umami flavor when the concentration exceeds 1.0 mg/mL [18]. Additionally, FAAs and flavor nucleotides have synergistic effects on umami and sweetness in swimming crabs and are influenced by heat treatment [19]. Therefore, it is necessary to explore the changes in protein degradation and flavor substances of swimming crab during steaming.

To examine how the swimming crab's flavor changed during steaming, high-performance liquid chromatography (HPLC) and an ultra-high-speed automatic amino acid analyzer were utilized to analyze nucleotides and FAAs, respectively; the electronic tongue was used to describe the taste modifications; taste quality indicators included the equivalent EUC and TAV. Protein degradation was also discussed. Furthermore, sensory evaluators graded swimming crabs steamed at different times. This study creates the ideal conditions for cooking swimming crab, establishes the link between protein degradation and flavor formation, and provides a theoretical basis for promoting industrial production.

2. Materials and Methods

2.1. Materials

Monopotassium phosphate, dipotassium phosphate, methanol, amino acid standards, and nucleotide standards were chromatographic reagents purchased from Shanghai Ample Scientistic Instrument Co., Ltd., (Shanghai, China). Analytical grades of sodium hydroxide, potassium hydroxide, perchloric acid, trichloroacetic acid, and other reagents were purchased from Sinopharm Chemical Reagent Co., (Shanghai, China).

2.2. Preparation of Sample

Swimming crabs (weight = 200 ± 20 g, n = 60) were bought from Luchaogang in November 2021 in Shanghai, China and transported to the laboratory within 2 h on ice. Crabs were washed clean with tap water and separated into 6 groups at random. Subsequently, the samples were steamed for 5, 10, 15, 20, and 25 min, respectively, in an induction cooker (1600 W power). The time started when the water temperature reached 100 °C, and fresh crabs were served as a control group. After cooling at room temperature, the body meat of each group was collected separately and stored in plastic bags at −80 °C for further analysis.

2.3. Moisture Content

The moisture content was measured according to AOAC [20]. Briefly, to measure the moisture content, 2.00 g of the sample was dried to a constant weight at 105 °C in the drying oven (Boxun, Shanghai, China).

2.4. Total Nitrogen (TN)

TN was measured according to AOAC. The sample of 0.20 g was weighed and digested for two hours with 12 mL concentrated sulfuric acid, 0.40 g copper sulfates, and 3.50 g potassium sulfates (kjeldahl8400, FOSS, Hillerød, Denmark). The absorbing liquid was 1% (m/v) boric acid solution, and the titrant was 0.1 mol/L hydrochloric acid. Both the 0.1% (m/v) bromocresol green and 0.1% (m/v) methyl red served as indicators.

2.5. Non-Protein Nitrogen (NPN)

NPN was determined based on the method of Cambero et al. [21]. The sample of 1.00 g was weighed and homogenized for 2 min in 10 mL of 10% (v/v) trichloroacetic acid. After standing for 2 h at 4 °C and centrifuged (10,000× g, 4 °C, 15 min), 2 mL supernatant was taken and adjusted. The following steps were the same as TN.

2.6. Protein Degradation Index (PI)

NPN stands for the non-protein nitrogen content and TN for the overall nitrogen concentration [12]. The percentage ratio between NPN and TN was used to determine PI.

$$PI = \frac{NPN}{TN} \tag{1}$$

2.7. FAAs

A minor adjustment was made based on Zhuang et al. [22]. Briefly, the sample of 0.50 g was weighed in a 50 mL tube and homogenized for 2 min with 15 mL of precooled trichloroacetic acid solution (5%, w/v). After ultrasonic extraction for 15 min, the sample was centrifuged at 10,000× g for 15 min at 4 °C before being stored at 4 °C for 2 h. The pH of 2.0 was then adjusted in 5 mL of supernatant using sodium hydroxide solution (3 M and 1 M). A 0.22-μm syringe filter was used to filter the sample solution after the supernatant had been diluted to 10 mL with ultrapure water. An ultra-high-speed automatic amino acid analyzer was used to determine and analyze FAAs (LA-8800, Hitachi, Japan). Each sample was examined three times.

2.8. Nucleotides

5′-nucleotides were analyzed according to Zhang et al. [23]. The sample weighed 3.00 g, was placed in a 50 mL tube and homogenized for 1 min in a 10 mL cold solution of 10% (v/v) perchloric acid before being centrifuged at 10,000× g for 15 min at 4 °C. The supernatant was separated from the precipitation, and remained precipitation was rinsed with 5 mL 5% (v/v) perchloric acid solution before being centrifuged in the same manner. The action mentioned above was carried out twice, and the pH of 5.75 was achieved in the combined supernatant by adding cold potassium hydroxide solution (6 M and 1 M). The supernatant was diluted to 50 mL and then filtered with a 0.22-μm syringe filter for the test after being kept at 4 °C. The entire procedure was carried out below 4 °C. Finally, the analysis was performed by HPLC (Waters Co., Milford, MA, USA). Each sample underwent a triple analysis, and the analytical conditions were performed by Tu, Wu, Wang and Shi [17].

2.9. TAV and EUC

The ratio of a flavor ingredient's concentration to its threshold value is known as the TAV. TAV can be used to determine how much a flavor substance contributes to the overall flavor of samples [24]. When the TAV value is less than one, the sense barely affects the taste. While the value is greater than one, the compound has a significant impact on flavor. The greater the TAV is, the more significantly it contributes to the overall flavor of samples.

EUC is the equivalent of monosodium glutamate (MSG), which represents the intensity of umami produced by the synergistic action of umami amino acids (Asp and Glu) and 5′-nucleotides (IMP, GMP, and AMP). The following formula is the equation for EUC [25]:

$$EUC = \Sigma a_i b_i + 1218 \times (\Sigma a_i b_i) \times (\Sigma a_j b_j) \tag{2}$$

where a_i is the content of umami amino acids (Asp and Glu); b_i is the relative umami concentration (RUC) of each umami amino acid (Asp, 0.077; Glu, 1); a_j is the flavor nucleotides content (IMP, AMP and GMP); b_j is the RUC of the taste nucleotide (IMP, 1; GMP, 2.3; AMP, 0.18) and the synergistic constant is 1218 [26].

2.10. Electronic Tongue

A sample (5.00 g) was homogenized for 2 min in ultrapure water (100 mL) using a high-speed homogenizer (Shanghai Fokker Equipment Co., Ltd., Shanghai, China). The sample solution was centrifuged at 10,000× g for 15 min at 4 °C after being kept in a cooling bath for 30 min. A 30 mL supernatant was prepared for an electronic tongue on-machine test (ISENSO, Super Tongue, New York, USA), and the entire experiment was conducted under 4 °C. A triplicate of each sample was used for analysis.

2.11. Sensory Assessment

Eight sensory assessors (four women and four men, aged 22–25) who are not partial to food or allergic graded the fresh and steamed samples according to the five aspects of order, taste, appearance, morphology, and texture. The specific standards referred to the method of Yang et al. [27] and are shown in Table 1. After tasting 1 sample, the sensory evaluator gargled with pure water and moved on to the next. Each item was rated between 1 and 20 points, and the highest sensory evaluation score for each sample was 100 points. The higher the point is, the more acceptable the sample will be. The Sensory Evaluation Laboratory served as the location for the entire procedure.

Table 1. Sensory evaluation criteria of swimming crab during steaming.

	Score				
Index	**1–4**	**5–8**	**9–12**	**13–16**	**17–20**
Order	Unacceptable	No inherent aroma	Slight inherent aroma	Inherent aroma	Strong inherent aroma
Taste	Unacceptable	No inherent umami	Slightly inherent umami	Inherent umami	Strong inherent umami
Appearance	Dim	Slightly dim	Fairly bright	Bright	Translucent
Morphology	Very loose	Loose	Partial loose	Tight	Very tight
Texture	Very soft or stiff	Soft	Slight elastic	Elastic	Firm

2.12. Statistical Analysis

All tests were performed three times. Three duplicates of each experiment were evaluated. SPSS26.0 (SPSS Inc., Chicago, IL, USA) was used to analyze the data, and the findings are presented as means and standard deviations (SD). The Duncan technique was used, and $p < 0.05$ was used to denote a significant difference in the one-way analysis of variance (ANOVA). The charts were created by Origin2021 (Origin Lab Corporation, Hampton, Northampton, MA, USA).

3. Results and Discussion

3.1. Moisture Content Analysis

The moisture content of swimming crab steamed at different times is displayed in Table 2. With the increase in steaming time, the moisture content increased and then decreased. The moisture content of the fresh sample was 79.85% and rose a little after steaming for 5 min (80.84%), which may be due to the accumulation of water vapor in the crab's meat, and the shell prevents the evaporation of moisture at the beginning of steam. Subsequently, the moisture content decreased slightly and was always lower than that of the fresh sample. The result could be attributed to the protein denaturation, especially the severe contraction of myofibrillar protein after heating, which lowered the muscle's water-holding ability [28]. In the later stage of steaming, the moisture content was almost unchanged, which might be because the external water vapor content was close to saturation, so the sample water loss was not apparent. Moreover, crab meat contains collagen, which can be turned into gelatin after high-temperature treatment and absorb water and reduce water loss [28].

Table 2. Changes of moisture, total nitrogen, and non-protein nitrogen content in swimming crab meat during steaming.

	0 min	5 min	10 min	15 min	20 min	25 min
Moisture content (%)	79.85 ± 0.35 ^{b}	80.84 ± 0.07 ^{a}	78.28 ± 0.17 ^{d}	78.64 ± 0.19 ^{cd}	78.68 ± 0.33 ^{cd}	78.82 ± 0.25 ^{c}
TN (g/100 g)	86.80 ± 2.22 ^{b}	90.12 ± 0.71 ^{a}	85.19 ± 0.90 ^{b}	85.97 ± 0.33 ^{b}	82.55 ± 0.91 ^{c}	82.06 ± 0.28 ^{c}
NPN (g/100 g)	8.88 ± 0.05 ^{c}	10.2 ± 0.51 ^{bc}	12.23 ± 1.05 ^{a}	11.39 ± 0.46 ^{ab}	9.19 ± 0.00 ^{c}	9.71 ± 0.28 ^{c}
PI (%)	10.23	11.32	14.36	13.25	11.13	11.83

Data in the same row with different letters differ significantly ($p < 0.05$).

3.2. Protein Degradation Analysis

The changes in TN, NPN, and PI reflect the protein degradation during steaming. The contents of TN and NPN on a dry basis of swimming crab during steaming are shown in Table 2. The TN content increased and then decreased during the steaming process, which reached the highest value (90.12 ± 0.71 g/100 g) after steamed for 5 min and fell as the steaming process continued, which may have resulted from the loss of volatile nitrogen or the dissolution of some of the water-soluble nitrogen into the exudate. The NPN content increased and then decreased during steaming, which was always higher than that of the fresh sample. The result indicated that steaming treatment increased the NPN content of swimming crabs, and the highest value was obtained at 10 min (12.23 ± 1.05 g/100 g). It was mainly due to the protein degradation in the steaming process, which formed a variety of small molecules of peptides, amino acids, and other non-protein substances [29]. These protein degradation products have a significant effect on flavor formation. After continuing to steam, NPN began to decrease. The concentration of NPN in the steamed sample was decreased mainly as a result of the aggregation of FAAs and polypeptides, among which water-soluble amino acids and peptides were transported to the exudate (Zhao et al., 2008).

3.3. FAAs Analysis

Aquatic products may have various taste perceptions depending on the components, concentration, and threshold of FAAs [30]. Due to the variations in moisture content during the steaming process, the composition proportion of crab meat was affected. To investigate the specific changes of flavor substances in crab meat with different steaming times, the content of a dry basis is more appropriate. A total of 17 FAAs, including umami amino acids (Glu and Asp), sweet amino acids (Gly, Ala, Thr, Ser, Pro and Arg), and bitter amino acids (Val, Lys, Leu, Ile, Phe, His, Tyr and Met), were displayed in Table 3. During steaming, the amounts of total FAAs increased and then declined, and the content was the maximum (10,996.21 ± 1167.47 mg/100 g) at 10 min, which was consistent with the highest NPN content in crabs steamed for 10 min. In both fresh and steamed samples, the concentration of sweet amino acids was the highest, followed by bitter and umami amino acids. The variance trend for the three different types of amino acids was essentially the same, rising at the start of steaming and then falling. The maximum concentration of sweet amino acids (9418.84 ± 1326.57 mg/100 g) and umami amino acids (520.24 ± 21.34 mg/100 g) were obtained at 10 and 15 min, respectively. In addition, there was no significant change in bitter amino acid content in fresh and steamed samples. This means that steaming doesn't change the bitter taste of the crab meat; however, the proper steaming time can increase the umami and sweetness. Thr, Gly, Ala, and Pro were the most abundant sweet amino acids in fresh crab, among which the content of Gly (2530.88 ± 89.82 mg/100 g) was the highest. The content of Gly reached the maximum (2622.46 ± 390.54 mg/100 g) after crabs were steamed for 10 min and then decreased after continuous steaming. The changes in Thr, Ala, Pro, and Arg contents were about the same as Gly during steaming. It was reported that the taste of sweetness was noted in Arg at low concentrations (less than twice the threshold) and the taste of bitterness at high doses. However, unlike those hydrophobic amino acids with branched chains that create disagreeable bitterness, Arg can increase flavor complexity and freshness [31]. Asp and Glu were the primary umami amino acids, and the composition of the fresh sample and the sample that had been steamed for five mins did not differ significantly. However, the contents of Asp and Glu reached the maximum at 10 min (144.37 ± 5.17 mg/100 g) and 15 min (409.59 ± 16.41 mg/100 g), respectively. After continuing to steam, the contents of Asp and Glu started to decline. Most bitter amino acids reached their maximum contents when crabs were steamed for 10 min; however, due to the high threshold, bitter amino acids contributed little to the taste quality of fresh or steamed crab. Additionally, several bitter amino acids help to improve sweetness at levels below the threshold [32]. According to Table 3, the primary flavor amino acids of steamed swimming crab are Glu, Gly, Ala, and Arg, and the maximum contents are achieved at 10 or 15 min. After continuous steaming, the content of FAAs began to decrease, which was in line with the trend of NPN during steaming. The increase of free amino acid content in the steaming process was mainly due to protein degradation, however, after continuous steaming, intense aggregation occurred in protein, and the amount of decomposition of Maillard reaction was larger than the amount of thermal degradation of protein, which led to a decrease in free amino acid content in the later period of steaming [33]. In the process of steaming, muscle fibers will contract after being heated, along with a loss of juice. The proteins in the lost fluid are mainly water-soluble proteins and the degradation products of some polypeptide compounds with small molecular weight or myofibrillar proteins. During the heating process, various covalent bonds and non-covalent bonds that maintain protein molecular structure in muscle tissue gradually break, and myofibrilla proteins lose their advanced structure and begin to dissolve. All of these will contribute to the loss of free amino acids in muscle during steaming. In addition, the Maillard reaction, as the main pathway for volatile flavor compound generation, also accelerates FAAs consumption in this process [34]. Therefore, the free amino acid content of swimming crab could be affected by steaming time, and the crab steamed for 10–15 min may have better taste quality.

3.4. Nucleotide Analysis

Flavor nucleotides are essential compounds of aquatic products, especially GMP, IMP, and AMP, which provide umami and sweetness taste profiles and synergy with free amino acids to improve the umami taste [18,35]. The dry base content of umami nucleotides (GMP, IMP, and AMP) is shown in Table 4. In the fresh sample, the content of IMP (307.49 ± 18.56 mg/100 g) was higher than the contents of AMP (119.85 ± 22.48 mg/100 g) and GMP (6.7 ± 0.15 mg/100 g). Kong et al. [36] made a similar observation in the Chinese mitten crab. However, the contents of GMP and IMP continued to decrease during steaming, which is contrary to the trend of AMP. The same result was found in Pacific white shrimp and Antarctic krill [37]. Mendes et al. [38] also reported that IMP was mainly found in fish and AMP in crustaceans. Moreover, the GMP content of steamed swimming crab meat decreased sharply and wasn't detected after 5 min. This result was consistent with the findings of Gorbatov and Lyaskovskaya [39] and Zhang, Qiu, Zhang, Ho Row, Cheng, and Jin [37], who observed that different nucleotides have varying degrees of heat stability, leading to different degrees of thermal degradation. The thresholds of GMP, IMP, and AMP are 12.5 mg/100 g, 25 mg/100 g, and 50 mg/100 g, respectively. In the steaming process, the TAV value of IMP decreased gradually and was always less than one after 5 min. On the contrary, the TAV value of AMP was always greater than one during steaming and also greater than that of the fresh sample, which indicated that AMP contributed much to the flavor of steamed swimming crab, especially the TAV of AMP reached the maximum (6.67) at 10 min. During the steaming process, the total amount of IMP and AMP didn't change much; however, it decreased obviously at 25 min, which indicated that long-time steaming wasn't conducive to the formation of flavor. Therefore, swimming crabs should be steamed at a reasonable time to ensure delicious taste.

Table 3. The contents of free amino acids on a dry basis of swimming crab during steaming (mg/100 g).

		Content (mg/100 g)					
Amino Acid	**Threshold (mg/100 g)**	**0 min**	**5 min**	**10 min**	**15 min**	**20 min**	**25 min**
Asp	100	109.93 ± 19.74^{b}	106.5 ± 18.08^{b}	144.37 ± 5.17^{a}	110.64 ± 5.24^{b}	91.24 ± 8.73^{bc}	73.57 ± 2.24^{c}
Thr	260	503.59 ± 44.39^{a}	297.36 ± 51.86^{b}	436.84 ± 114.29^{a}	221.49 ± 33.89^{b}	187.15 ± 29.86^{b}	295.65 ± 2.31^{b}
Ser	150	44.72 ± 3.36^{b}	54.76 ± 3.85^{a}	25.87 ± 6.43^{c}	22.17 ± 2.08^{cd}	17.8 ± 1.53^{d}	42.59 ± 4.2^{b}
Glu	30	261.45 ± 10.64^{bc}	253.01 ± 30.13^{bc}	329.41 ± 92.37^{ab}	409.59 ± 16.41^{a}	409.02 ± 75.62^{a}	166.92 ± 4.12^{c}
Gly	130	2530.88 ± 89.82^{a}	2422.71 ± 168.64^{a}	2622.46 ± 390.54^{a}	1757.4 ± 62.35^{b}	1652.2 ± 243.43^{b}	1940.14 ± 51.97^{b}
Ala	60	683.76 ± 27.2^{a}	432.38 ± 52.15^{c}	551.5 ± 73.52^{b}	432.65 ± 44.95^{c}	671.45 ± 74.86^{a}	484.54 ± 12.28^{bc}
Cys	-	14.88 ± 0.36^{b}	9.72 ± 1.35^{bc}	12.6 ± 3.89^{bc}	14.35 ± 1.92^{b}	23.64 ± 3.99^{a}	8.24 ± 0.54^{c}
Val	40	129.99 ± 5.19^{ab}	96.62 ± 8.44^{c}	145.61 ± 27.54^{a}	113.96 ± 18.68^{bc}	146.41 ± 10.05^{a}	133.41 ± 3.59^{ab}
Met	30	182.7 ± 7.19^{a}	93.61 ± 11.23^{c}	127.65 ± 24.12^{bc}	112.24 ± 27.33^{bc}	182.42 ± 31.89^{a}	140.94 ± 5.17^{ab}
Ile	90	55.59 ± 3.04^{a}	34.1 ± 5.19^{b}	69.26 ± 19.43^{a}	53.22 ± 12.79^{ab}	58.03 ± 4.42^{a}	59.49 ± 1.09^{a}
Leu	190	106.35 ± 5.11^{a}	94.23 ± 9.84^{a}	137.14 ± 41.15^{a}	104.06 ± 25.35^{a}	123.73 ± 11.99^{a}	113.89 ± 3.37^{a}
Tyr	-	115.35 ± 3.27^{b}	159.66 ± 19.27^{a}	168.45 ± 37.47^{a}	135.29 ± 20.61^{ab}	132.5 ± 16.88^{ab}	150.46 ± 1.02^{ab}
Phe	90	168.66 ± 2.8^{a}	167.34 ± 38.02^{a}	167.51 ± 26.28^{a}	151.74 ± 17.33^{a}	153.21 ± 16.41^{a}	159.11 ± 0.62^{a}
Lys	50	183.5 ± 7.65^{a}	152.91 ± 30.99^{a}	188.84 ± 30.43^{a}	181.98 ± 24.39^{a}	172.03 ± 26.51^{a}	147.18 ± 2.35^{a}
His	20	92.01 ± 5.89^{a}	63.4 ± 7.81^{b}	86.54 ± 16.03^{a}	65.12 ± 12.11^{b}	73.91 ± 9.95^{ab}	88.34 ± 0.23^{a}
Arg	50	3327.15 ± 71.78^{bc}	3482.12 ± 261.82 bc	4336.19 ± 925.95 ab	2772.98 ± 88.97^{c}	3572.72 ± 779.01 bc	4695.08 ± 217.29^{a}
Pro	300	1132.15 ± 120.49^{b}	612.5 ± 123.95^{c}	1445.99 ± 194.25^{a}	689.76 ± 205.87^{c}	1024.22 ± 155.7^{b}	1431.03 ± 75.79^{a}
Total		9642.67 ± 349.39 ab	8532.94 ± 653.29 bc	$10{,}996.21 \pm 1167.47^{a}$	7348.65 ± 576.23^{c}	8691.67 ± 1249.04 bc	10130.57 ± 236.49 ab
SFAs		8222.25 ± 312.44 abc	7301.83 ± 606.14 bcd	9418.84 ± 1326.57 a	5896.45 ± 424.85^{d}	7125.54 ± 1202.59 cd	8889.03 ± 255.27 ab
BFAs		1034.15 ± 39.09^{a}	861.87 ± 109.28^{a}	1090.99 ± 220.55^{a}	917.61 ± 157.33^{a}	1042.22 ± 11.34^{a}	992.81 ± 17.44^{a}
UFAs		371.39 ± 30.38^{bc}	359.52 ± 46.27^{c}	473.77 ± 89.07^{ab}	520.24 ± 21.34^{a}	500.26 ± 84.32^{a}	240.49 ± 1.88^{d}

SFAs: the sweet amino acid content (Thr, Ser, Gly, Ala, Arg, and Pro); BFAs: the bitter amino acid content (Val, Met, Ile, Leu, Phe, Tyr, Lys, and His) and UFAs: the umami amino acid content (Asp and Glu). Each provided number corresponds to the mean and standard deviation. Significant differences ($p < 0.05$) are shown by different letters in the same row.

Table 4. The contents and TAVs of flavor nucleotides on a dry basis of swimming crab during steaming (mg/100 g).

	Content (mg/100 g)			TAV		
Steaming Time (min)	**GMP**	**IMP**	**AMP**	**GMP**	**IMP**	**AMP**
0	6.7 ± 0.15 [a]	307.49 ± 18.56 [a]	119.85 ± 22.48 [d]	0.54	12.3	2.40
5	0.83 ± 0.05 [b]	105.78 ± 0.05 [b]	227.99 ± 0.1 [c]	0.07	4.23	4.56
10	-	18.51 ± 2.76 [c]	333.54 ± 9.47 [a]	-	0.74	6.67
15	-	15.26 ± 0.51 [c]	296.26 ± 2.06 [b]	-	0.61	5.93
20	-	8.58 ± 0.38 [e]	332.83 ± 18.88 [a]	-	0.34	6.66
25	-	9.82 ± 0.05 [d]	280.78 ± 0.19 [b]	-	0.39	5.62

Significant differences ($p < 0.05$) are shown by different letters in the same line.

3.5. EUC and TAV Analysis

The EUC value has been universally acknowledged as the primary taste evaluation measure for assessing the umami level of aquatic items [40]. It reveals the direct synergistic action of flavor nucleotides and FAAs of umami and sweet taste and evaluates the flavor of umami enhancement in aquatic products [41]. Figure 1 shows the EUC and TAV values of swimming crab meat throughout steaming. The EUC and TAV values of the fresh sample were significantly higher than steamed ones ($p < 0.05$), and the EUC value was the highest at 4.65 g MSG/100 g in fresh meat. In other words, 100 g of fresh swimming crab meat equals 4.65 g of MSG. During the steaming process, the EUC value declined, and there was no significant change during 5 to 20 min, while it decreased to the lowest (0.16 g MSG/100 g) at 25 min ($p < 0.05$), which meant that steaming for too long had a harmful effect on the formation of flavor. Throughout the procedure, all TAVs of EUC values were significantly greater than one, which indicated that the interaction of nucleotides and FAAs had a significant role in umami taste qualities of swimming crab. This finding is in line with the work of Feng, et al. [42].

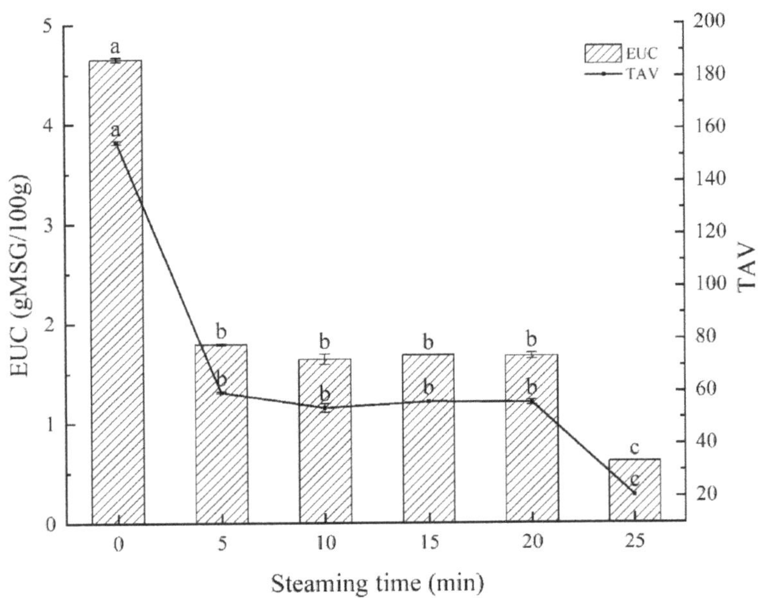

Figure 1. Comparison of EUC and TAV in swimming crab meat during steaming. Different letters are significantly different at $p < 0.05$.

3.6. Electronic Tongue Analysis

The electronic tongue (e-tongue) is a modern qualitative and quantitative analysis instrument, primarily made up of an interactive sensitive sensor array, a signal acquisition circuit, and a pattern recognition-based data processing approach [43]. E-tongue is now widely used to identify the taste profiles of foods such as livestock and poultry meat, marine products, fruit and vegetable products, etc. Figure 2 shows the PCA chart for the taste profiles of swimming crab meat during different steaming times. According to Garcia Hernandez et al. [44], the two primary components have enough information to accurately represent the sample's overall characteristics if their cumulative contribution rate exceeds 85%. The first principal component (PC1) and the second principal component (PC2) accounted for 90.1785% and 8.3820%, respectively; the total proportion was 98.56% (>85%), which indicates that each sample's flavor change information could be accurately recorded. According to Figure 2, there was no crossover among the six samples distributed in the region, which means that the taste of each sample significantly differs from the others, and the e-tongue can distinguish the tastes of swimming crab meat at different steaming times.

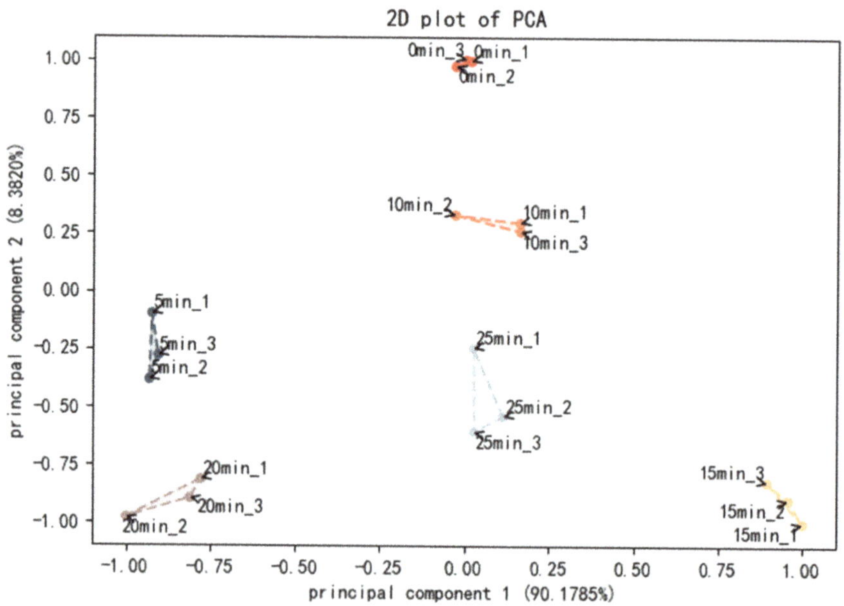

Figure 2. The PCA plot of electronic tongue data of swimming crab steamed at different times.

3.7. Sensory Evaluation Analysis

Figure 3 shows the sensory score of swimming crabs depending on steaming time. The sensory score of steamed samples increased initially and then decreased, but it was always higher than the fresh sample (62.5 $\pm$ 3.55). The maximum sensory score (88.9 $\pm$ 7.6) was obtained at 15 min and then reduced to 78.8 $\pm$ 9.7 at 25 min. From the results, we held that steamed crab is more acceptable than the fresh sample, while over time, steaming reduces the sensory properties. The reason may be that in the steaming process, the protein was degraded continuously, resulting in the accumulation of free amino acids and nucleotides. Some of these umami and sweet amino acids, such as Asp, Glu, Thr, Gly, Arg, and Pro, peaked at 10 or 15 min. Thus, the sensory properties of sweet and umami tastes were significantly influenced. As the steaming continued, the quality suffered from decreased water-holding capacity, cook loss, and muscle shrinkage, which resulted in a decrease in overall acceptance [45]. The longer the steaming time is, the lesser the acceptance will be.

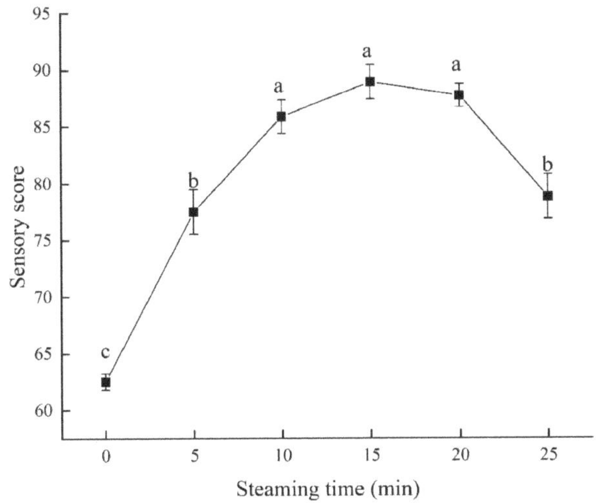

Figure 3. Sensory evolution of swimming crab meat during steaming. Different letters are significantly different at $p < 0.05$.

4. Conclusions

The changes in protein degradation and non-volatile flavor compounds in steamed swimming crab muscle were analyzed. Generally, throughout the steaming process (0–25 min), the protein continuously degraded to many flavor precursor substances resulting in changes in taste properties. The levels of NPN and total FAAs peaked at 10 min; the amount of sweet and umami amino acids peaked at 10 min and 15 min, respectively. AMP contributed significantly to the taste of steamed crab, and the content of AMP reached the highest at 10 min. Additionally, the flavor profile of the swimming crab was distinguished through the PCA plot of the e-tongue, and the sensory score reached the highest at 15 min. As the steaming time extends, the taste quality deteriorates. In conclusion, 10–15 min is an ideal steaming time for swimming crab (weight = 200 $\pm$ 20 g). The findings provide a theoretical foundation for the taste changes in swimming crab during steaming and serve as a reference for further process of swimming crab. However, the dynamic changes of volatile flavor compounds in the steaming process need further study.

Author Contributions: Q.C.: Data curation (lead); investigation (lead); validation (lead); writing—original draft (lead). Y.Z.: Investigation (equal); writing—review and editing (equal). L.J.: Investigation (equal); writing—review and editing (equal). N.X.: writing—review and editing (equal). X.W.: Conceptualization (equal); writing—review editing (equal). W.S.: Conceptualization (equal); funding acquisition (lead); administration (lead); writing—review and editing (equal). All authors have read and agreed to the published version of the manuscript.

Funding: The study was supported by the National Key R&D Program (NO. 2018YFD0900100) from the Ministry of science and technology of China.

Institutional Review Board Statement: Not applicable.

Informed Consent Statement: Informed consent was obtained from all subjects involved in the study.

Data Availability Statement: Data is contained within the article.

Acknowledgments: The study was supported by the National Key R&D Program (No. 2018YFD0900100) from the Ministry of science and technology of China.

Conflicts of Interest: The authors declare no conflict of interest.

Ethical Guidelines: Ethics approval was provided by the Shanghai Ocean University for this research (Approval No. SHOU-DW-2021-086).

References

1. Jin, M.; Wang, M.; Huo, Y.; Huang, W.; Mai, K.; Zhou, Q. Dietary lysine requirement of juvenile swimming crab, *Portunus trituberculatus*. *Aquaculture* **2015**, *448*, 1–7. [CrossRef]
2. Sun, P.; Jin, M.; Ding, L.; Lu, Y.; Ma, H.; Yuan, Y.; Zhou, Q. Dietary lipid levels could improve growth and intestinal microbiota of juvenile swimming crab, *Portunus trituberculatus*. *Aquaculture* **2018**, *490*, 208–216. [CrossRef]
3. FAO. *The State of World Fisheries and Aquaculture 2022*; Towards Blue Transformation; FAO: Rome, Italy, 2022. [CrossRef]
4. Hackett, S.; Krachey, M.; Dewees, C.; Hankin, D.; Sortais, K. An economic overview of dungeness carb (*Cancer magister*) processing in California. *Calif. Coop. Ocean. Fish. Investig. Rep.* **2003**, *44*, 86–93.
5. Risso, S.J.; Carelli, A.A. Nutrient composition of raw and cooked meat of male southern King Crab (*Lithodes santolla* Molina, 1782). *J. Aquat. Food Prod. Technol.* **2012**, *21*, 433–444. [CrossRef]
6. Luo, X.; Xiao, S.; Ruan, Q.; Gao, Q.; An, Y.; Hu, Y.; Xiong, S. Differences in flavor characteristics of frozen surimi products reheated by microwave, water boiling, steaming, and frying. *Food Chem.* **2022**, *372*, 131260. [CrossRef]
7. Li, X.; Xie, W.; Bai, F.; Wang, J.; Zhou, X.; Gao, R.; Xu, X.; Zhao, Y. Influence of thermal processing on flavor and sensory profile of sturgeon meat. *Food Chem.* **2022**, *374*, 131689. [CrossRef]
8. Wang, S.; Hu, M.; Zhao, L.; Liu, Q.; Cao, R. Changes in lipid profiles and volatile compounds of shrimp (*Penaeus vannamei*) submitted to different cooking methods. *Int. J. Food Sci. Technol.* **2022**, *57*, 4234–4244. [CrossRef]
9. Yu, H.-Z.; Chen, S.-S. Identification of characteristic aroma-active compounds in steamed mangrove crab (*Scylla serrata*). *Food Res. Int.* **2010**, *43*, 2081–2086. [CrossRef]
10. Shi, S.; Wang, X.; Wu, X.; Shi, W. Effects of four cooking methods on sensory and taste quality of *Portunus trituberculatus*. *Food Sci. Nutr.* **2020**, *8*, 1115–1124. [CrossRef]
11. Dima, J.B.; Barón, P.J.; Zaritzky, N.E. Mathematical modeling of the heat transfer process and protein denaturation during the thermal treatment of Patagonian marine crabs. *J. Food Eng.* **2012**, *113*, 623–634. [CrossRef]
12. Wang, Y.; Wu, H.; Shi, W.; Huang, H.; Shen, S.; Yang, F.; Chen, S. Changes of the flavor substances and protein degradation of black carp (*Mylopharyngodon piceus*) pickled products during steaming. *J. Sci. Food Agric.* **2021**, *101*, 4033–4041. [CrossRef] [PubMed]
13. Song, J.; Wang, H.; Wu, X.; Wang, X.; Shi, W. The flavor of gonad and meat of female *Portunus Trituberculatus* cultured in indoor and outdoor. *J. Food Biochem.* **2019**, *43*, e12743. [CrossRef] [PubMed]
14. Wu, X.; Zhou, B.; Cheng, Y.; Zeng, C.; Wang, C.; Feng, L. Comparison of gender differences in biochemical composition and nutritional value of various edible parts of the blue swimmer crab. *J. Food Compos. Anal.* **2010**, *23*, 154–159. [CrossRef]
15. Luo, J.; Monroig, O.; Zhou, Q.; Tocher, D.R.; Yuan, Y.; Zhu, T.; Lu, J.; Song, D.; Jiao, L.; Jin, M. Environmental salinity and dietary lipid nutrition strategy: Effects on flesh quality of the marine euryhaline crab *Scylla paramamosain*. *Food Chem.* **2021**, *361*, 130160. [CrossRef] [PubMed]
16. Yuan, Y.; Wang, X.; Jin, M.; Jiao, L.; Sun, P.; Betancor, M.B.; Tocher, D.R.; Zhou, Q. Modification of nutritional values and flavor qualities of muscle of swimming crab (*Portunus trituberculatus*): Application of a dietary lipid nutrition strategy. *Food Chem.* **2020**, *308*, 125607. [CrossRef] [PubMed]
17. Tu, L.; Wu, X.; Wang, X.; Shi, W. Effects of fish oil replacement by blending vegetable oils in fattening diets on nonvolatile taste substances of swimming crab (*Portunus trituberculatus*). *J. Food Biochem.* **2020**, *44*, e13345. [CrossRef]
18. Chen, D.; Zhang, M. Non-volatile taste active compounds in the meat of Chinese mitten crab (*Eriocheir sinensis*). *Food Chem.* **2007**, *104*, 1200–1205. [CrossRef]
19. Liu, C.; Meng, F.; Tang, X.; Shi, Y.; Wang, A.; Gu, Z.; Pan, Z. Comparison of nonvolatile taste active compounds of wild and cultured mud crab *Scylla paramamosain*. *Fish. Sci.* **2018**, *84*, 897–907. [CrossRef]
20. *Official Methods of Analysis of AOAC International*; 4th Revision; 1998; Volume Edition 16; Volume I and II.
21. Cambero, M.I.; Jaramillo, C.J.; Ordonez, J.A.; Cobos, A.; pereira-Lima, C.I. Effect of cooking conditions on the flavour compounds and composition of shrimp (*Parapenaeus longirostris*) broth. *Z. Lwbensm Unters. A* **1998**, *206*, 311–322. [CrossRef]
22. Zhuang, K.; Wu, N.; Wang, X.; Wu, X.; Wang, S.; Long, X.; Wei, X. Effects of 3 feeding modes on the volatile and nonvolatile compounds in the edible tissues of female Chinese mitten crab (*Eriocheir sinensis*). *J. Food Sci.* **2016**, *81*, S968–S981. [CrossRef]
23. Zhang, L.; Yin, M.; Zheng, Y.; Xu, C.; Tao, N.; Wu, X.; Wang, X. Brackish water improves the taste quality in meat of adult male Eriocheir sinensis during the postharvest temporary rearing. *Food Chem.* **2021**, *343*, 128409. [CrossRef]
24. Rotzoll, N.; Dunkel, A.; Hofmann, T. Quantitative studies, taste reconstitution, and omission experiments on the key taste compounds in morel mushrooms (*Morchella deliciosa* Fr.). *J. Agric. Food Chem.* **2006**, *54*, 2705–2711. [CrossRef] [PubMed]
25. Wang, S.; He, Y.; Wang, Y.; Tao, N.; Wu, X.; Wang, X.; Qiu, W.; Ma, M. Comparison of flavour qualities of three sourced *Eriocheir sinensis*. *Food Chem.* **2016**, *200*, 24–31. [CrossRef] [PubMed]
26. Yamaguchi, S.; Yoshikawa, T.; Ikeda, S.; Ninomiya, T. Measurement of the relative taste intensity of some L-α-amino acids and 5′-nucleotides. *J. Food Sci.* **1971**, *36*, 846–849. [CrossRef]
27. Yang, W.; Shi, W.; Zhou, S.; Qu, Y.; Wang, Z. Research on the changes of water-soluble flavor substances in grass carp during steaming. *J. Food Biochem.* **2019**, *43*, e12993. [CrossRef]
28. Combes, S.; Lepetit, J.; Darche, B.; Lebas, F. Effect of cooking temperature and cooking time on Warner–Bratzler tenderness measurement and collagen content in rabbit meat. *Meat Sci.* **2003**, *39*, 91–96. [CrossRef]
29. Sun, L.; Xia, W. Effect of steam cooking on muscle and protein heat-denature of tuna. *Food Mach.* **2010**, *26*, 22–25. [CrossRef]

30. Rabie, M.A.; Peres, C.; Malcata, F.X. Evolution of amino acids and biogenic amines throughout storage in sausages made of horse, beef and turkey meats. *Meat Sci.* **2014**, *96*, 82–87. [CrossRef]
31. Fuke, S.; Konosu, S. Taste-active components in some foods: A review of Japanese research. *Physiol. Behav.* **1991**, *49*, 863–868. [CrossRef]
32. Lioe, H.N.; Apriyantono, A.; Takara, K.; Wada, K.; Yasuda, M. Umami taste enhancement of MSG/NaCl mixtures by subthreshold L-α-Aromatic amino acids. *J. Food Sci.* **2005**, *70*, s401–s405. [CrossRef]
33. Li, X.; Luan, A.; Li, X.; Wang, F.; Huang, Y.; Li, A.; Liu, Y. Protein degradation and aggregation in silver carp (*Hypophthalmichthys molitrix*) muscle during hot air drying. *LWT-Food Sci. Technol.* **2022**, *163*, 113540. [CrossRef]
34. Xie, Q.; Xu, B.; Xu, Y.; Yao, Z.; Zhu, B.; Li, X.; Sun, Y. Effects of different thermal treatment temperatures on volatile flavour compounds of water-boiled salted duck after packaging. *LWT* **2022**, *154*, 112625. [CrossRef]
35. Hong, H.; Regenstein, J.M.; Luo, Y. The importance of ATP-related compounds for the freshness and flavor of post-mortem fish and shellfish muscle: A review. *Food Sci. Nutr.* **2017**, *57*, 1787–1798. [CrossRef] [PubMed]
36. Kong, L.; Cai, C.; Ye, Y.; Chen, D.; Wu, P.; Li, E.; Chen, L.; Song, L. Comparison of non-volatile compounds and sensory characteristics of Chinese mitten crabs (*Eriocheir sinensis*) reared in lakes and ponds: Potential environmental factors. *Aquaculture* **2012**, *364–365*, 96–102. [CrossRef]
37. Zhang, R.; Qiu, W.; Zhang, M.; Ho Row, K.; Cheng, Y.; Jin, Y. Effects of different heating methods on the contents of nucleotides and related compounds in minced Pacific white shrimp and Antarctic krill. *LWT-Food Sci Technol* **2018**, *87*, 142–150. [CrossRef]
38. Mendes, R.; Quinta, R.; Nunes, M.L. Changes in baseline levels of nucleotides during ice storage of fish and crustaceans from the Portuguese coast. *Eur. Food Res. Technol.* **2001**, *212*, 141–146. [CrossRef]
39. Gorbatov, V.M.; Lyaskovskaya, Y.N. Review of the flavour-contributing volatiles and water-soluble non-volatiles in pork meat and derived products. *Meat Sci.* **1980**, *4*, 209–225. [CrossRef]
40. Zhang, N.; Wang, W.; Li, B.; Liu, Y. Non-volatile taste active compounds and umami evaluation in two aquacultured pufferfish (*Takifugu obscurus* and *Takifugu rubripes*). *Food Biosci.* **2019**, *32*, 100468. [CrossRef]
41. Liu, T.; Xia, N.; Wang, Q.; Chen, D. Identification of the non-volatile taste-active components in crab sauce. *Foods* **2019**, *8*, 324. [CrossRef]
42. Feng, C.; Tian, L.; Jiao, Y.; Tan, Y.; Liu, C.; Luo, Y.; Hong, H. The effect of steam cooking on the proteolysis of pacific oyster (*Crassostrea gigas*) proteins: Digestibility, allergenicity, and bioactivity. *Food Chem.* **2022**, *379*, 132160. [CrossRef]
43. Haddi, Z.; Mabrouk, S.; Bougrini, M.; Tahri, K.; Sghaier, K.; Barhoumi, H.; El Bari, N.; Maaref, A.; Jaffrezic-Renault, N.; Bouchikhi, B. E-Nose and e-Tongue combination for improved recognition of fruit juice samples. *Food Chem.* **2014**, *150*, 246–253. [CrossRef] [PubMed]
44. Garcia Hernandez, C.; Salvo Comino, C.; Martin Pedrosa, F.; Garcia Cabezon, C.; Rodriguez Mendez, M.L. Analysis of red wines using an electronic tongue and infrared spectroscopy. Correlations with phenolic content and color parameters. *LWT-Food Sci. Technol.* **2020**, *118*, 108785. [CrossRef]
45. Deng, Y.; Luo, Y.; Wang, Y.; Yue, J.; Liu, Z.; Zhong, Y.; Zhao, Y.; Yang, H. Drying-induced protein and microstructure damages of squid fillets affected moisture distribution and rehydration ability during rehydration. *J. Food Eng.* **2014**, *123*, 23–31. [CrossRef]

foods

MDPI

Review

Endogenous Proteases in Sea Cucumber (*Apostichopus japonicas*): Deterioration and Prevention during Handling, Processing, and Preservation

Xinru Fan [1,3,4], Ke Wu [1], Xiuhui Tian [2], Soottawat Benjakul [5], Ying Li [1,3,4], Xue Sang [1,3,4], Qiancheng Zhao [1,3,4,*] and Jian Zhang [2,*]

1 College of Food Science and Engineering, Dalian Ocean University, Dalian 116023, China
2 Yantai Key Laboratory of Quality and Safety Control and Deep Processing of Marine Food, Shandong Marine Resource and Environment Research Institute, Yantai 264006, China
3 Dalian Key Laboratory of Marine Bioactive Substances Development and High Value Utilization, Dalian 116023, China
4 Liaoning Provincial Marine Healthy Food Engineering Research Centre, Dalian 116023, China
5 International Center of Excellence in Seafood Science and Innovation, Faculty of Agro-Industry, Prince of Songkla University, Hat Yai 90110, Songkhla, Thailand
* Correspondence: qczhao@dlou.edu.cn (Q.Z.); zjsd408@163.com (J.Z.)

Abstract: The sea cucumber is an essential nutrient source and a significant economic marine resource associated with successful aquaculture. However, sea cucumbers are highly susceptible to autolysis induced by endogenous protease after postmortem, and the phenomenon of body wall "melting" occurs, which seriously affects the food quality of products and the degree of acceptance by consumers. To satisfy the growing demand for fresh or processed sea cucumbers, we must clarify the autolysis mechanism of sea cucumbers and the methods to achieve autolysis regulation. In this paper, the factors leading to the quality deterioration and texture softening of sea cucumbers are reviewed, with emphasis on enzymatic characteristics, the autolysis mechanism, the effects of autolysis on the physicochemical properties of the body wall of the sea cucumber, and the development of potential natural protease inhibitors. We aim to provide some reference in future preservation and processing processes for sea cucumbers, promote new processing and preservation technologies, and advance the sea cucumber industry's development.

Keywords: sea cucumber; autolysis mechanism; protease inhibitors; quality deterioration; microstructure

Citation: Fan, X.; Wu, K.; Tian, X.; Benjakul, S.; Li, Y.; Sang, X.; Zhao, Q.; Zhang, J. Endogenous Proteases in Sea Cucumber (*Apostichopus japonicas*): Deterioration and Prevention during Handling, Processing, and Preservation. *Foods* **2024**, *13*, 2153. https://doi.org/10.3390/foods13132153

Academic Editor: Ali Bougatef

Received: 24 May 2024
Revised: 28 June 2024
Accepted: 5 July 2024
Published: 8 July 2024

1. Introduction

With the increasing global demand for fresh and processed aquatic products, food and products with prime quality and safety are required for competitiveness and consumer acceptability. Deterioration, particularly in terms of quality attributes such as texture and flavor, has a detrimental effect on the palatability and market demand for processed aquatic products. Processors are focusing on quality control from the farm to the table. Consequently, storage stability, transportation, and logistics have become crucial issues.

Sea cucumber, especially Japanese sea cucumber (*Apostichopus japonicas*), a marine echinoderm species with abundant nutritional ingredients, has been recognized as a low-cholesterol, low-fat, and low-carbohydrate marine animal [1]. Sea cucumber also comprises a variety of amino acids and vitamins [2,3]. The bioactive compounds found in sea cucumber, including polysaccharides, saponins, and cerebrosides, have demonstrated multiple bioactivities, such as antioxidant effects, anti-hyperlipidemic properties, potential anti-tumor activity, and anticoagulant properties. Additionally, it may contribute to the prevention of a fatty liver [4–6]. Currently, there are approximately 58 species of sea cucumbers recognized as suitable for consumption [7], with the Japanese sea cucumber

(*Apostichopus japonicas*) being a significant economic aquaculture and fishing species in China, Japan, and various other Asian countries and regions [8]. According to FAO [9], the global quantity of cultured Japanese sea cucumbers (*Apostichopus japonicus*) reached approximately 256,000 tons in 2022, representing a significant portion (10.1%) of other types of aquatic animal production. In China, the amount of cultivated sea cucumbers reached 248,508 tons, an increase of 11.59% compared to that reported in the previous year [10]. The growing demand for sea cucumber has increased competition within the aquaculture and food manufacturing industries. Consumer preferences for fresh and alive sea cucumbers have been expanding.

After postmortem, fresh sea cucumber is prone to self-dissolution, known as "autolysis", which significantly affects its storage, transportation, and preservation methods. To counter this undesirable phenomenon, blanching is typically implemented to inactivate enzymes, especially proteases, to preserve their structural integrity [8]. Sea cucumber has been subjected to several processing techniques. Over 80% of processed sea cucumber products are dried, frozen, pre-boiled, salted, and ready-to-eat [11]. Nevertheless, the continued popularity of traditional Chinese deep-processed cuisine has effectively driven innovation to maintain the freshness of raw sea cucumbers for processing sectors.

Autolysis induced by endogenous proteases plays a crucial role in lowering the edible quality, involving texture, mouthfeel and flavor properties of post-harvested sea cucumbers. These endogenous proteases are mainly released from sea cucumbers' body walls or intestines via the disruption of collagen fibers or the interfibrillar proteoglycan bridge in dermis tissues [12]. Subsequently, the liberated proteases can induce texture softening, epidermis damage, and tissue melting, resulting in white spots formed at the dermis. These white spots significantly impact the processing and quality of sea cucumber [13]. Autolysis typically occurs during fishing, transportation, storage and other processing steps due to several factors, such as ultraviolet irradiation (UV), cutting, mechanical damage, and other environmental parameters, such as high temperatures [12,14]. Cathepsin B, D, E, and rMMP-2 have been documented to primarily catalyze the autolysis of sea cucumber through the cleavage of collagen fibers [15,16]. Moreover, an increasing number of novel endogenous enzymes in sea cucumbers have been unearthed, and various endogenous enzymes have been found to exhibit synergistic effects on the hydrolysis of sea cucumbers [17]. Consequently, further exploration is warranted to elucidate the classification, origin, enzymatic properties, and other fundamental information pertaining to internal sea cucumber enzymes. Other unresolved causes need to be further investigated. For instance, the systematic evaluations of the proteases involved in sea cucumber autolysis are limited. Furthermore, it is not clear that multiple endogenous proteases in a sea cucumber's autolysis and the mode of action of those proteases have not been elucidated. Therefore, further research on the sea cucumber autolytic process is still necessary.

This review mainly discusses the origins of endogenous proteases and autolysis on the ultrastructure, textural properties, and chemical composition of the sea cucumber's body wall. The classification, characteristics, possible mechanisms of enzymatic hydrolysis mediated by endogenous proteases, and the protease inhibitors derived from natural resources were explored during sea cucumber processing and preservation. Furthermore, some future research gaps are proposed.

2. Autolysis Phenomenon in Sea Cucumber

2.1. Autolysis in Sea Cucumber during Processing and Storage

Generally, the autolysis of sea cucumbers can be activated by external inducing factors within a short period. The culture environment (including the pH, water temperature, salt concentration, and nutritional condition), exposure to UV light, and mechanical damage (such as injury and cutting) have been known to accelerate autolysis [12,18]. The hydrolysis of structural proteins by aquatic endogenous proteases and UV-induced hydrolysis contribute to the autolysis of sea cucumbers [19,20]. One of the most significant manifestations of autolysis is "melting" [21]. There are two critical stages in the autolysis reactions of sea cucumbers during commercial production. One occurs after postharvest when raw sea cucumbers are subjected to prolonged transportation, harsh manual handling, or undesirable storage conditions. These can lead to the enhanced hydrolysis of structural proteins induced by endogenous proteases [22]. The other arises from inappropriate heating conditions. Sea cucumbers are preboiled (60 °C for 3–5 min, or 90–100 °C for 10–15 min) immediately after being caught to suddenly inactivate endogenous proteases and immobilize their shape [8]. Nonetheless, inadequate processing temperatures could retain some proteases, which still promote hydrolysis, damaging collagen fibers, causing collagen gelation, and destroying the texture [23]. However, based on the temperature-dependent autolysis (Table 1), the tenderization of the texture can be performed, and quality control has been monitored during the thermal treatment of the sea cucumber body wall (SCBW), which has been applied to achieve product specifications in the sea cucumber processing industry [24].

Table 1. A summary of the key endogenous enzymes, related inhibitors, and their effects on physicochemical properties of sea cucumber (*Apostichopus japonicas*) during processing and preservation.

Enzymes Species	Endogenous Enzymes	Heating/Storage Condition	Target Protein	Inhibitors	Metal ions (Activation (↑)/Inactivation (↓))	Molecule Weight	Textural Properties Decrease (↓)	References
Metalloprotease (MMP)	Matrix metalloproteinase-2 (MMP-2)	25 °C for 72 h	Type I collagen	NaN_3 (0.03%), PMSF (5 mM), E-64 (0.1 mM) and $CaCl_2$ (5 mM)	-	-	-	[16]
	Gelatinolytic metalloproteinase (GMP)	37 °C for 18 h	Collagen bands (α1 and β)	EDTA (10 mM); EGTA (10 mM); 1,10-phenanthroline (10 mM)	Ca^{2+} (↑)	45 kDa	-	[25]
	Collagenase type I	30 °C for 72 h	Collagen fibers	-	-	-	-	[13]
	Matrix metalloproteinase (MMP)	Boiled (10 min) sea cucumber stored at 4 °C for 60 d	Extracellular matrix in the dermis; Interfibrillar proteoglycan bridges	EDTA Na_2 (10 mM); 1,10-phenanthroline (10 mM)	Ca^{2+} (↑)	-	Shear force (↓); hardness (↓); elasticity (↓); cohesiveness (↓); chewiness (↓); coverability (↓)	[21,26]
Cysteine protease (CP)	Cysteine proteinases	37 °C for 2~72 h	Collagen fibers	PMSF (1 mM), 1,10-phenanthroline (1 mM), and E-64 (0.5 mM)	-	-	-	[19]
	Cathepsin L-like proteinase	20~70 °C for 30 min	The epidermis of the body wall	E-64 (0.1 mM); Indoacetic acid (1 mM); Antipain (1 mM)	Zn^{2+} (↓); Cu^{2+} (↓); Fe^{2+} (↓)	30.9 kDa	-	[27,28]
	Cathepsin L and Cathepsin B	4 °C for 8 days	Body wall	-	-	-	Hardness (↓); chewiness (↓); springiness (↓); adhesiveness (↓)	[29]
Serine proteinase (SP)	High alkaline protease (serine protease-like)	4~65 °C for 1 h	-	EDTA (10 mM); PMSF (5 mM)	Ca^{2+} (↑); Mg^{2+} (↑); Cu^{2+} (↑); Zn^{2+} (↓); Hg^{2+} (↓);	20.6 kDa; 39.1 kDa; 114.1 kDa;	-	[30]
	Serine proteinase	37 °C for 12 h	Collagen bands (α, β and γ)	Pefabloc SC (2 mM); Benzamidine (5 mM)	-	34 kDa	-	[15]
Mixed protease system	Cysteine (cathepsins, calpains, caspases and proteasomes), serine protease and MMP	37 °C for 1~36 h	Body wall	E-64 (0.1 mM); PMSF (1 mM); 1,10-phenanthroline (1 mM)	-	-	Hardness (↓); chewiness (↓)	[17,24]

2.2. *Role of UV Light in the Autolysis of Sea Cucumber*

The process of apoptosis, also known as programmed cell death, is genetically regulated and involves receptor recognition and signal transduction [24]. It plays a significant role in the autolytic process of sea cucumbers. During the early stage of UVA-induced (400–320 nm) autolysis in sea cucumbers, physiological damage occurs in their body wall, leading to the accumulation of reactive oxygen species (ROS) generated through cellular respiration [24]. This triggers an oxidative stress response within the extracellular matrix, which activates the mitogen-activated protein kinase (MAPK) signaling pathway and accelerates the phosphorylation of extra-cellular regulated protein kinase (ERK), c-Jun N-terminal kinase (JNK), and p38 MAPK [20]. Furthermore, calcium imbalance occurs along with the abnormal morphology of the endoplasmic reticulum. Changes are observed in the mitochondrial membrane potential, resulting in cytochrome c release and the subsequent activation of caspase-3 and caspase-9 enzymes, which lead to DNA degradation and cell death [18,31]. A subsequent study confirmed that the injection of cytosolic calcium chelator, namely 1,2-bis (2-aminophenoxy) ethane-N,N,N′,N′-tetraacetic acid acetoxymethyl ester (pan- caspase inhibitor), into the body cavity effectively regulates the Ca^{2+} release to control apoptosis in coelomic cells [22].

Autophagy can be induced during the apoptosis process of sea cucumber cells. Essential proteins expressed by LC3-II, Atg5, PI3K, AKT, and mTOR autophagosomes cease their expression within 2 h after UV stimulation. Additionally, apoptosis-related signaling pathways such as PI3K/AKT/mTOR are also involved in sea cucumber autophagy [32]. Sea cucumber autolysis typically occurs after 6 h of UV treatment. Therefore, autophagy does not participate in the entire process of sea cucumber autolysis. Moreover, an injection of autophagy inducers can effectively delay UV-induced sea cucumber autolysis, demonstrating that autophagy plays a positive role in maintaining the body wall homeostasis of sea cucumbers [32]. However, current research on regulating sea cucumber autolysis through both apoptosis and autophagy is still experimental and has yet to be implemented in actual practice. However, further pertinent studies are warranted to effectively minimize UV-induced autolysis effectively.

2.3. Physicochemical Changes in the SCBW Induced by Autolysis

2.3.1. Microscopic Molecular Structure

The SCBW primarily comprises collagen, glycoprotein, proteoglycan, and other soluble/insoluble compounds in the reticular formation (considered as 'mutable collagenous tissue or MCT'). The components above constitute the essential edible part of a sea cucumber [13]. Sea cucumber collagen is classified as type I collagen, representing approximately 70% of the SCBW protein content [15]. Collagen fibrils form collagen molecules through covalent helical self-assembly and non-covalent binding via internal electrostatic interactions connected via glycosaminoglycan (GAG) [33]. The collagen fiber structure consists of a collagen fibril network with transverse connections and a proteoglycan bridging structure that collectively constitutes the sea cucumber MCT and determines its mechanical properties [34].

Generally, fresh sea cucumbers can undergo complete autolysis within 72 h. Ultrastructural changes in the body wall of the sea cucumber can be observed through the apparent visualization, tissue staining, and electron microscopy (Figure 1). Following the action of endogenous protease, the stratum corneum on the SCBW exfoliates and disappears, leading to the disorganization of the cell arrangement and fibrous network structure [35]. The aggrecan bridging structure between collagen fibrils was compromised, causing depolymerization into bundles or single fibrils and increasing the space between them [36]. Subsequently, the organization of collagen fibrils is loosened, while the periodic striae becomes blurred or even disappeared [13]. This process was accompanied by the degradation and dissolution of certain insoluble components in the extracellular matrix. This coincides with an increased content of water-soluble protein, soluble collagen, and glycosaminoglycan [12].

2.3.2. Mechanical Properties

The mechanical properties of SCBWs, especially in textural profiles, were estimated during autolysis or the postmortem period (Table 1). Hardness and chewiness were identified as the most prominent indicators of changes in the mechanical properties during the postharvest storage of sea cucumber. The hardness was approximately half of the initial value after the first day of iced storage. By the second day, the softening process was almost complete, in which the hardness value was reduced to only one-fifth of its initial value. With increased storage time, complete autolysis typically occurs by the third day, leading to the loss of the mastication feature and a significant reduction in apparent viscosity due to complete protein hydrolysis in the SCBW [29]. Rheological evaluation has been applied to assess the mechanical properties of ready-to-eat sea cucumber and thermally treated sea cucumber. The apparent viscosity, storage modulus and loss modulus were utilized as assessment parameters. The three indices above were reduced with a prolonged refrigerated storage time, particularly the ready-to-eat sea cucumber injected with inhibitors. It exhibited varying degrees of adverse deterioration during storage [38].

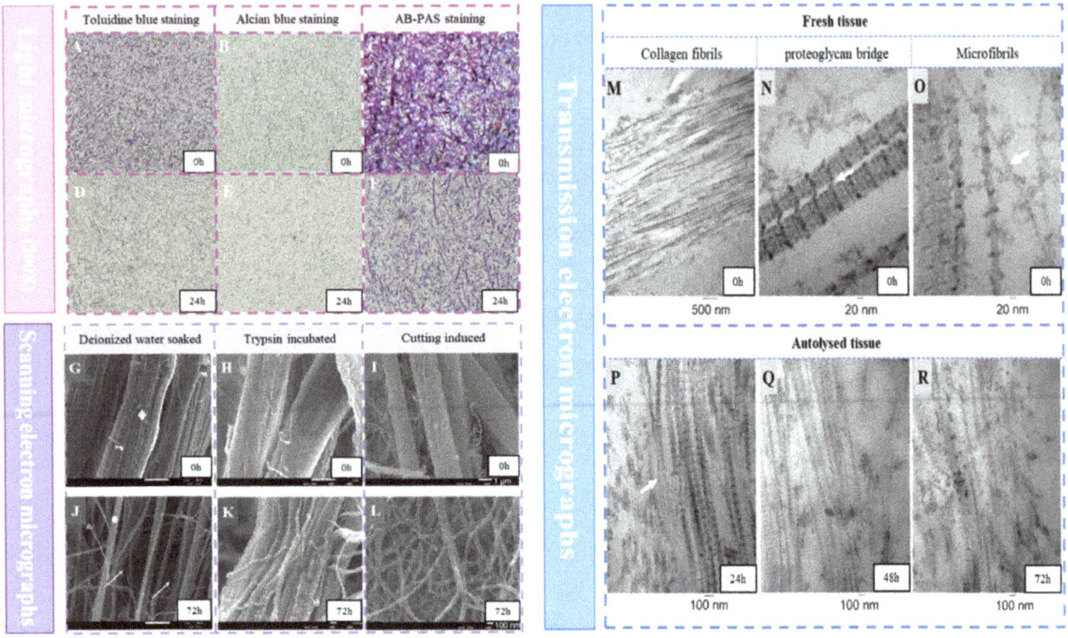

Figure 1. Histological analysis by light microscopy (**A–F**), scanning electron microscopy (**G–L**), and transmission electron microscopy (**M–R**) of fresh and autolyzed sea cucumber (*Apostichopus japonicas*) body walls (SCBWs) at room temperature (20–25 °C) for 72 h [12,37,38]. Note: AB-PAS: alcian blue-periodic acid schiff. White arrow: the location where collagen fibers fracture.

The degree of cross-linking between the body wall fibers was determined by the hardness of the SCBW, which is generally believed to be regulated by the release of endogenous matrix metalloproteinase (MMP) inhibitor [21]. The cross-linking and depolymerization of disulfide bonds and hydrophobic interactions govern the rheological properties. The change in mechanical properties of MCT is due to alterations in the interaction between fibrils (interfibrous matrix), rather than changes in the mechanical properties of the fibrils themselves [38,39]

2.3.3. Chemical Compositions

The autolysis of SCBW can be monitored by the degree of soluble fractions and oxidative indicators [13]. Due to the disintegration of the original microstructure of SCBW collagen fibers during autolysis, the GAG components were precipitated and the content of soluble collagen, such as TCA-soluble peptides, water/acid-soluble collagen, hydroxyproline, and protein fragments increased obviously [40,41]. Sea cucumber typically possesses a stable triple helical collagen structure resistant to extraction or degradation [13]. Endogenous proteases can hydrolyze collagen. The release of calcium ions activated MMP activity in response to external factors, releasing free water. Released water further promoted the occurrence of biochemical reactions [21]. However, the cysteine-containing hydrolyzed non-collagenous proteins, mainly yolk protein and actin, significantly contributed to SCBW deterioration [17,24,42].

Currently, there are two crucial processing methods for manufacturing sea cucumber products. One is the use of high-temperature and high-pressure boiling or steaming (over 100 °C, 0.15–0.21 MPa) to achieve the ripening of the SCBW followed by a water-swollen ripening process for ready-to-eat sea cucumbers [8,43]. The other is to tenderize the SCBW rich in endogenous enzymes at low temperatures and using long-term boiling technology [17]. Each type of thermal treatment usually accompanies an oxidation reaction in the SCBW [23]. Although the autolysis of fresh sea cucumber or pre-boiled frozen cucumber can affect the quality of subsequent processing, it is generally believed that intense processing leads to severe protein oxidation [24]. In particular, the process of free radical oxidation mediated by ROS has been of concern [44]. Hydroxyl radicals were generated in the SCBW heated at 37 °C for 3 h, in which protein oxidation occurred during the mild heating process [45].

3. The Characteristics and Mechanism of Endogenous Enzymes in the Sea Cucumber

3.1. Endogenous Proteases: Characteristics and Mode of Action

Endogenous proteases causing autolysis in the sea cucumber have been isolated and purified from various tissues of sea cucumber carcasses (Figure 2). Other enzymes have also been isolated and characterized. Recently, cathepsin B [46], cathepsin D [47], β-1,3-glucanase [48], alkaline phosphatase (ALP) [30,49], acetylcholine esterase (AChE) [25], serine protease (SP) [15], serine endopeptidases (SEP) [50], and superoxide dismutase (SOD) [51] have been identified in the intestinal tract of sea cucumbers (Table 2). In addition, cathepsin L-like proteinase [27,28], cathepsin D [52], cathepsin E [52], cathepsin K [53], acid phosphatase (ACP) [54], gelatinolytic metalloproteinase (GMP) [55], cysteine-like protease [56], collagenase [57], and α-1,4-amylase [58] have been identified in the SCBW. Furthermore, a novel superoxide dismutase known as a specific deep-sea sea cucumber (*Psychropotes verruciaudatus*) copper-zinc superoxide dismutase (PVCuZnSOD) has been discovered, which has broad application potentials in the fields of bioengineering, cosmetics, and medicine [59].

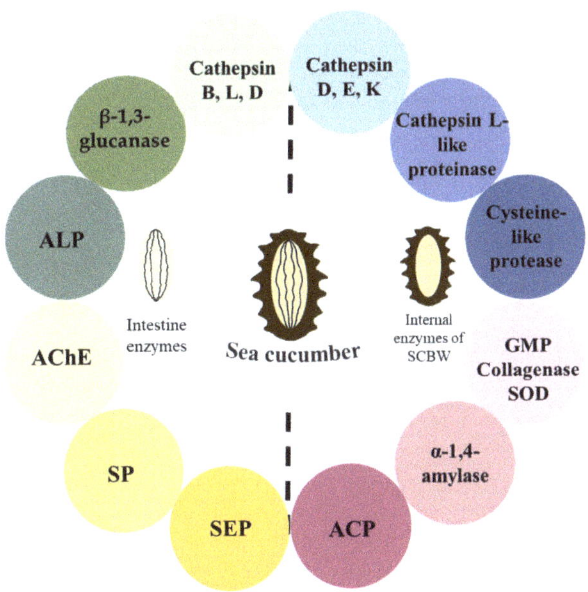

Figure 2. Endogenous enzymes in the intestine and body wall causing the autolysis of the sea cucumber. Note: SCBW: sea cucumber body wall, ALP: acid phosphatase, AChE: acetylcholine esterase, SP: serine protease SEP: serine endopeptidases, GMP: gelatinolytic metalloproteinase, ACP: acid phosphatase, SOD: superoxide dismutase.

Table 2. Characteristics of various endogenous enzymes in sea cucumber.

Sea Cucumber Species	Enzyme Types	Extracted Position	Molecule Weight (kDa)	Optimum Temperature (°C)	Optimum pH	Activator	Inhibitors	References
Janpanese sea cucumber (*Apostichopus japonicus*)	Cathepsin B	Intestine	49	45	5.5	DTT, L-Cys, EGTA, EDTA	E-64, IAA, Antipain, Cu^{2+}, Ni^{2+}, Zn^{2+}	[46]
	Cathepsin L	Intestine	-	48	4.4	-	E-64, IAA, Antipain	[60]
	Cathepsin D	Intestine	-	50	3.0	Cd^{2+}	E-64, IAA, Pepstatin A, Fe^{3+}, Fe^{2+}	[47]
		Body wall		60	5.0	DTT	Pepstatin A, PMSF, Zn^{2+},Cu^{2+}, Fe^{2+},Fe^{3+}, Mn^{2+}	[51]
	β-1,3-glucanase	Intestine	37.5	40	5.5	Mn^{2+}	Cu^{2+}, Ag^{+}, Zn^{2+}, Fe^{2+}, Ca^{2+}	[48]
	ACP	Body wall	148	40	4.0	Mg^{2+}	Zn^{2+}, Cu^{2+}, Fe^{2+}, Fe^{3+}	[54]
	ALP	Intestine	166 ± 9	40	11.0	Mg^{2+}	Zn^{2+}, Ca^{2+}, EDAT	[49]
	AChE	Intestine	68	35	7.5	-	Eserine, BW284C51	[25]
	SP	Intestine	34	35-40	6.0–9.0	EDTA	Leupeptin, Cu^{2+}, Zn^{2+}, Mg^{2+}, Mn^{2+}, Ca^{2+}, Fe^{2+}	[15]
	SEP	Intestine	-	40	9.0	-	PMSF	[50]
	Cathepsin L-like proteinase	Body wall	30.9	50	5.0–5.5	DTT, L-Cys, EDTA	E-64, IAA, Antipain, Zn^{2+}, Fe^{2+}, Cu^{2+}	[27]
	Cathepsin E	Body wall	-	40	4.0	DTT	Pepstatin A, PMSF, Fe^{3+}, Fe^{2+}, Cu^{2+}	[52]
	Cathepsin K	Body wall	-	50	5.0	Mg^{2+}, DTT, L-Cys,	E-64, IAA, Antipain, PMSF, EDTA, Zn^{2+}, Fe^{2+}, Fe^{3+}, Cu^{2+}	[53]
	Cysteine-like protease	Body wall	35.5	50	7.0	L-cysteine hydrochloride	Antipain, leupeptin, Cu^{2+}, Mg^{2+}, Fe^{2+}, Fe^{3+}	[56]
	GMP	Body wall	45	40-45	8.0–9.0	Ca^{2+}, Ba^{2+}	EDTA, EGTA, 1,10-phenanthroline	[55]
	α-1,4-amylase	Body wall	420	80	9.0	Cu^{2+}, Mg^{2+}	Mn^{2+}, K^{+}, Fe^{3+}	[58]
	Collagenase	Body wall	45	40	8.0	Mn^{2+}	EDTA, 1,10-phenanthroline	[57]
	SOD	Body wall	-	30-60	4.0	Ca^{2+}, Zn^{2+}	H_2O_2, Fe^{2+}	[51]
Deep-sea sea cucumber (*Psychoropotes verruciaudatus*)	PVCuZnSOD	-	15	20	4–11	Mg^{2+}, Ni^{2+}	Mn^{2+}, Cu^{2+}, Zn2+, Co^{2+}	[59]

Note: ACP: acid phosphatase, ALP: acid phosphatase, AChE: acetylcholine esterase, SP: serine protease, SEP: serine endopeptidases, GMP: gelatinolytic metalloproteinase, SOD: superoxide dismutase, PVCuZnSOD: Psychropotes verruciaudatus copper-zinc superoxide dismutase, DTT: Dithiothreitol, L-Cys: L-Cysteine, EDTA: ethylenediamine tetraacetic acid, EGTA: ethylene glycol tetraacetic acid, E-64: trans-epoxysuccinyl-L-leucyl-amido (4-guanidino) butane, IAA: iodoacetic acid, PMSF: phenylmethylsulfonyl fluoride, BW284C51: 1,5-bis(4-allyldimethylammonium phenyl)-pentan-3-one dibromide.

These enzymes have been typically isolated and purified using anion exchange chromatography, gel filtration chromatography, the specific fluorescent substrate method, the ammonium sulfate precipitation method, and other techniques [49]. It has been found that sea cucumber primarily contains four types of proteases. Type 1 includes cysteine proteases (CP) like cathepsins B and L, which play a role in cellular protein transformation processes [61]. Moreover, cathepsin B can enhance apoptosis when the sea cucumber is under pathogenic attack [62]. The optimal activity for cathepsin B was found at pH 5.5 and 45 °C. Cathepsin L exhibits its highest activity at pH 5.0–5.5 and 50 °C. Both of these two enzymes can be inhibited by trans-epoxysuccinyl-L-leucyl-amido (4-guanidino) butane (E-64), iodoacetic acid (IAA), antipain, Cu^{2+}, and Zn^{2+} [46,52]. In a fresh SCBW, cathepsin L is in cellular vacuoles released from intracellular stores and diffuses into tissues after UV stimulation for involvement in sea cucumber autolysis [35]. Type 2 consists of serine proteinase (SP). SP is separated from the intestinal tract of the sea cucumber, which effectively hydrolyzes gelatin at pH 6.0–9.0 and 30–40 °C, and is inhibited by leupeptin, Cu^{2+}, Zn^{2+}, Mg^{2+}, Mn^{2+}, Ca^{2+}, and Fe^{2+} [50]. An enzyme extracted from the sea cucumber intestine, regarded as SEP, showed the maximal enzymatic activity at pH 9.0 and 40 °C [51]. Type 3 is mainly aspartic protease. Cathepsin D, the most abundant aspartic protease in lysosomes, is widely expressed in sea cucumber tissues and has been implicated in sea cucumber autolysis. It plays a crucial role in protein degradation, apoptosis, and autophagy [63,64]. Type 4 includes matrix metalloproteinases (MMPs). MMPs are the primary endogenous proteases facilitating sea cucumber autolysis by catalyzing the hydrolysis of collagen, gelatin and membrane proteins [65]. The MMPs identified in sea cucumber autolysis include tensilin, GMP, MMP, and recombinant matrix metalloproteinase-2 (rMMP-2) [55,66]. The enzyme activity can be almost completely inhibited under the combined action of enzyme inhibitors, especially EDTA and 1,10-phenanthroline [60].

3.2. ROS Mediated Oxidative Stress: Occurrence and Mode of Action

When sea cucumbers experience an imbalance in ROS homeostasis, a reduced ROS scavenging ability and subsequent enrichment of ROS occur, resulting in oxidative stress (OS) reactions [24]. The stress can activate endogenous proteases via two main mechanisms (Figure 3) [9]. First, through direct effects, when ROS attack, lysosomal membrane destruction takes place. As a result, permeability increases, causing the release of small molecular weight cathepsin B, cathepsin L, and cathepsin D (approximately 43 kDa) directly into the body wall of sea cucumbers, in which hydrolysis can occur. Secondly, by controlling the release of metal ions, highly oxidizing hydroxyl radicals from ROS disrupt Ca^{2+} regulation by folding proteins on the endoplasmic reticulum. This leads to an excessive cytoplasmic Ca^{2+} release, activating endogenous proteases, such as calpain and MMP [21]. Consequently, cytoskeletal protein hydrolysis occurs, resulting in cell death, while caspase activity is further accelerated, ultimately inducing apoptosis [24]. Moreover, via the regulation of cysteine release, ROS can activate MMP activities, thus achieving a bidirectional control of apoptosis [21].

3.3. Degradation of the SCBW

There is a complex enzyme system in sea cucumbers. Synergistic or competitive effects of various endogenous enzymes contribute to the deterioration of sea cucumber processing and storage stability [13]. Cysteine protease is involved in the hydrolysis of collagen and non-collagen proteins in the SCBW [19,42]. By disrupting the proteoglycan bridge between fibers, cysteine protease causes the partial depolymerization of collagen fibers in the sea cucumber body wall into collagen fibrils, which release glycosaminoglycan, hydroxyproline, and collagen fragments, and increase the degree of structural disorder of collagen fibers [19]. Cathepsin B, L, and K are lysosomal cysteine proteases. When the UV-induced autolysis of sea cucumber body wall occurs, the gene expression of cathepsin L is up-regulated, and the enzyme activities of cathepsin B and cathepsin L increase [14,60]. Cathepsin B has both endopeptidase and exopeptidase activities capable of degrading collagen, connective tissue proteins, and certain natural enzymes [67]. The content of cathepsin L is higher in the sea cucumber epidermis than in the dermis. Its density and distribution are proportional to the sea cucumber autolysis rate and extent [27,35]. Under UV induction, cathepsin L is released from the cell and makes contact with the extracellular substrate collagen fiber, which decomposes into collagen fibrils, thus weakening the sea cucumber texture [53]. Cathepsin K can induce multi-site degradation on stable collagen triple helix structures while degrading the body's elastic fiber [68,69]. In the autolysis process of sea cucumber via the dissolution of lysosomes and the release of cathepsin, cathepsin K can hydrolyze the collagen of the SCBW under acidic conditions and the non-collagen part under neutral conditions [53].

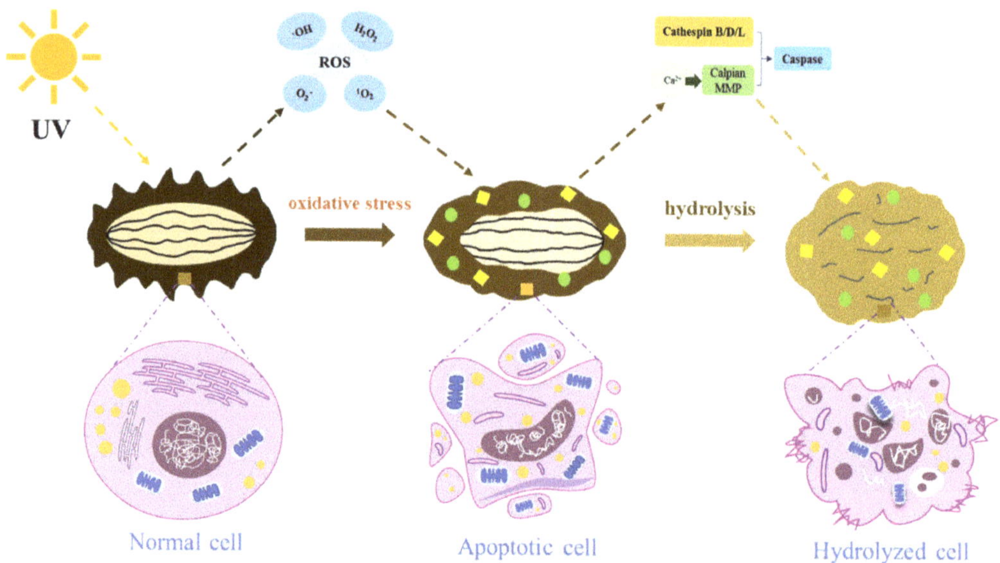

Figure 3. The mechanism of apoptosis participating in the autolysis process of a sea cucumber under UV induction. Note: ROS: reactive oxygen species, MMP: matrix metalloproteinase, "■": glycosaminoglycan (GAG), "●": soluble collagen.

Serine protease (SP) is a kind of protease with a serine group in active sites, which can cleave protein chains and activate MMPs [15]. Nevertheless, SP can inactivate MMPs as tissue inhibitors [15]. SP extracted from the intestine of sea cucumber will destroy serine residual sites in crude collagen fibers, reduce the cross-linking of sea cucumber collagen fibers, and soften the SCBW [15]. Trypsin is an SP, which can decompose the proteoglycan bridge structure between collagen fibrils during the autolysis process of sea cucumber, resulting in the release of water-soluble glycosaminoglycan and protein fragments of sea cucumber [37].

Cathepsin D and cathepsin E are found in sea cucumbers and belong to the aspartic protease family [70]. Cystatin D has been identified in lysosomes and is widely distributed in various tissues and cells of mammals, especially in sea cucumbers, with the highest content in the intestine followed by the muscle and body wall [71]. Cathepsin E is a non-secretory intracellular protease with a relatively limited distribution. It exists only in some sea cucumber tissues [52]. Cystatin D participates in protein degradation under the strong acid conditions of lysosomes with strong fibrin and fibrinogen degradation activity and the hydrolysis of the carboxyl COOH end peptide of collagen [63]. Cathepsin D and E have serine and cysteine residues that regulate their activity, which may be involved in protein degradation in the autolysis process of the sea cucumber [52]. In addition, the relative expression level of the cathepsin D gene is significantly up-regulated during the autolysis of sea cucumber, and the purified recombinant sea cucumber cathepsin D has been confirmed to promote the degradation of sea cucumber muscle [64].

MMP primarily functions to degrade various proteins, such as collagen, mucin, glycoprotein and other components of the extracellular matrix [72]. Tensilin was isolated, purified, and identified as a member of the MMP family from the body wall of the North Atlantic sea cucumber (*Cucumaria frondosa*) [66]. Tensilin is a collagen fiber-binding protein that interacts with stiparin, inhibiting its ability to bond collagen fibers [73]. Gelatinase has also been isolated and identified from the SCBW. Its inhibition can effectively terminate the dissolution of soluble proteins in the autolysis process of sea cucumber [55]. The proteoglycan bridge can be destroyed by MMPs, resulting in the complete depolymerization of collagen fibers into smaller collagen fibril bundles and fibrils. Additionally, MMPs cause partial degradation of collagen fibers by releasing soluble hydroxyproline and pyridine crosslinking products [26]. Among the members of the MMP family, rMMP-2 is a distinctive protease that efficiently catalyzes the hydrolysis of type I collagen and gelatin, leading to the decomposition of collagen fibers into fibrils in the SCBW [41].

4. Inhibitors for Sea Cucumber Endogenous Enzyme Proteases

Endogenous proteases, especially MMP, are pivotal enzymes in sea cucumber autolysis, particularly at the body wall, causing the quality loss during processing, transportation, and short-term storage [41]. Therefore, the inhibition of SCBW autolysis is mainly achieved through chelators of metal ions, which are required for MMP activity. Moreover, some natural extracts and organic acids are considered to be sea cucumber endogenous protease inhibitors. EDTA and 1,10-phenanthroline, as bivalent metal chelating agents, have demonstrated a positive inhibitory effect toward SCBW hydrolysis [57]. Under the stimulation of external conditions, the content of Ca^{2+} in the dermis of sea cucumber increases, leading to the synchronous activation of the MMP function. Adding an MMP inhibitor (1,10-phenanthroline) allows the chelation of Ca^{2+}, thus inhibiting MMP activation, and delaying the autolysis of the sea cucumber [21]. Previous studies have revealed that UV irradiation induces the release of Ca^{2+} from the endoplasmic reticulum and activates the apoptosis pathway in sea cucumbers [20]. This results in the upregulation of casase-3 and casase-9 associated with apoptosis-related metabolic enzymes, ultimately leading to coelom cell apoptosis [17]. However, injecting a cytoplasmic calcium chelator into a sea cucumber effectively maintains cellular Ca^{2+} homeostasis and inhibits sea cucumber autolysis until 48 h have passed by controlling the cell apoptosis processes [22].

Tea polyphenol has the chelating ability of metal ions [74]. Oxalic acid is an organic acid with strong metal-chelating characteristics [75]. When fresh sea cucumber was soaked in 0.2% (w/v) tea polyphenol and 0.2% (w/v) oxalic acid solution, the endogenous enzyme activity could be considerably inhibited by chelating calcium ions, which could dramatically impede the degradation and oxidation of sea cucumber proteins during refrigeration storage, thus reducing the release of soluble proteins, glycosaminoglycans and hydroxyproline in sea cucumber. Consequently, the textural deterioration (softening) of sea cucumber quality can be delayed 48 h (total shelf life of 72 h) [38]. Furthermore, there is growing interest in brown algae feedstocks rich in polyphenols sourced from marine resources as potential inhibitors. Extracts of phlorotannins obtained from brown algae (*Ascophyllum nodosum*) have been demonstrated to promote the cross-linking and aggregation of endogenous proteases within sea cucumbers while reducing the thermal stability of proteases. Those extracts could inhibit enzymes in a dose-dependent manner [76]. During subsequent heating processes (37 °C for 3 h), these polyphenol and phlorotannin extracts can effectively hinder oxidation generation by eliminating free radicals, mitigate protein side chain modification, enhance thermal stability properties of the body wall texture, and delay protein degradation during processing [77,78]. The polyphenol oxidase-mediated (-)-epigallocatechin gallate has been verified to inhibit CP activities by modifying the enzyme structure and promoting protein oxidation [79]. The coelomic fluid isolate from sea cucumber has also been demonstrated to enhance the inhibition rate of SP and CP in sea cucumber up to 56.2–87.2%, which can serve as a potential source for the preparation of inhibitors [80].

5. Conclusions

Sea cucumbers are susceptible to self-degradation and tissue damage, ultimately causing evisceration and significant economic losses. Endogenous enzymes, particularly proteases, in sea cucumbers are primarily involved in autolysis. Additionally, UV-induced stress also triggers the autolysis of sea cucumbers. To tackle these problems, the specific inhibitors, which can maintain cellular homeostasis and inhibit proteolysis, can be employed to retard the deterioration of sea cucumber quality. Various natural products and chemical compounds, such as protease inhibitors, can suppress the activity of autolytic enzymes in sea cucumbers. Future research should focus on developing effective protease inhibitors to stabilize sea cucumber quality during storage and processing. Optimizing inhibitor application methods, understanding the underlying biochemical mechanisms, and integrating sustainable processing technologies are crucial steps toward enhancing sea cucumber production's economic viability and competitiveness. These efforts will advance scientific knowledge and support the seafood industry in meeting growing market demands sustainably.

Author Contributions: Conceptualization, X.F.; methodology, X.F.; investigation, Y.L. and X.S.; data curation, X.F.; writing—original draft preparation, X.F. and K.W.; writing—review and editing, S.B.; supervision, S.B.; project administration, X.T. and J.Z.; funding acquisition, Q.Z. All authors have read and agreed to the published version of the manuscript.

Funding: This research was funded by the Foundation of the Liaoning Educational Committee (grant number JYTQN2023127), the Fund of the Yantai Key Laboratory of Quality and Safety Control and Deep Processing of Marine Food (grant number QSCDP202306), the Star of Science and Technology of Dalian (grant number 2023RQ048), and the Natural Science Foundation of Liaoning Province (grant number 2022-BS-270).

Institutional Review Board Statement: Not applicable.

Informed Consent Statement: Not applicable.

Data Availability Statement: No new data were created or analyzed in this study. Data sharing is not applicable to this article.

Conflicts of Interest: The authors declare no conflicts of interest.

References

1. Pangestuti, R.; Arifin, Z. Medicinal and health benefit effects of functional sea cucumbers. *J. Tradit. Complement. Med.* **2018**, *8*, 341–351. [CrossRef] [PubMed]
2. Haider, M.S.; Sultana, R.; Jamil, K.; Lakht-e-Zehra, O.M.T.; Shirin, K.G.; Afzal, W. A study on proximate composition, amino acid profile, fatty acid profile and some mineral contents in two species of sea cucumber. *J. Anim. Plant Sci.* **2015**, *25*, 168–175.
3. Yin, P.P.; Jia, A.R.; Heimann, K.; Zhang, M.S.; Liu, X.; Zhang, W.; Liu, C.H. Hot water pretreatment-induced significant metabolite changes in the sea cucumber *Apostichopus japonicus*. *Food Chem.* **2020**, *314*, 126211. [CrossRef] [PubMed]
4. Liu, X.; Sun, Z.L.; Zhang, M.S.; Meng, X.M.; Xia, X.K.; Yuan, W.P.; Xue, F.; Liu, C.H. Antioxidant and antihyperlipidemic activities of polysaccharides from sea cucumber *Apostichopus japonicus*. *Carbohydr. Polym.* **2012**, *90*, 1664–1670. [CrossRef] [PubMed]
5. Zhao, Y.C.; Xue, C.H.; Zhang, T.T.; Wang, Y.M. Saponins from sea cucumber and their biological activities. *J. Agric. Food Chem.* **2018**, *66*, 7222–7237. [CrossRef]
6. Xu, J.; Wang, Y.M.; Feng, T.Y.; Zhang, B.; Sugawara, T.; Xue, C.H. Isolation and anti-fatty liver activity of a novel cerebroside from the sea cucumber *Acaudina molpadioides*. *Biosci. Biotechnol. Biochem.* **2011**, *75*, 1466–1471. [CrossRef]
7. Sánchez-Tapia, I.A.; Slater, M.; Olvera-Novoa, M.A. Evaluation of the growth and survival rate of the Caribbean Sea cucumber, *Isostichopus badionotus* (Selenka, 1867), early juveniles produced in captivity. *J. World Aquac. Soc.* **2019**, *50*, 763–773. [CrossRef]
8. Fan, X.R.; Ma, Y.S.; Li, M.; Li, Y.; Sang, X.; Zhao, Q.C. Thermal treatments and their influence on physicochemical properties of sea cucumbers: A comprehensive review. *Int. J. Food Sci. Technol.* **2022**, *57*, 5790–5800. [CrossRef]
9. FAO. *The State of World Fisheries and Aquaculture: Blue Transformation in Action*; FAO: Rome, Italy, 2024; p. 33.
10. *China Fishery Statistics Yearbook*; China Agriculture Press: Beijing, China, 2023.
11. Duan, X.; Zhang, M.; Mujumdar, A.S.; Wang, S.J. Microwave freeze drying of sea cucumber (*Stichopus japonicus*). *J. Food. Eng.* **2010**, *96*, 491–497. [CrossRef]

12. Liu, Y.X.; Zhou, D.Y.; Liu, Z.Q.; Lu, T.; Song, L.; Li, D.M.; Dong, X.P.; Qi, H.; Zhu, B.W.; Shahidi, F. Structural and biochemical changes in dermis of sea cucumber (*Stichopus japonicus*) during autolysis in response to cutting the body wall. *Food Chem.* **2018**, *240*, 1254–1261. [CrossRef]
13. Liu, Y.X.; Liu, Z.Q.; Song, L.; Ma, Q.R.; Zhou, D.Y.; Zhu, B.W.; Shahidi, F. Effects of collagenase type I on the structural features of collagen fibres from sea cucumber (*Stichopus japonicus*) body wall. *Food Chem.* **2019**, *301*, 125302. [CrossRef] [PubMed]
14. Zhu, B.W.; Zheng, J.; Zhang, Z.S.; Dong, X.P.; Zhao, L.L.; Mikiro, T. Autophagy plays a potential role in the process of sea cucumber body wall "Melting" Induced by UV irradiation. *Wuhan Univ. J. Nat. Sci.* **2008**, *13*, 232–238. [CrossRef]
15. Yan, L.J.; Zhan, C.L.; Cai, Q.F.; Weng, L.; Du, C.H.; Liu, G.M.; Su, W.J.; Cao, M.J. Purification, characterization, cDNA cloning and in vitro expression of a serine proteinase from the intestinal tract of sea cucumber (*Stichopus japonicus*) with collagen degradation activity. *J. Agric. Food Chem.* **2014**, *62*, 4769–4777. [CrossRef] [PubMed]
16. Yan, L.J.; Sun, L.C.; Cao, K.Y.; Chen, Y.L.; Zhang, L.J.; Liu, G.M.; Jin, T.C.; Cao, M.J. Type I collagen from sea cucumber (*Stichopus japonicus*) and the role of matrix metalloproteinase-2 in autolysis. *Food Biosci.* **2021**, *41*, 100959. [CrossRef]
17. Dong, X.F.; He, B.Y.; Jiang, D.; Yu, C.X.; Zhu, B.W.; Qi, H. Proteome analysis reveals the important roles of protease during tenderization of sea cucumber *Apostichopus japonicus* using iTRAQ. *Food Res. Int.* **2019**, *131*, 108632. [CrossRef]
18. Yang, J.F.; Gao, R.C.; Wu, H.T.; Li, P.F.; Hu, X.S.; Zhou, D.Y.; Zhu, B.W.; Su, Y.C. Analysis of apoptosis in ultraviolet-induced sea cucumber (*Stichopus japonicus*) melting using terminal deoxynucleotidyl-transferase-mediated dutp nick end-labeling assay and cleaved caspase-3 immunohistochemistry. *J. Agric. Food Chem.* **2015**, *63*, 9601–9608. [CrossRef] [PubMed]
19. Liu, Y.X.; Zhou, D.Y.; Ma, D.D.; Liu, Z.Q.; Liu, Y.F.; Song, L.; Dong, X.P.; Li, D.M.; Zhu, B.W.; Konno, K.; et al. Effects of endogenous cysteine proteinases on structures of collagen fibres from dermis of sea cucumber (*Stichopus japonicus*). *Food Chem.* **2017**, *232*, 10–18. [CrossRef]
20. Qi, H.; Fu, H.; Dong, X.f.; Feng, D.D.; Li, N.; Wen, C.R.; Nakamura, Y.; Zhu, B.W. Apoptosis induction is involved in UVA-induced autolysis in sea cucumber *Stichopus japonicus*. *J. Photochem. Photobiol. B Biol.* **2016**, *158*, 130–135. [CrossRef]
21. Liu, Z.Q.; Zhou, D.y.; Liu, Y.X.; Liu, X.Y.; Liu, Y.; Liu, B.; Song, L.; Shahidi, F. In vivo mechanism of action of matrix metalloprotease (MMP) in the autolysis of sea cucumber (*Stichopus japonicus*). *J. Food Process. Preserv.* **2020**, *44*, e14383. [CrossRef]
22. Tan, Z.F.; Ding, Y.T.; Tian, J.Y.; Liu, Z.Q.; Bi, J.R.; Zhou, D.Y.; Song, L.; Chen, G.B. Inhibition of ultraviolet-induced sea cucumber (*Stichopus japonicus*) autolysis by maintaining coelomocyte intracellular calcium homeostasis. *Food Chem.* **2022**, *368*, 130768. [CrossRef]
23. Wang, J.; Lin, L.; Sun, X.; Hou, H. Mechanism of sea cucumbers (*Apostichopus japonicus*) body wall changes under different thermal treatment at micro-scale. *LWT-Food Sci. Technol.* **2020**, *130*, 109461. [CrossRef]
24. Dong, X.F.; Qi, H.; Feng, D.D.; He, B.Y.; Zhu, B.W. Oxidative stress involved in textural changes of sea cucumber *Stichopus japonicus* body wall during low-temperature treatment. *Int. J. Food Prop.* **2018**, *21*, 2646–2659. [CrossRef]
25. Wu, H.T.; Li, D.M.; Zhu, B.W.; Du, Y.; Chai, X.Q.; Murata, Y. Characterization of acetylcholinesterase from the gut of sea cucumber *Stichopus japonicus*. *Fish. Sci.* **2013**, *79*, 303–311. [CrossRef]
26. Sun, L.M.; Wang, T.T.; Zhu, B.W.; Niu, H.L.; Zhang, R.; Hou, H.M.; Zhang, G.L.; Murata, Y. Effect of matrix metalloproteinase on autolysis of sea cucumber *Stichopus japonicus*. *Food Sci. Biotechnol.* **2013**, *22*, 1–3. [CrossRef]
27. Zhou, D.Y.; Chang, X.N.; Bao, S.S.; Song, L.; Zhu, B.W.; Dong, X.P.; Zong, Y.; Li, D.M.; Zhang, M.M.; Liu, Y.X.; et al. Purification and partial characterisation of a cathepsin L-like proteinase from sea cucumber (*Stichopus japonicus*) and its tissue distribution in body wall. *Food Chem.* **2014**, *158*, 192–199. [CrossRef] [PubMed]
28. Zhu, B.W.; Zhao, L.L.; Sun, L.M.; Li, D.M.; Murata, Y.; Yu, L.; Zhang, L. Purification and characterization of a cathepsin L-like enzyme from the body wall of the sea cucumber *Stichopus japonicus*. *Biosci. Biotechnol. Biochem.* **2008**, *72*, 1430–1437. [CrossRef] [PubMed]
29. Xiong, X.; He, B.Y.; Jiang, D.; Dong, X.F.; Yu, C.X.; Hang, Q. Postmortem biochemical and textural changes in the sea cucumber *Stichopus japonicus* body wall (SJBW) during iced storage. *LWT-Food Sci. Technol.* **2020**, *118*, 108705. [CrossRef]
30. Fu, X.Y.; Xue, C.H.; Miao, B.C.; Li, Z.J.; Yang, W.G.; Wang, D.F. Study of a highly alkaline protease extracted from digestive tract of sea cucumber (*Stichopus japonicus*). *Food Res. Int.* **2005**, *38*, 323–329. [CrossRef]
31. Su, L.; Yang, J.F.; Fu, X.; Dong, L.; Zhou, D.Y.; Sun, L.M.; Gong, Z.W. Ultraviolet-ray-induced sea cucumber (*Stichopus japonicus*) melting is mediated by the caspase-dependent mitochondrial apoptotic pathway. *J. Agric. Food Chem.* **2018**, *66*, 45–52. [CrossRef]
32. Sun, J.H.; Zheng, J.; Wang, Y.N.; Yang, S.L.; Yang, J.F. The exogenous autophagy inducement alleviated the sea cucumber (*Stichopus japonicus*) autolysis with exposure to stress stimuli of ultraviolet light. *J. Sci. Food Agric.* **2022**, *102*, 3416–3424. [CrossRef]
33. Scott, J.E. Supramolecular organization of extracellular matrix glycosaminoglycans, in vitro and in the tissues. *FASEB J.* **1992**, *6*, 2639–2645. [CrossRef]
34. Scott, J.E.; Orford, C.R.; Hughes, E.W. Proteoglycan-collagen arrangements in developing rat tail tendon. An electron microscopical and biochemical investigation. *Biochem. J.* **2019**, *195*, 573–581. [CrossRef] [PubMed]
35. Liu, Y.X.; Zhou, D.Y.; Ma, D.D.; Liu, Y.F.; Li, D.M.; Dong, X.P.; Tan, M.Q.; Du, M.; Zhu, B.W. Changes in collagenous tissue microstructures and distributions of cathepsin L in body wall of autolytic sea cucumber (*Stichopus japonicus*). *Food Chem.* **2016**, *212*, 341–348. [CrossRef]
36. Liang, X.W. Changes of Collagenous Tissue in Body Wall of Autolytic Sea Cucumber Sitcgopus japonicus. Master's thesis, Dalian Polytechnic University, Dalian, China, 2016. (In Chinese)
37. Liu, Z.Q.; Tuo, F.Y.; Song, L.; Liu, Y.X.; Dong, X.P.; Li, D.M.; Zhou, D.Y.; Shahidi, F. Action of trypsin on structural changes of collagen fibres from sea cucumber (*Stichopus japonicus*). *Food Chem.* **2018**, *256*, 113–118. [CrossRef] [PubMed]

38. Liu, Z.Q.; Zhou, D.Y.; Liu, Y.X.; Yu, M.M.; Liu, B.; Song, L.; Dong, X.P.; Qi, H.; Shahidi, F. Inhibitory effect of natural metal ion chelators on the autolysis of sea cucumber (*Stichopus japonicus*) and its mechanism. *Food Res. Int.* **2020**, *133*, 109205. [CrossRef]
39. Bi, J.R.; Yong, L.; Cheng, S.S.; Dong, X.P.; Kamal, T.; Zhou, D.Y.; Li, D.M.; Jiang, P.F.; Zhu, B.W.; Tan, M. Changes in body wall of sea cucumber (*Stichopus japonicus*) during a two-step heating process assessed by rheology, LF-NMR, and texture profile analysis. *Food Biophys.* **2016**, *11*, 257–265. [CrossRef]
40. Liu, Z.Q.; Li, D.Y.; Song, L.; Liu, Y.X.; Yu, M.M.; Zhang, M.; Rakariyatham, K.; Zhou, D.Y.; Shahidi, F. Effects of proteolysis and oxidation on mechanical properties of sea cucumber (*Stichopus japonicus*) during thermal processing and storage and their control. *Food Chem.* **2020**, *330*, 127248. [CrossRef]
41. Liu, Z.Q.; Liu, Y.X.; Zhou, D.Y.; Liu, Y.X.; Dong, X.P.; Li, D.M.; Shahidi, F. The role of matrix metalloprotease (MMP) to the autolysis of sea cucumber (*Stichopus japonicus*). *J. Sci. Food Agric.* **2019**, *99*, 5752–5759. [CrossRef]
42. Wu, H.T.; Li, D.M.; Zhu, B.W.; Sun, J.J.; Zheng, J.; Wang, F.L.; Konno, K.; Xi, J. Proteolysis of noncollagenous proteins in sea cucumber, *Stichopus japonicus*, body wall: Characterisation and the effects of cysteine protease inhibitors. *Food Chem.* **2013**, *141*, 1287–1294. [CrossRef] [PubMed]
43. Gu, P.; Qi, S.Z.; Zhai, Z.Y.; Liu, J.; Liu, Z.Y.; Jin, Y.; Qi, Y.X.; Zhao, Q.C.; Wang, F.J. Comprehensive proteomic analysis of sea cucumbers (*Stichopus japonicus*) in thermal processing by HPLC-MS/MS. *Food. Chem.* **2022**, *373*, 131368. [CrossRef]
44. Dong, X.F.; Shen, P.; Yu, M.Q.; Yu, C.X.; Zhu, B.W.; Hang, Q. (-)-Epigallocatechin gallate protected molecular structure of collagen fibers in sea cucumber *Apostichopus japonicus* body wall during thermal treatment. *LWT-Food Sci. Technol.* **2020**, *123*, 109076. [CrossRef]
45. Xiong, X.; Xie, W.C.; Xie, J.W.; Qi, H.; Yang, X.H.; Chen, H.X.; Song, L.; Dong, X.F. Protein oxidation results in textural changes in sea cucumber (*Apostichopus japonicus*) during tenderization. *LWT-Food Sci. Technol.* **2021**, *144*, 111231. [CrossRef]
46. Sun, L.M.; Zhu, B.W.; Wu, H.T.; Yu, L.; Zhou, D.Y.; Dong, X.P.; Yang, J.F.; Li, D.M.; Ye, W.X.; Murata, Y. Purification and characterization of cathepsin B from the gut of the sea cucumber (*Stichopus japonicas*). *Food Sci. Biotechnol.* **2011**, *20*, 919–925. [CrossRef]
47. Guo, X.K.; Li, A.T.; Han, J.R.; Du, Y.N.; Guo, T.M.; Yu, C.P.; Tang, Y.; Wu, H.T. Extraction and characterization of cathepsin D from sea cucumber (*Stichopus japonicus*) guts. *Sci. Technol. Food Ind.* **2017**, *18*, 135–139. (In Chinese) [CrossRef]
48. Zhu, B.W.; Zhao, J.G.; Yang, J.F.; Mikiro, T.; Zhang, Z.S.; Zhou, D.Y. Purification and partial characterization of a novel β-1,3-glucanase from the gut of sea cucumber *Stichopus japonicus*. *Process. Biochem.* **2008**, *43*, 1102–1106. [CrossRef]
49. Wu, H.T.; Li, D.M.; Zhu, B.W.; Cheng, J.H.; Sun, J.J.; Wang, F.L.; Yang, Y.L.; Song, Y.K.; Yu, C.X. Purification and characterization of alkaline phosphatase from the gut of sea cucumber *Stichopus japonicus*. *Fish. Sci.* **2013**, *79*, 477–485. [CrossRef]
50. Xu, S.Q.; Zhang, Z.Y.; Nie, B.; Du, Y.N.; Tang, Y.; Wu, H.T. Characteristics of the intestine extracts and their effect on the crude collagen fibers of the body wall from sea cucumber *Apostichopus japonicus*. *Biology* **2023**, *12*, 705. [CrossRef] [PubMed]
51. Fu, H.; Zhang, J.Y.; Zhao, X.T.; Liu, Q.; Xu, Z.; Feng, D.D.; Wei, J.Y.; Hang, Q. Extraction and characterization of superoxide dismutase from the body wall of sea cucumber *Stichopus japonicus*. *Food Sci. Technol.* **2015**, *40*, 150–154. (In Chinese) [CrossRef]
52. Wang, L.; Nian, Y.Y.; Xu, P.; Ji, X.T.; Cui, Y.T.; Zhang, G.L.; Hou, H.M.; Sun, L.M. Two aspartic proteases in sea cucumber (*Stichopus japonicus*): Enzymatic properties and effect on autolysis. *Food Sci.* **2018**, *39*, 99–105. (In Chinese) [CrossRef]
53. Ji, X.T.; Wang, L.; Xue, P.; Nian, Y.Y.; Zhou, W.R.; Zhang, Y.F.; Zhang, G.L.; Hou, H.M.; Sun, L.M. Characteristics of cathepsin K of sea cucumber *Stichopus japonicus* and its effects on autolysis. *J. China Agric. Univ.* **2017**, *22*, 72–77. (In Chinese)
54. Zhu, B.W.; Yu, J.W.; Zhang, Z.S.; Zhou, D.Y.; Yang, J.F.; Li, D.M.; Murata, Y. Purification and partial characterization of an acid phosphatase from the body wall of sea cucumber *Stichopus japonicus*. *Process. Biochem.* **2009**, *44*, 875–879. [CrossRef]
55. Wu, H.L.; Hu, Y.Q.; Shen, J.D.; Cai, Q.F.; Liu, G.M.; Su, W.J.; Cao, M.J. Identification of a novel gelatinolytic metalloproteinase (GMP) in the body wall of sea cucumber (*Stichopus japonicus*) and its involvement in collagen degradation. *Process. Biochem.* **2013**, *48*, 871–877. [CrossRef]
56. Qi, H.; Dong, X.P.; Gao, Y.; Liu, L.; Mikiro, T.; Zhu, B.W. Purification and characterization of a cysteine-like protease from the body wall of the sea cucumber *Stichopus japonicus*. *Fish Physiol. Biochem.* **2007**, *33*, 181–188. [CrossRef]
57. Zhong, M.; Hu, C.Q.; Ren, C.H.; Luo, X.; Cai, Y.M. Characterization of a main extracellular matrix autoenzyme from the dermis of sea cucumber *Stichopus monotuberculatus*: Collagenase. *Int. J. Food Prop.* **2016**, *19*, 2495–2509. [CrossRef]
58. Zhang, J.; Zhang, Y.Q.; Luo, C.H.; Liu, Z.D.; Cheng, Y.F.; Lu, F.F.; Zhang, T. Purifi cation and characterization of α-1,4-amylase in body wall of sea cucumber. *Food Sci.* **2015**, *36*, 137–141. (In Chinese) [CrossRef]
59. Li, Y.N.; Chen, Z.F.; Zhang, P.; Gao, F.; Wang, J.F.; Lin, L.; Zhang, H.B. Characterization of a novel superoxide dismutase from a deep-sea sea cucumber (*Psychoropotes verruciaudatus*). *Antioxidants* **2023**, *12*, 1227. [CrossRef]
60. Yan, J.N.; Guo, X.K.; Tang, Y.; Li, A.T.; Zhu, Z.M.; Chai, X.Q.; Duan, X.H.; Wu, H.T. Contribution of cathepsin L to autolysis of sea cucumber *Stichopus japonicus* intestines. *J. Aquat. Food Prod. Technol.* **2019**, *28*, 233–240. [CrossRef]
61. Martin, S.L.; Moffitt, K.L.; McDowell, A.; Greenan, C.; Bright-Thomas, R.J.; Jones, A.M.; Webb, A.K.; Elborn, J.S. Association of airway cathepsin B and S with inflammation in cystic fibrosis. *Pediatr. Pulmonol.* **2010**, *45*, 860–868. [CrossRef]
62. Jiang, P.Z.; Gao, S.; Chen, Z.F.; Sun, H.J.; Li, P.P.; Yue, D.M.; Pan, Y.J.; Wang, X.D.; Mi, R.; Dong, Y.; et al. Cloning and characterization of a phosphomevalonate kinase gene that is involved in saponin biosynthesis in the sea cucumber *Apostichopus japonicus*. *Fish Shellfish Immunol.* **2022**, *128*, 67–73. [CrossRef]
63. Benes, P.; Vetvicka, V.; Fusek, M. Cathepsin D-Many functions of one aspartic protease. *Crit. Rev. Oncol./Hematol.* **2008**, *68*, 12–28. [CrossRef]

64. Yu, C.P.; Cha, Y.; Wu, F.; Xu, X.B.; Qin, L.; Du, M. Molecular cloning and functional characterization of cathepsin D from sea cucumber *Apostichopus japonicus*. *Fish Shellfish Immunol.* **2017**, *70*, 553–559. [CrossRef] [PubMed]
65. Singh, A.; Nelson-Moon, Z.L.; Thomas, G.J.; Hunt, N.P.; Lewis, M.P. Identification of matrix metalloproteinases and their tissue inhibitors type 1 and 2 in human masseter muscle. *Arch. Oral Biol.* **2000**, *45*, 431–440. [CrossRef] [PubMed]
66. Tipper, J.P.; Lyons-Levy, G.; Atkinson, M.A.L.; Trotter, J.A. Purification, characterization and cloning of tensilin, the collagen-fibril binding and tissue-stiffening factor from *Cucumaria frondosa* dermis. *Matrix Biol.* **2002**, *21*, 625–635. [CrossRef] [PubMed]
67. Coriicchiato, O.; Cajot, J.-F.; Abrahamson, M.; Chan, S.J.; Keppler, D.; Sordat, B. Cystatin C and cathepsin B in human colon carcinoma: Expression by cell lines and matrix degradation. *Int. J. Cancer* **1992**, *52*, 645–652. [CrossRef] [PubMed]
68. Garnero, P.; Borel, O.; Byrjalsen, I.; Ferreras, M.; Drake, F.H.; McQueney, M.S.; Foged, N.T.; Delmas, P.D.; Delaissé, J.-M. The collagenolytic activity of cathepsin K is unique among mammalian proteinases. *J. Biol. Chem.* **1998**, *273*, 32347–32352. [CrossRef] [PubMed]
69. Helske, S.; Syväranta, S.; Lindstedt, K.A.; Lappalainen, J.; Öörni, K.; Lommi, J.; Turto, H.; Werkkala, K.; Kupari, M.; Kovanen, P. Increased expression of elastolytic cathepsins S, K, and V, and their inhibitor cystatin C in stenotic aortic valves. *Arterioscler. Thromb. Vasc. Biol.* **2006**, *7*, 242–243. [CrossRef] [PubMed]
70. Davies, D.R. The structure and function of the aspartic proteinases. *Annu. Rev. Biophys. Biophys. Chem.* **1990**, *19*, 189–215. [CrossRef]
71. Takashi, S.; Hideaki, S.; Yasuaki, S.; Yuzo, K.; Kenji, Y. An immunocytochemical study on distinct intracellular localization of Cathepsin E and Cathepsin D in human gastric cells and various rat cells. *J. Biochem.* **1991**, *110*, 956–964. [CrossRef]
72. Laronha, H.; Caldelra, J. Structure and function of human matrixmetalloproteinases. *Cell* **2020**, *9*, 1076. [CrossRef]
73. Trotter, J.A.; Lyons-Levy, G.; Chino, K.; Koob, T.J.; Keene, D.R.; Atkinson, M.A.L. Collagen fibril aggregation-inhibitor from sea cucumber dermis. *Matrix Biol.* **1999**, *18*, 569–578. [CrossRef]
74. Zhang, Y.T.; Zhao, B.L. Green tea polyphenols enhance sodium nitroprusside-induced neurotoxicity in human neuroblastoma SH-SY5Y cells. *J. Neurochem.* **2003**, *86*, 1189–1200. [CrossRef] [PubMed]
75. Fomina, M.; Hillier, S.; Charnock, J.M.; Melville, K.; Alexander, L.J.; Gadd, G.M. Role of oxalic acid overexcretion in transformations of toxic metal minerals by *Beauveria caledonica*. *Appl. Environ. Microbiol.* **2005**, *71*, 371–381. [CrossRef] [PubMed]
76. Ming, Y.; Wang, Y.Z.; Xie, Y.Q.Q.; Sun, C.H.; Dong, X.F.; Nakamura, Y.; Chen, X.; Dong, X.P.; Qi, H. Phlorotannin extracts from *Ascophyllum nodosum* inhibited proteases activities and structural changes from *Apostichopus japonicus*. *ACS Food Sci. Technol.* **2022**, *2*, 1586–1596. [CrossRef]
77. Ming, Y.; Wang, Y.Z.; Xie, Y.Q.Q.; Dong, X.F.; Nakamura, Y.; Chen, X.; Qi, H. Polyphenol extracts from Ascophyllum nodosum protected sea cucumber (*Apostichopus japonicas*) body wall against thermal degradation during tenderization. *Food Res. Int.* **2023**, *164*, 112419. [CrossRef] [PubMed]
78. Guo, Y.C.; Ming, Y.; Li, X.; Sun, C.H.; Dong, X.P.; Qi, H. Effect of phlorotannin extracts from *Ascophyllum nodosum* on the textural properties and structural changes of *Apostichopus japonicus*. *Food Chem.* **2024**, *437*, 137918. [CrossRef] [PubMed]
79. Guo, Y.C.; Ming, Y.; Sun, K.L.; Dong, X.F.; Nakamura, Y.; Dong, X.P.; Qi, H. Polyphenol oxidase mediates (−)-epigallocatechin gallate to inhibit endogenous cathepsin activity in *Apostichopus japonicus*. *Food Chem.* **2024**, *449*, 139166. [CrossRef]
80. Du, Y.N.; Li, A.T.; Yan, J.N.; Jiang, X.Y.; Wu, H.T. Inhibitory effect of coelomic fluid isolates on autolysis of minced muscle tissue from sea cucumber Stichopus japonicus. *J. Food Meas. Charact.* **2021**, *15*, 4575–4581. [CrossRef]

MDPI AG
Grosspeteranlage 5
4052 Basel
Switzerland
Tel.: +41 61 683 77 34

Foods Editorial Office
E-mail: foods@mdpi.com
www.mdpi.com/journal/foods

www.ingramcontent.com/pod-product-compliance
Lightning Source LLC
Chambersburg PA
CBHW061616210926
51850CB00009B/2345
9783725822997